T. A. Dettlaff · A. S. Ginsburg
O. I. Schmalhausen

Sturgeon Fishes

Developmental Biology and Aquaculture

Translated by G. G. Gause and S. G. Vassetzky

With 81 Figures and 20 Plates

Springer-Verlag
Berlin Heidelberg New York
London Paris Tokyo
Hong Kong Barcelona
Budapest

Authors:
Professor Dr. Tatiana A. Dettlaff
Dr. Anna S. Ginsburg
Dr. Olga I. Schmalhausen

N.K. Koltsov Institute of Developmental Biology
Russian Academy of Sciences
26 Vavilov Street, Moscow 117808, Russia

Translators:
Dr. George G. Gause
Dr. Sergei G. Vassetzky

N.K. Koltsov Institute of Developmental Biology
Russian Academy of Sciences
26 Vavilov Street, Moscow 117808, Russia

ISBN-13: 978-3-642-77059-3 e-ISBN-13: 978-3-642-77057-9
DOI: 10.1007/978-3-642-77057-9

Library of Congress Cataloging-in-Publication Data. Dettlaff, Tatiana Antonovna. [Razvitie osetrovykh ryb. English] Sturgeon fishes: developmental biology and aquaculture / T.A. Dettlaff, A.S. Ginsburg, O.I. Schmalhausen: translated by G.G. Gause and S.G. Vassetzky. p. cm. Translation of: Razvitie osetrovykh ryb. Includes bibliographical references and index. (alk. paper) 1. Sturgeons – Development. 2. Embryology – Fishes. 3. Sturgeon fisheries. 4. Sturgeons – Larvae. I. Ginsburg, Anna Samoilovna. II. Schmalhausen, O.I. (Olga Ivanovna) III. Title. QL638.A25D4813 1992 639.3'744–dc20 92-6807 CIP

Softcover reprint of the hardcover 1st edition 1993

Typesetting: Macmillan, Bangalore
31/3145-5 4 3 2 1 0 – Printed on acid-free paper

Foreword

This book is the first comprehensive description of development of the Acipenserid fish published in the English language. It contains the results of more than 40 years of studies by the authors and their colleagues. My own life in science has been intimately related both with the authors and the fish, which are the subject of this book. Therefore, it gives me a great pleasure to present to the English reader an expanded version of the book.

Those interested in the history of biology must be well aware of the fact that genetics in the USSR was practically demolished by Lysenko at the session of the Lenin All-Union Academy of Agricultural Sciences in 1948. However, it is much less well-known that other fundamental branches of biology were also persecuted at that time, experimental embryology (developmental mechanics) among them. As a result, many embryologists, including the authors of this book, were forced to turn to more applied problems, this being the only way to continue research. They had to abandon amphibians and concentrate their efforts on sturgeon.

At that time the conservation of Acipenserid fish, a unique group of fish, most of which (about 90%) occur in the USSR, became very urgent. Construction of hydroelectric stations on the spawning rivers markedly reduced the natural reproduction of sturgeon. Hence, it was important to replenish the natural populations on a large scale with juveniles obtained under artificial conditions. The authors of this book were among the first scientists who started doing research in this direction.

Hence, this, partly involuntary, change in the topic of studies proved to be fruitful from different viewpoints. On the one hand, a new experimental model was introduced in developmental biology, which had a number of advantages. Studies of development of the Acipenserid species made it possible to uncover new patterns of embryonic and postembryonic development which were very interesting for fundamental science. On the other hand, since these studies were carried out at the surgeon hatcheries and the authors were well aware of the problems related to obtaining and collecting mature gametes, artificial insemination, incubation of embryos, and rearing of prelarvae, they could offer recommendations aimed at the improvement of sturgeon culture on the basis of their own fundamental studies.

This way of combining basic research and application of its results in fish culture proved to be very worthwhile. I shall provide a few examples in order not to be accused of offering unsubstantiated statements.

T.A. Dettlaff developed, in her studies on Acipenserid fish, a basically novel approach to studying the temporal patterns of development based on the use of relative, dimensionless units of the development duration. This approach was later used on a number of animal species. Recommendations for determining the time of obtaining mature gametes from sturgeon spawners and the time of sampling in order to assess the conditions of development and analyze the causes of embryonic mortality and defects during incubation were the practical results of these studies. Recommendations for artificial insemination of sturgeon eggs at the hatcheries were based on profound studies of fertilization and mechanisms of polyspermy block carried out by A.S. Ginsburg. These recommendations allow fish culturists to obtain the highest possible rate of fertilization with the minimal number of polyspermic (non-viable) embryos. Experimental studies of the effects of various water pollutants on sturgeon prelarvae allowed O.I. Schmalhausen to uncover several interesting defects. The results of these studies were used to exclude materials dangerous for prelarvae from use in the construction of holding facilities.

These and other practical recommendations of the authors were tested by the safest criterion, time, and are now successfully used at the sturgeon hatcheries in the USSR.

However important the applied aspects of these studies were, the authors were, first of all, embryologists and they succeeded in studying throughout this time the whole process of development of these peculiar fish very carefully and successively, from the formation of a mature egg to the transition of prelarvae to active (exogenous) feeding. The authors have used both classical descriptive approaches and physiological and experimental-embryological methods. The book contains not only results obtained by the authors. It includes practically all important studies of their students and researchers from other laboratories in this field. The book is provided with excellent illustrations, including drawings made by the authors. These illustrations give a detailed picture of the development of sturgeon, which is sometimes inaccessible not only to inexperienced researcher, but also to modern photography.

A few words about the "heroes" of this book, Acipenserid fish. The taste quality of the meat and, especially, caviar which are recognized all over the world are not the only advantages of these fish. Their taxonomic status is no less important for a biologist.

This group of ancient fish, fairly limited in the species composition, has many peculiar features. No phylogenetic studies in developmental biology, physiology, biochemistry, and, nowadays, molecular biology could be complete without studies of these fish. Specific features of sturgeons, such as enormous fecundity, large size of eggs, and the possibility of obtaining a great number of synchronously developing embryos, will always attract the developmental biologist.

Soviet scientists until recently had a monopol in studies of sturgeon. In some countries of Western Europe, America, and the Far East programs are now developed and realized, which are aimed at restoring the local populations of Acipenserid fish. In addition, commercial sturgeon culture is being developed in an ever increasing number of countries. These commercial enterprises are based on the local, imported (*Acipenser baeri* became very popular in Western Europe), or hybrid forms. This activity called for research which, on one hand, was aimed at development of breeding and rearing of sturgeon and, on the other hand, gave a possibility to research fundamental scientific problems.

The 1st International Symposium on Sturgeon Fish held in 1989 in France showed the great interest of scientists from various countries towards joint studies in this area of research. Publication of this fundamental treatise on development of the Acipenserid fish in English will help all those who started or intend to start working with sturgeon to acquaint themselves with knowledge accumulated in this area by the Soviet scientists.

Since this book contains both the results of fundamental studies and practical recommendations, it will be interesting for a wide readership, including scientists, teachers, and students who deal with the problems of comparative and experimental embryology and, of course, practical fish culturists involved in breeding and rearing of sturgeon.

B. F. Goncharov
Head of Group for Regulation of Gametogenesis
Institute of Developmental Biology
Russian Academy of Sciences

Contents

Introduction

Sturgeon, fish of extremely great commercial value, belong to one of the most ancient groups of the *Osteichthyes.* They are widely distributed in the Northern Hemisphere above the 30th parallel, and within this zone they are found everywhere along the coasts of the Atlantic and Pacific Oceans, the Mediterranean and Black seas, as well as in rivers, lakes and inland seas. Their abundance in this huge range differs greatly: in many parts of the range they are found only as isolated, sparse populations and occur in significant numbers only in a few regions. The highest concentration of sturgeon is in the seas of the Ponto-Caspian depression, -the Caspian, Azov and Black seas. The Caspian represents a unique reservoir: in recent years it has produced up to 92% of the sturgeon catch in the USSR (Barannikova 1987).

The sharp fall in the abundance of sturgeon in many regions, and their complete disappearance in some places has been taken by many researchers as an indication that they are a relic group of fish doomed to extinction as a result of competition with higher fish (cf. Magnin 1959; Gerbil'skii 1962). Although the *Acipenseridae* indeed are an ancient group of the *Osteichthyes,* they are beautifully adapted to contemporary environmental conditions and even have great advantages as compared with the *Teleostei.* These advantages include: wide range of spawning temperatures, longer retention of fertilizability by sperm and eggs in water, early euryhalinity of juvenile fish and the broad spectrum of their feeding, protection from predators afforded juvenile fish by their scutes, ecological plasticity as expressed in the existence of anadromous, semianadromous and freshwater forms, etc. The decrease in the abundance of sturgeon throughout the world is a result of human activity: fishing over thousands of years, and, at present, the regulation of the flow of rivers and their pollution, and mass destruction of juvenile sturgeon as a result of their being incidentally caught in small-mesh nets intended for teleost fish, and other factors (Gerbil'skii 1962; Marti 1964). Therefore, the problem of conserving sturgeon and replenishing their stocks in the entire range has become urgent for scientists of various countries. The authors of this book hope that the materials given here will prove useful in solving this difficult but important problem.

The family of sturgeon, *Acipenseridae,* includes four genera containing 24 species; 17 of them belong to the genus *Acipenser* (sturgeon), two to the genus *Huso* (giant sturgeon), two to the genus *Scaphirhynchus* (shovel-nosed sturgeon) and three to the genus *Pseudoscaphirhynchus* (Aral shovelnoses). Thirteen species reside in USSR territory (Berg 1948). These are the two species of the genus *Huso*: *H. dauricus* (Georgi) (kaluga, giant Siberian sturgeon) in the Amur basin and the Amur estuary and *H. huso* (L.) (beluga, giant sturgeon) in the Caspian and Black seas from whence they migrate into the rivers. The genus *Acipenser* includes: *A. gueldenstaedti* Brandt (Russian sturgeon) living in the Caspian from whence they migrate into rivers to spawn, and its subspecies, *A. g. colchicus* V. Marti (Black-Azov-sea sturgeon), as well as *A. persicus* Borodin (Persian sturgeon) in the Caspian Sea. There are also *A. baeri* Brandt (Siberian sturgeon) in the rivers of Siberia, beginning with the Ob' and extending to Kolyma, and its subspecies, *A. baeri chatys* Drijagin (little Siberian sturgeon) living in the Lena and Kolyma basins; *A. schrencki* Brandt (Amur sturgeon) in the Amur basin; *A sturio* L. (Baltic sturgeon) in the Baltic and Black seas and their basins; *A. medirostris* Ayres (Sakhalin sturgeon in Asia or green sturgeon in north America) in the seas of Japan and Okhotsk. To the same genus belong also *A. stellatus* Pallas (sevryuga, stellate sturgeon) inhabiting the Caspian, Azov and Black Seas and their basins; *A nudiventris* Lovetzky (bastard sturgeon) in the Black, Caspian and Aral Seas and their basins; *A. ruthenus* (L.) (sterlet) in the rivers of the Black and Caspian Sea basins, in the basins of lakes Ladoga and Onega, in the Severnaya Dvina, Ob' and Enisei. There are also three species belonging to the genus *Pseudoscaphirhynchus*, which inhabit the Aral sea basin: *P. kaufmanni* (Bogdanow) (big Amu Darja shovelnose) as well as the rare *P. hermanni* (Kessler) (little Amu Darja shovelnose) and *P. fedtschenkoi* (Kessler) (Syr-Darja shovelnose).

The majority of sturgeon species are not abundant outside the USSR. According to data presented by the French ichthyologist Magnin (1959), sturgeon had practically disappeared in the Mediterranean and Adriatic by the middle of the 20th century and only rare individuals of *A. sturio*, *A. naccarii* Bonoparte and *H. huso* are found. Only *A. sturio* are found on the Atlantic coast of Europe. *H. huso*, *A. gueldenstaedti colchicus*, *A. stellatus*, and *A. nudiventris* migrate from the Black sea into the Danube. *A. ruthenus* also occurs in the Danube. *H. huso*, *A. persicus*, *A. stellatus*, and *A. nudiventris* which are found in the southern part of the Caspian spawn in rivers of the southern coast.

In America there are five species of the genus *Acipenser*: on the Pacific coast, *A. medirostris* and *A. transmontanus* Richardson (white sturgeon); on the Atlantic, *A. oxyrhynchus* Mitchill (Atlantic sturgeon), *A. brevirostrum* Lesuer (short-nosed sturgeon) and, in fresh waters, *A. fulvescens* Rafinesque (lake sturgeon), as well as two species of the genus *Scaphirhynchus*: *S. platorhynchus* (Rafinesque) (shovel-nosed sturgeon) and *S. albus* (Forbes et Richardson) (pallid sturgeon).

Finally, five species of sturgeon occur on the Asian coast of the Pacific: *A. medirostris*; in China, rarely *A. sinensis* Gray (Chinese sturgeon) and *A. dabryanus* Dumeril (Korean sturgeon), and, in Japan, still more rarely, *A. micadoi* Hilgendorf and *A. kikuchii* Jordan et Snyder.

A large part of the entire world catch of sturgeon (about 90%) is in the USSR (Barannikova 1987). Of greatest commercial importance are: *A. gueldenstaedti* and *A. gueldenstaedti colchicus*, *A. stellatus*, and *H. huso*. Catches include also *A. baeri*, *A. nudiventris*, *H. dauricus*, and *A. ruthenus*.

Outside the USSR the largest catch belongs to Iran (in the southern Caspian). In Europe sturgeon (in small quantities) are caught in Rumania, Bulgaria and Portugal. In the USA there is fishing for *A. sturio*, *A. medirostris*, *A. transmontanus*, and *A. fulvescens*, as well as *S. platorhynchus* and *S. albus*.

Sturgeon include anadromous species (living in the sea and migrating into rivers to spawn), semianadromous species (which spend most of their lifetime in rivers and make migrations into pro-estuarine regions, but do not go out into the open sea), and river-resident (freshwater) species which spend their entire lifetime in rivers. Some species, as for example *A. baeri* and *A. ruthenus*, have both semianadromous and river-resident forms.

At the present time, in the USSR the flow of most rivers in which sturgeon used to spawn has been regulated. This deprived sturgeon of the major part of their natural spawning grounds. In view of this, great attention is given to artificial propagation in order to preserve the diversity of sturgeon species and maintain their numbers. Since the decrease in abundance of sturgeon is not due to insufficient adaptation or competitive relationships with teleosteans, replenishment of their stocks by means of hatchery propagation of various sturgeon species seems quite practical.

The first experiment of artificial propagation of sturgeon was carried out in 1869 when Owsjannikow (1871) artificially inseminated *A. ruthenus* eggs, incubated them, and obtained prelarvae. In subsequent years, insemination and incubation were also performed for some other sturgeon species: *A. fulvescens* by Seth Green in 1875, *A. sturio* by Frauen in 1881, *A. stellatus* in 1884 and *A. gueldenstaedti* in 1899 by Borodin (Borovik 1916). These and similar experiments were aimed mainly at obtaining materials for investigation. The first experiment specially for fish culture was carried out in 1886 by Mohr who released 500 000 *A. sturio* larvae in the Elbe. However, despite the relatively successful results of the first experiments, most of the work on artificial propagation remained small-scale for a long time. This was mainly due to difficulties in obtaining mature spawners which were caught on spawning grounds.

Derzhavin was the first to stimulate maturation of sturgeons (Derzhavin 1938, 1939, 1947). He placed them in conditions close to those found in nature during the migration to spawning grounds: they were kept in a basin with strong circular flow and pebbly bottom. In these conditions a few females matured, but this could not ensure the mass propagation of sturgeon.

The scientific principles of hatchery propagation of sturgeon were laid down by long-term investigations of Nikolai L. Gerbil'skii and his coworkers. They developed a method of pituitary injections for stimulating maturation of sturgeon (Gerbil'skii 1941, 1947). This method allows the obtaining of large amounts of eggs suitable for fish breeding from females caught in prespawning condition. Such a method is indispensable for hatchery propagation of sturgeon.

Contemporary sturgeon culture in the USSR widely uses hatchery reproduction and refined biotechnology.

Work on artificial propagation of sturgeon has recently been started in many countries: USA, France, Italy, Hungary, Poland, China, Japan, etc. In the USA successful attempts to reproduce *A. transmontanus*, *A. oxyrhynchus*, and *A. brevirostrum* have been made. A major advance has been made with *A. transmontanus* (Doroshov et al. 1983).

However, a marked difficulty has arisen recently in obtaining suitable brood stock for fish culture: the number of females not responding to gonadotropic hypophysial hormones by reaching maturity has sharply increased. As a result of flow regulation in rivers and change in their hydrological regime, the migratory behavior of sturgeon, dynamics of their movements and the structure of their populations have changed, as well as the physiological condition of spawners (Barannikova 1979). At the present time, sturgeon fishery managers are faced with new problems, which a wide circle of scientists and practical fish culturists are trying to solve. These problems include the development of fish hatchery methods for various species and populations, development of methods for in vivo evaluation of gonad maturity, and selection of brood stock, as well as for eliminating losses of eggs, embryos, prelarvae, larvae and juveniles. The problem of water is critical, since pollution of water by oil and inadequately purified sewage water causes the condition of spawners to deteriorate and periodically produces mass extinction of prelarvae at the hatcheries. Also, the preservation of natural spawning, the improvement of existing and creation of artificial spawning grounds are very important tasks.

The efficiency of hatchery propagation is a well established fact: individuals of *H. huso* migrating from the Caspian into the Volga and Ural belong to those generations when natural spawning grounds have already been lost, i.e., they were raised at the hatcheries. At the same time a sharp increase in the abundance of *H. huso* juveniles has been noted in the northern Caspian. Stocks of sturgeon in the Azov sea are also mainly due to hatchery reproduction, since natural spawning in the Don is almost non-existent.

At present, there are 26 sturgeon hatcheries in the USSR (Barannikova 1987), most of which are located in river estuaries or nearby dams: in the Caspian sea basin — eight on the Volga, three on the Kura, one on the Terek, and one on the Sulak; in the Azov sea basin — three on the Don and four on the Kuban': in the Black sea basin — one on the Dniepr and one on the Rion; in Siberia — four hatcheries. Sturgeon hatcheries release annually 100–140 million juveniles in natural water bodies. In all, since 1975 more than a billion juveniles of *H. huso*, *A. gueldenstaedti*, *A. stellatus*, and *A. nudiventris* have been released.

Hatcheries provide for various stages in the biotechnological chain of fish culture: work with spawners, insemination and incubation of eggs, rearing of prelarvae, raising juvenile fish in ponds, keeping account of and distributing juvenile fish, and also the growing of live food. In many hatcheries three species of sturgeon are raised, in some, two. For keeping brood stock, hatcheries have isolation areas, and some hatcheries have temperature-controlled production lines designed by B. N. Kazanskii for long-term keeping of brood stock. These lines make it possible to keep fish until they complete stage IV of maturity (cf. Trusov 1964a), i.e., until they acquire the ability to respond to pituitary injection by maturation. Also, it is possible to keep them in such a condition at low temperature for two-three months and thus extend the period, during which they can be used for fish culture (Kazanskii 1963; Kazanskii and Molodtsov 1974).

The first information about embryology of sturgeon was given by Kowalewsky et al. (1870) who described development of the embryos and prelarvae in *A. ruthenus*. These observations were later supplemented, mainly by Salensky (1878, 1880, 1881). Later embryological studies, which were carried predominantly on *A. ruthenus*, touched upon development of individual organs. However, studies of sturgeon development are of great scientific interest in comparative-embryological aspect. The *Acipenseridae* are an ancient group of fish that preserved primitive structural features relating them to the *Chondrostei*. At the same time, in the structure of eggs they are very similar to the *Amphibia*, since in the eggs of these groups, unlike most of the *Teleostei*, yolk inclusions are distributed over the cytoplasm. Hence, these groups are similar in many features of morphogenesis, in which the *Acipenseridae* differ from the *Teleostei*. Sturgeon can be used easily in developmental, physiological, and biochemical studies due to the high viability of embryos and possibility of obtaining large numbers of the embryos of the same age at the hatcheries. At present, sturgeon are similar to amphibians in the degree of our knowledge about their development. It will be seen later that many current problems of developmental biology are being studied using sturgeon embryos.

On the other hand, studies of sturgeon development are important for scientific improvement of fish culture methods. These studies include investigations into mechanisms of hormonal regulation of the transition of fish from the prespawning condition to the spawning one, mechanisms of oocyte maturation, fertilization, development of embryos and prelarvae. Another subject is sources of losses and causes of various defects during embryonic and prelarval development.

The present book gives an account of the contemporary state of our knowledge of these questions with special emphasis on practical sturgeon-culture problems. The book summarizes results of many years of investigation by the authors and their coworkers at the D. P. Filatov Laboratory of Experimental Embryology of the N. K. Kol'tsov Institute of Developmental Biology, USSR Academy of Sciences. Also summarized are appropriate data from the literature.

Studies were carried out predominantly on *A. gueldenstaedti*, *A. gueldenstaedti colchicus*, *A. stellatus*, and *H. huso*. In addition, the development of *A. ruthenus* and prelarval development of *P. kaufmanni* were studied.

Structure of embryos of different Acipenserid species is very similar. Therefore, developmental stages can be characterized by a sum of features common for all studied species. We have identified 35 successive stages of embryonic development. They were first described for *A. stellatus* and illustrated with photographs of the embryos of this species and, for comparison, embryos of *A. gueldenstaedti* and *H. huso* at some stages (Dettlaff and Ginsburg 1954). These stages were later described for *A. gueldenstaedti colchicus* and illustrated with drawings given in this book. These illustrations have already been published, partly (Ginsburg and Dettlaff 1955, 1969) and in full (Ginsburg and Dettlaff 1975; Dettlaff et al. 1981).

In the development of sturgeon prelarvae ten stages were identified which are characterized by a complex of common features. However, during this period there are more species-specific differences increasing with the age of prelarvae. These stages were described previously for embryos of *H. huso* and illustrated with photographs (Schmalhausen 1968). In this book the development of prelarvae, as well as embryos, is described for *A. gueldenstaedti colchicus* in comparison with *H. huso*, *A. stellatus*, *A. ruthenus*, and *P. kaufmanni*. Drawings of successive stages of prelarval development include those previously published for *A. gueldenstaedti colchicus* (Schmalhausen 1975) and *H. huso* (Dettlaff et al. 1981) and unpublished drawings of consecutive stages in the development of *P. kaufmanni*.

The book is complete with eight appendices containing recommendations for practical use at sturgeon hatcheries. Temperature is given everywhere in degrees Celsius.

The authors collected the materials and carried out experiments annually, since 1949, at various fishery stations and sturgeon hatcheries: for many years on the Don (at the village Rogozhkino and the Rogozhkino sturgeon hatchery) and on the Volga (at the Volgograd and the Ikryanino sturgeon hatcheries and at the Saratov division of All-Union Scientific Institute of Maritime Fisheries and Oceanography, and, in separate years, — on the Kuban' (in Kadushkino and Temryuk), the Kama (in Rybnaya Sloboda) and on the Kura (in Mingechaur, at the Kura experimental and Ust'-Kura sturgeon hatcheries). Some of these stations (Kadushkino, Saratov, Mingechaur, Rybnaya Sloboda) and the Volgograd hatchery are located in the area of spawning grounds, while the others — in the lower river flow and in the delta areas. Investigations were carried out with the constant aid of workers at the Ministry of Fisheries and its divisions, and with the active assistance of fish culturists. The authors express their deep thanks to them all.

Chapter 1 Oocyte Maturation and Ovulation

1.1 Maturation of Sturgeon Oocytes in Nature and at the Hatcheries

The females of anadromous sturgeon species migrate to rivers from the sea with their germ cells still immature. Their oocytes are located in the ovaries. If river flow is not regulated, some fish such as vernal *H. huso*, *A. gueldenstaedti*, *A. persicus*, and *A. stellatus* migrate into rivers with their gonads at the completed stage IV of maturity; their ovaries contain large fully grown oocytes. These female spawn during the same year. Other fish enter the rivers with gonads at stages III–IV and incomplete stage IV of maturity. These fish, which include the hiemal summer-migrating *A. gueldenstaedti* and the hiemal *H. huso* and *A. stellatus*, nest in holes for several months and spawn only during the following year.

The sturgeon with gonads at the completed stage IV of maturity migrate to the spawning grounds where the oocytes mature. With ripe males present, oocytes ovulate (come out of the follicles) and the female fish releases the ovulated oocytes in water.

The transition of a fish to the spawning condition can be visualized as follows (Fig. 1): via the sense organs, favorable environmental conditions (temperature, flow, pebbly bottom, etc.) act on the central nervous system of the fish, i.e., on the region of hypothalamus which in turn stimulates the cells of the hypophysis to release gonadotropic hormones into the blood. These hormones stimulate the maturation of oocytes and their ovulation. Ovulated oocytes that can be fertilized are called eggs. The female releases the eggs into water as they ovulate in small batches, and, therefore, the spawning of all the ripe eggs continues for a long time. Simultaneously, the male fish release sperm into the water. The spawning takes place in conditions favorable to both the maturation of the oocytes, and their subsequent development.

Under hatchery conditions, oocyte maturation and ovulation in sturgeon proceed, as for natural spawning, under the influence of the gonadotropic hypophysial hormones. For oocyte maturation and ovulation to take place, the spawners are injected with a suspension of powdered acetone-dried sturgeon pituitaries. In addition, a method of stimulation of the maturation of gametes by injecting a synthetic analog of mammalian LH–RH has recently been developed. The method has already successfully been used at some sturgeon hatcheries (Goncharov 1984a, b, 1985, 1990). The difference is in the fact that the

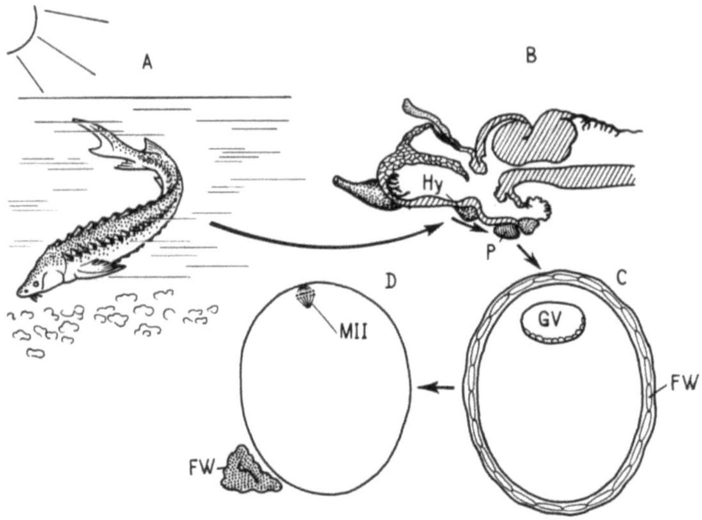

Fig. 1A–D. Schematic representation of the processes leading to maturation of the sturgeon oocytes (Ginsburg and Dettlaff 1969). **A** Ecological conditions at the spawning ground; **B** the sturgeon's brain responding to the spawning conditions by secretion of the hypophysial hormone in blood; **C** ovarian oocyte at the completed stage IV of maturity; **D** ovulated oocyte (mature egg). *FW* Follicle wall; *GV* germinal vesicle; *Hy* hypothalamus; *MII* metaphase of the II meiotic division; *P* pituitary

increase in the concentration of the hypophysial hormone in the blood of a female fish is affected not by the spawning conditions which stimulate the secretion of the hormone from the hypophysis, but by injection of the female with a suspension of pituitaries taken from other fish or with LH–RH. This is a very significant difference. It means that under hatchery conditions, the fish culturist himself must provide the optimal conditions for oocyte maturation in the body of the female fish. In order to accomplish this task, it is necessary to know what conditions are required for the transition of the ovaries to the completed stage IV of maturity, the main patterns of oocyte maturation and its dependence on environmental conditions.

1.2 Changes in the Oocytes During Their Maturation

In the course of maturation, the nucleus of an oocyte and its cytoplasm undergo profound changes. Normally, oocytes mature in the female ovary. However, under experimental conditions, this process can also take place in vitro, what greatly expands the possibility of studying the mechanisms of oocyte maturation.

1.2.1 Structure and Properties of the Fully Grown Oocyte

An ovary that has reached the stage IV of maturity (Fig. 2) contains large pigmented oocytes of definitive size. Apart from these, the naked eye is able to distinguish small unpigmented translucent oocytes at the initial stages of growth, i.e., oocytes of the next generation which will begin to grow rapidly after the first-generation oocytes are spawned, and will themselves be spawned during the following spawning period.

The fully grown oocyte is enveloped by egg membranes and encased in a three-layered follicle (see 1.2.6 and Fig. 5C). The egg membranes include: external jelly coat and underlying zona radiata consisting of zona radiata externa and zona radiata interna (Figs. 3, 4). The internal layer of the follicle wall (follicular epithelium) is in close contact with the jelly coat. At this stage,

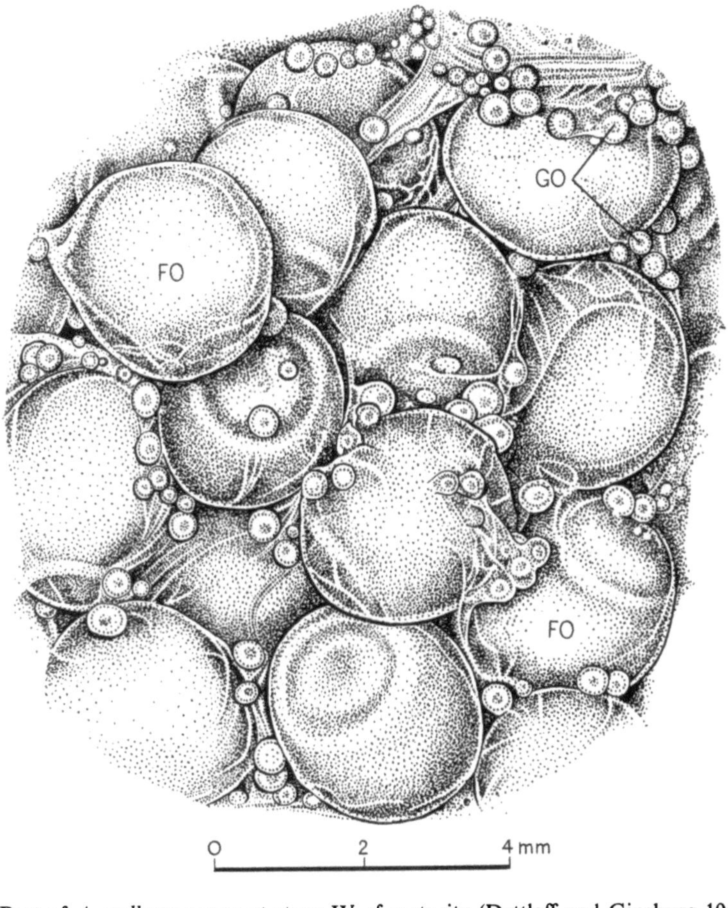

Fig. 2. Part of *A. stellatus* ovary at stage IV of maturity (Dettlaff and Ginsburg 1954). *FO* Large full-grown oocytes in the follicles; *GO* small oocytes at the beginning of rapid growth

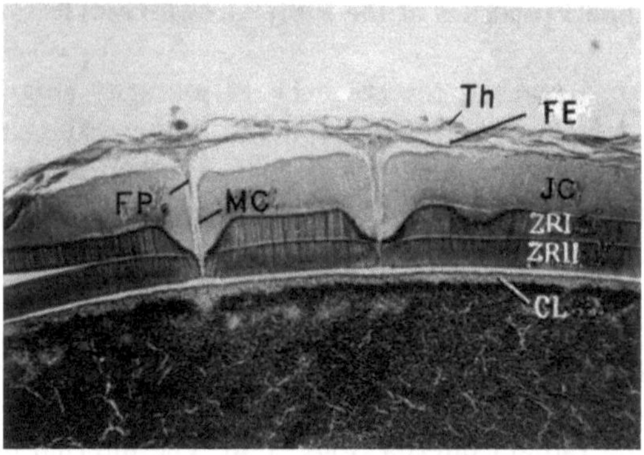

Fig. 3. A section through the animal area of *A. gueldenstaedti* oocyte (courtesy of N.E. Peseridi). *CL* Cortical layer of the oocyte; *FE* follicular epithelium; *FP* process of a follicle cell filling the future micropylar canal; *JC* jelly coat; *MC* micropylar canal filled by the process of a closing follicle cell; *Th* thecas; *ZR I* zona radiata externa; *ZR II* zona radiata interna

close connections are retained between the follicular epithelium and the oocyte: cytoplasmic processes of follicle cells penetrate the egg membranes, reach the surface of ooplasm and interdigitate with microvilli extending from the cytoplasmic cortex. There are canals in the egg membranes in the center of the animal area, over the oocyte's nucleus. They are filled with processes of a few large follicle cells and represent prospective micropylar canals (Fig. 3).

The oocyte (Fig. 4) is a large cell, the cytoplasm of which contains a lot of reserve nutrients used during embryonic development. The oocyte has a distinctly polarized structure: its vegetal part is filled with yolk granules (which attain, after fixation, upto 15 μm in diameter) and large lipid inclusions, while the animal part contains the bulk of the cytoplasm which includes, at a much lower concentration, only minute yolk granules and small lipid inclusions. Yolk granules contain a compact central body of crystalline structure surrounded by a thin granular layer (Grodzińsky 1963; Ginsburg 1968).

The nucleus of the oocyte is displaced toward the animal region; its position shows whether the gonads have reached the completed stage IV of maturity at which the follicles acquire the ability to respond to the effect of the hypophysial hormone by maturation. As the gonads pass from the incomplete to the completed stage IV of maturity, the nucleus moves towards the animal pole and finds itself completely or almost completely surrounded by the fine-grained yolk of the animal region (Trusov 1964a, b, 1975; Kazanskii et al. 1978; see Fig. 4 and Appendix 1). The nucleus is a large oval vesicle. As in the majority of other animals, during the period of rapid growth and in fully grown sturgeon oocytes the nucleus is many times larger than the nuclei of somatic cells. It also has a peculiar structure and is conventionally called "germinal vesicle" (GV); this name was given by Purkinje 150 years ago.

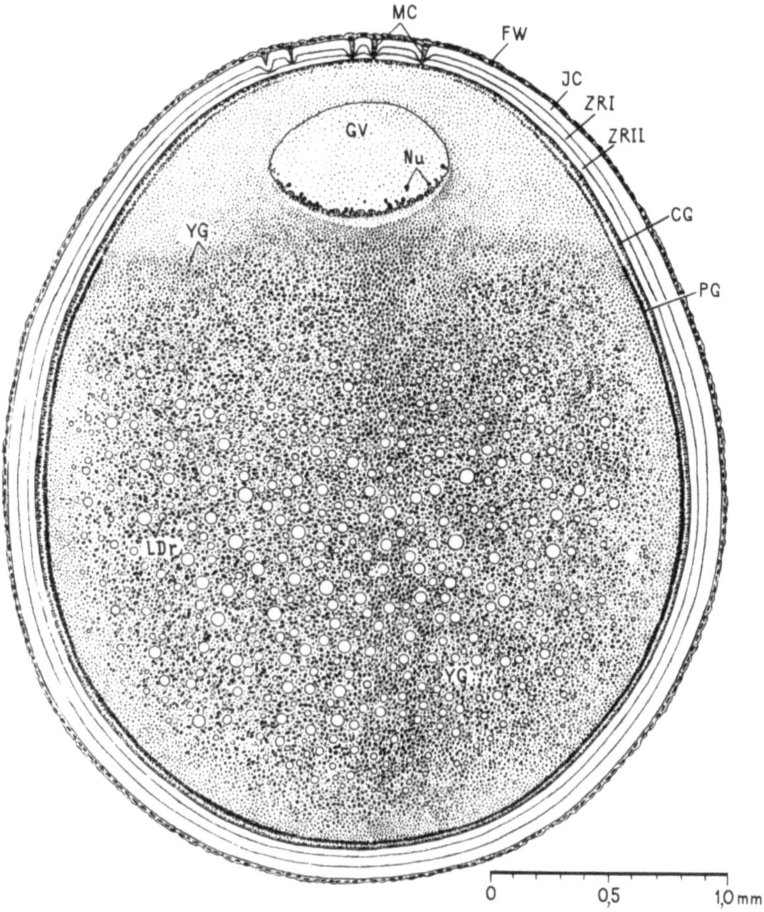

Fig. 4. A full-grown oocyte (section along the animal-vegetal axis) (Ginsburg and Dettlaff 1969). *CG* Cortical granules; *FW* follicle wall; *GV* germinal vesicle; *JC* jelly coat; *LDr* lipid droplets; *MC* micropylar canals; *Nu* nucleoli; *PG* pigment granules; *YG* yolk granules; *Zr I* zona radiata externa; *ZR II* zona radiata interna

The GV is filled with a nuclear sap (karyoplasm), and suspended in it is a small solid body containing chromosomes. At this stage, the oocyte contains a diploid number of chromosomes joined in tetrads consisting of two temporarily connected (conjugated) homologous chromosomes of the paternal and maternal genomes. Each chromosome is in turn divided into two sister chromatids. The chromosome apparatus remains in this condition (at diplotene of I meiotic division) until the beginning of the period of maturation inertia. At this time the chromosomes under the microscope look like lengthened bivalents with several chiasmata (cf. Fig. 8A, B). Numerous intensely staining nucleoli are found in the karyoplasm of the GV. They are located closer to the nuclear membrane in that part of the nucleus which faces the vegetal pole. Here, the nuclear membrane forms many evaginations. In this region, a small amount of karyoplasm leaks into the cytoplasm; the karyoplasm does not spread and, on sections, resembles a crescent-like lacuna adjacent to the nucleus and filled with material devoid of yolk inclusions (Fig. 7A).

There is a space between the oocyte's surface and zona radiata interna. It is less than 1 μm wide and contains microvilli and numerous outgrowths of the cytoplasmic cortex (Fig. 5A) (Aizenshtadt and Dettlaff 1972). Microvilli enter the ductules of zona radiata.

Cortical granules are located under the plasma membrane all over the oocyte's surface (Fig. 4). These organelles are lysosome-like structures and present in female gametes only (Aizenshtadt 1984). We will see that they play an important part in fertilization. Cortical granules are located in one-two rows (Fig. 5A). Their shape can be rounded, ellipsoid, or irregular; often, they have outgrowths. The size of profiles of the cortical granules on ultrathin sections in *A. gueldenstaedti* and *A. stellatus* vary from a few tenths to 3–4 μm (in the greater diameter). They contain protein-bound acid mucopolysaccharide (Dettlaff 1962). Correspondingly, the contents of the granules on electron micrographs has a fibrillar structure characteristic of mucopolysaccharides (Ginsburg 1968, 1972, 1987). In sturgeon, two components can be distinguished in the contents of the granules: numerous inclusions with densely arranged fibers are embedded in a matrix with looser fibrillar material (Fig. 5A, B). The granules are limited by a membrane.

There are numerous rod-like mitochondria under the layer of cortical granules and, at a depth of 3–5 to 20 μm from the surface, pigment granules whose concentration determines the characteristic pigment pattern of the egg. In addition, this zone contains cisternae of the Golgi complex and numerous smooth-surfaced vesicles.

The fully grown oocyte is characterized by the presence, in the animal region, of many annulate lamellae which form large aggregates, up to 10 μm in diameter (Aizenshtadt and Dettlaff 1972). Individual annulate lamellae are always strictly parallel to each other and often form concentric rows (Fig. 6A).

The surface layer of ooplasm at this stage possesses neither elasticity, nor contractility. After an incision is made on the surface, the wound does not close. Oocytes are permeable by water, and in hypotonic solutions or water they undergo plasmolysis and degeneration. The turgor of ooplasm is low. At the early stages of maturation the GV can easily be removed from the oocyte by making a small incision on the surface in the center of the animal region. After the incision is made, the GV increases slightly in volume, rises to the surface, pushes apart the wound edges, and emerges. If the GV membrane is ruptured with a needle inside the oocyte, a large lacuna of nucleoplasm remains in place of the GV. However, this contact between the nucleoplasm and cytoplasm does not lead to maturation of the cytoplasm (Dettlaff 1961; Dettlaff et al. 1964; Davydova 1968).

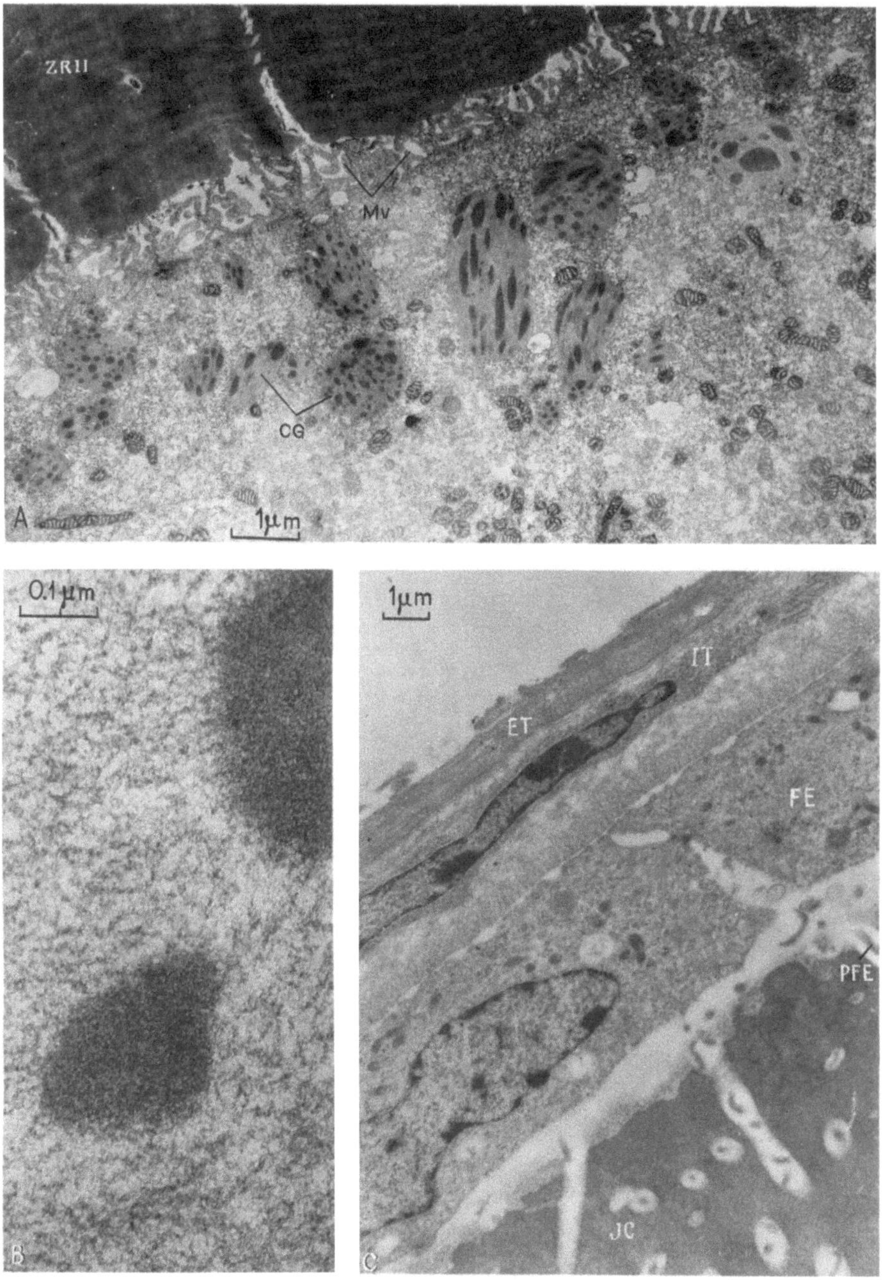

Fig. 5A–C. Structure of the cortical layer of a full-grown ovarian oocyte of *A. stellatus* (before the pituitary injection) and the follicle wall. **A** Cortical layer of the oocyte; **B** fine structure of the cortical granule contents; **C** follicle wall. *CG* Cortical granules; *ET* external theca; *FE* follicular epithelium cell; *IT* internal theca; *JC* jelly coat; *Mv* microvilli; *PFE* process of the follicular epithelium cell; *ZR II* zona radiata interna

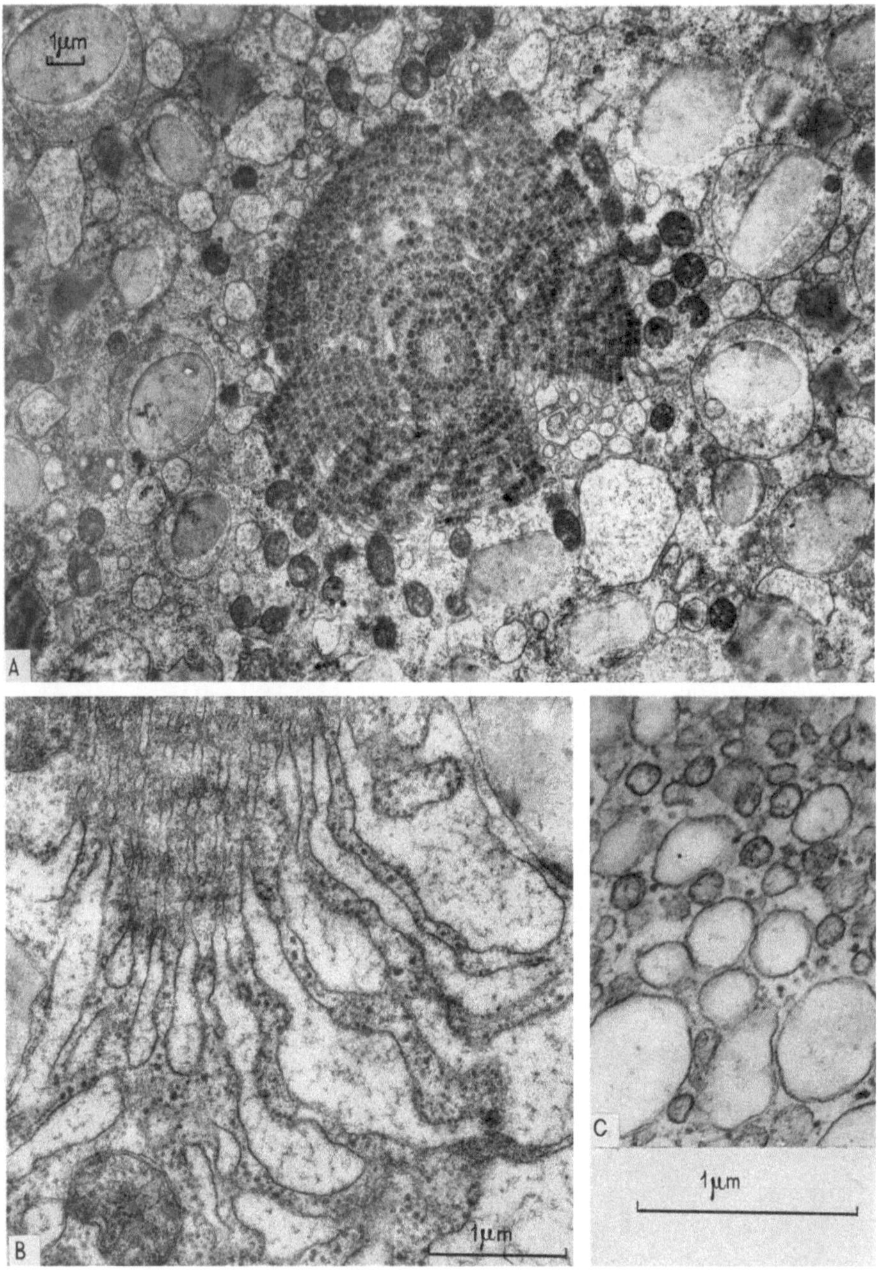

Fig. 6A–C. Changes in annulate lamellae during oocyte maturation in *A. stellatus* (courtesy of T.B. Aizenshtadt). **A** Fully grown oocyte (before pituitary injection) – a spherical aggregate of annulate lamellae forming concentric rows. **B** Beginning of the period of maturation inertia – terminal extensions of annulate lamellae. **C** Oocyte with the GV broken down – small smooth-surfaced vesicles arising as a result of disintegration of annulate lamellae

1.2.2 Changes in Oocytes During Their Maturation in Vivo

Changes occurring in oocytes were studied by taking them from an ovary at certain intervals [with the help of a probe (test rod) or through a cut in the body wall], fixing them and then examining the histological slides (Votinov 1947; Kazanskii 1957a, 1962; Dettlaff and Davydova, see Dettlaff and Skoblina 1969; Detlaff 1988). In the experiments of Dettlaff and Davydova, some oocytes at the time of fixation were placed in Ringer's solution, and their ability to mature in vitro studied. These experiments have proved that there are two periods which can be distinguished during oocyte maturation in sturgeon, just as in amphibians (Tchou-Su and Wang Yu-lan 1958), a hormone-dependent period and a hormone-independent period, or the period of maturation inertia. Oocytes transferred to a hormone-free Ringer solution during the hormone-dependent period do not mature. After a certain time, during the hormone-independent period, the removal of oocytes from the female body no longer interrupts their maturation which may be completed in saline. Oocytes which have matured in vitro can be fertilized after being transferred to water. They develop normally and produce viable prelarvae which begin active feeding (Dettlaff 1970a).

The duration of the hormone-dependent period varies significantly depending on the initial condition of the females (Goncharov 1976). In the experiments of Dettlaff and Davydova, the hormone-dependent period in females with gonads at the completed stage IV of maturity took up approximately one-third of the whole period of maturation, i.e., the period from the time the female was injected with a pituitary suspension up to the time of oocyte ovulation and obtaining ripe eggs. By the end of the hormone-dependent period, the oocyte acquires maturation inertia, but its structure does not change very much yet (cf Fig. 7A, B); the GV shifts towards the animal pole and gradually approaches the surface of the oocyte (Fig. 7B, C). At the same time the number and dimensions of the nuclear membrane evaginations on the vegetal side of the GV increase and the amount of nucleoplasm that has passed into the cytoplasm also increases. The nucleoli in the GV gradually become vacuolized and fragmented. As the GV moves close to the oocyte's surface, it flattens slightly (Fig. 7C). Subsequently, the GV breakdown takes place (Fig. $7D_{1-3}$). At the lower border of the GV in the area where the nucleoplasm passes into the cytoplasm and, later, in the lacuna of nucleoplasm, many cytasters appear (Fig. 8, 2). The nucleoplasm gradually spreads in the animal cytoplasm, forming a network of small lacunae (Fig. 7E). In addition to cytasters, globules of hydrophilic colloid may be detected in these lacunae (Fig. 8). The area where the nucleoplasm mixes with the cytoplasm, has a high glycogen content; in the nucleoplasm of intact GV glycogen has not been found (Dettlaff 1961).

When the GV breaks down, a spindle forms with strongly but not yet maximally condensed chromosomes, located in a small nucleoplasmic lacuna, alongside the incompletely disintegrated nuclear membrane. Chromosome condensation takes place when the chromosomes pass into the cytoplasm, since

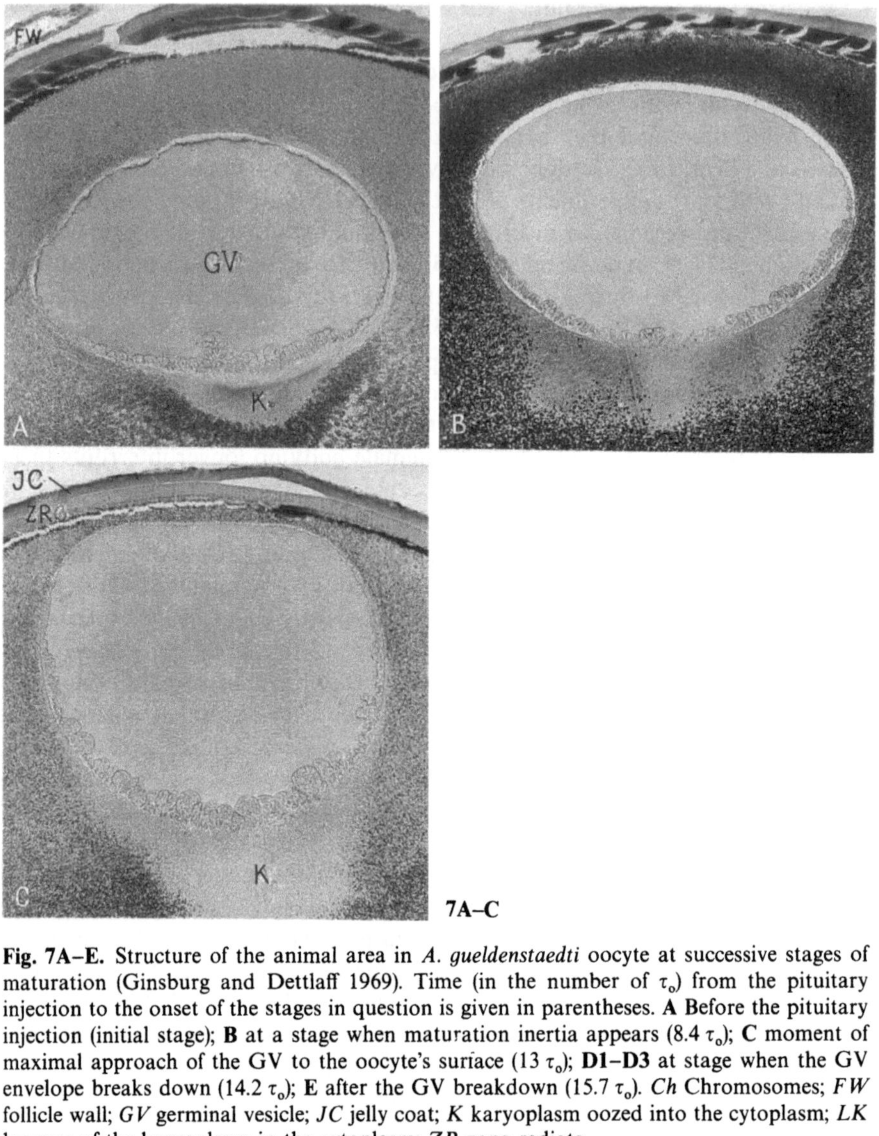

7A–C

Fig. 7A–E. Structure of the animal area in *A. gueldenstaedti* oocyte at successive stages of maturation (Ginsburg and Dettlaff 1969). Time (in the number of τ_0) from the pituitary injection to the onset of the stages in question is given in parentheses. **A** Before the pituitary injection (initial stage); **B** at a stage when maturation inertia appears (8.4 τ_0); **C** moment of maximal approach of the GV to the oocyte's surface (13 τ_0); **D1–D3** at stage when the GV envelope breaks down (14.2 τ_0); **E** after the GV breakdown (15.7 τ_0). *Ch* Chromosomes; *FW* follicle wall; *GV* germinal vesicle; *JC* jelly coat; *K* karyoplasm oozed into the cytoplasm; *LK* lacunae of the karyoplasm in the cytoplasm; *ZR* zona radiata

before the GV breakdown they appear as bivalents with several chiasmata (Fig. 8) (Dettlaff and Skoblina 1969; Ginsburg and Dettlaff 1969).

During the disintegration of the nuclear membrane (Fig. 7D), a certain part of the ovary will contain oocytes with a large GV and oocytes with a sharply diminished vesicle, or with parts of a collapsed but not yet decomposed membrane with chromosomes in the cytoplasm beside it.

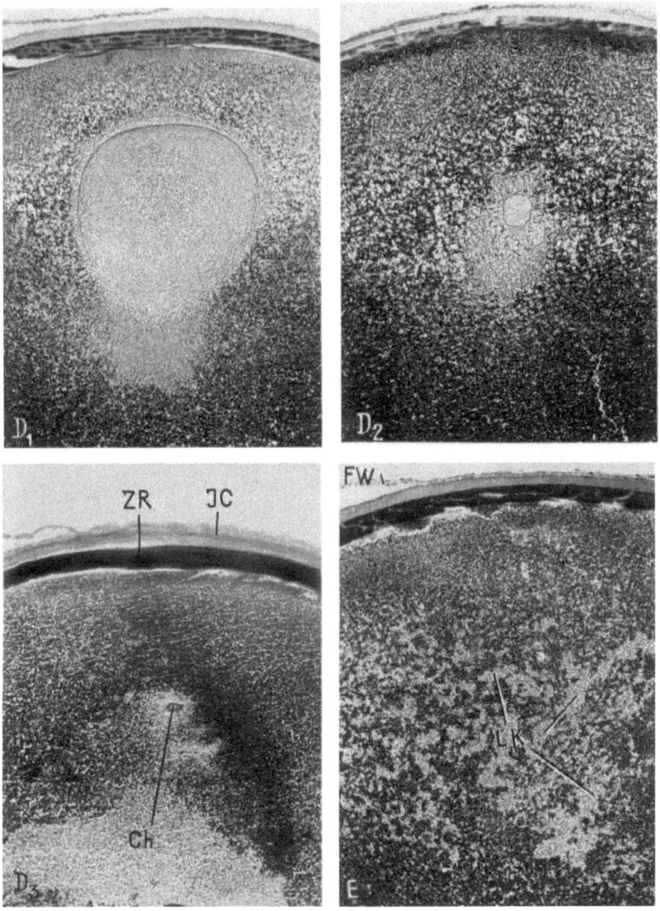

7D–E

As the GV breaks down, annulate lamellae disappear in the cytoplasm. Dilatations filled with a loose material first appear at the ends of annulate lamellae. Then these dilatations bud off and form aggregates of small vesicles and flat cisternae (Fig. 6) (Aizenshtadt and Dettlaff 1972; Aizenshtadt 1984). Aggregates of smooth-surfaced vesicles, which have originated from annulate lamellae and attain 20 μm in diameter, are displaced to the periphery of the cytoplasm. Comparison of light and electron microscopy observations suggests that aggregates of small smooth-surfaced vesicles correspond to globules of hydrophilic colloid (Fig. 8, 3) which are released to the perivitelline space after egg activation (Dettlaff 1962; see Sect. 2.3.2).

If we need to determine quickly how far the maturation of the oocytes has advanced, i.e., whether the GV has disintegrated, we can resort to the very

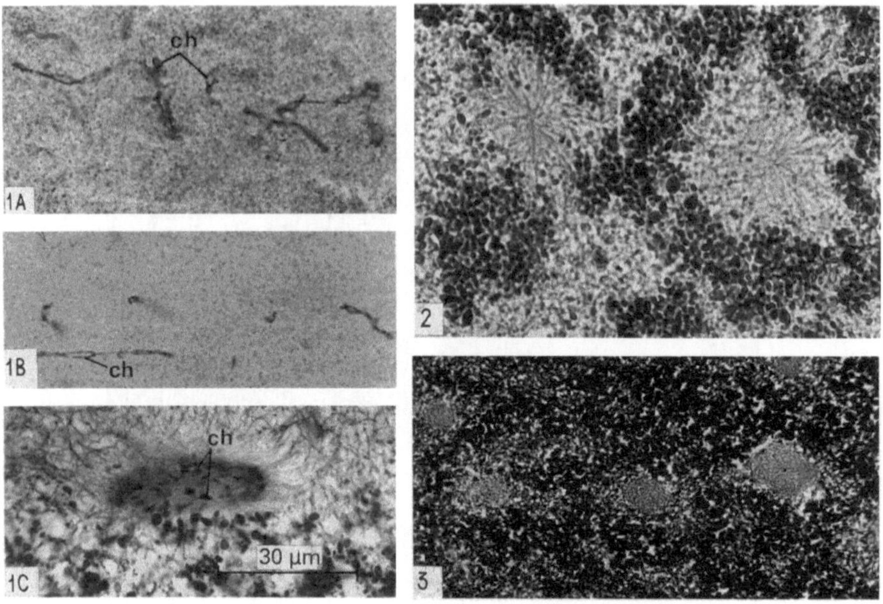

Fig. 8. Nuclear and cytoplasmic structures in a maturing oocyte of *A. gueldenstaedti*. **1** Chromosomes at: **A** the hormone-dependent period (4 τ_0) **B** the period of maturation inertia (12 τ_0), **C** the stage of GV envelope disintegration (14.4 τ_0); **2** cytasters in the cytoplasm; **3** globules of hydrophilic colloid (*1* and *3* from Dettlaff 1988). *Ch* Chromosomes

simple procedure of placing the oocyte in boiling water for 1–2 min and then cutting it in half along the animal-vegetal axis with a razor blade. If the GV is still intact, it will be clearly seen even with a naked eye.

During the second half of the maturation period, the oocyte enters a succession of meiotic phases; this leads to the reduction of the chromosome number in the egg. Meiosis begins and proceeds (until metaphase II) when the oocyte is still in the ovary; however, as in other vertebrates, this process in sturgeon is completed only after fertilization.

A spindle of the I meiotic division forms in the cytoplasm of the oocyte and moves to the surface (Fig. 9A, B). The chromosome (tetrads) on it are at first distributed randomly (prometaphase of the I meiotic division – prometaphase I); they later take up the equatorial plane of the spindle and form the metaphase plate (Fig. 9B). After that, the spindle begins to elongate, and the pairs of chromatids from each tetrad diverge to opposite poles of the spindle (anaphase I, Fig. 9C). When they reach their final location (telophase I), the peripheral group of chromosomes together with a small amount of cytoplasm separates from the oocyte (Fig. 9D) forming the I polar body. The group of remaining chromosomes begins without delay the II meiotic division which reaches metaphase II (Fig. 9E). At this stage, nuclear division is arrested (blocked).

Fig. 9A–E. Successive stages of nuclear transformations in the *A. gueldenstaedti* oocyte during the period of maturation inertia (Ginsburg and Dettlaff 1969). **A** Early prometaphase; **B** metaphase; **C** anaphase; **D** late telophase of the I meiotic division; **E** metaphase of the II meiotic division (mature egg). *MF* Maturation funnel; *MSI* spindle of the I meiotic division; *MSII* spindle of the II meiotic division; *PB* polar body; *ZR II* zona radiata interna

Changes in the properties of the oocyte's surface layer commence just before the GV breakdown and then proceed quickly. These changes include the appearance of contractility in the cortical layer of the cytoplasm and alteration of its permeability, followed somewhat later by the appearance of capacity for cortical reaction. During this period hydration and heat sensitivity of the oocyte increase. At metaphase II heat sensitivity is many times higher than at metaphase I (Vassetzky 1966, 1988). Permeability of the egg membranes during the maturation period drops markedly, and any changes in the ionic composition of the environment have a lower effect on the oocytes.

The capacity for cortical reaction (capacity for propagated exocytosis of the cortical granules' contents) appears in oocytes soon after the GV breakdown (Dettlaff 1961). It has been studied in detail on *A. stellatus* (Ginsburg and

Nikiforova 1979; Ginsburg 1987). Two specific features have been revealed. One of them is that at the stage when the capacity for cortical reaction is detected in a part of the oocytes for the first time, at prometaphase I, the reaction in some of them proceeds much more slowly than in mature eggs. Second, in many maturing oocytes exocytosis of the cortical granules in the center of the animal region is inhibited. As a result, the contact between the ooplasm and zona radiata interna in the region of micropylar canals is observed for a longer time than normally.

If the oocytes enclosed in the follicle wall are activated by needle pricking at meta- or anaphase I in Ringer solution, the contents of cortical granules discharged under the zona radiata does not swell. Accordingly, perivitelline space does not form and the jelly coat does not become adhesive. If, however, these oocytes are freed from the follicle wall, such changes do occur (Dettlaff 1961). Apparently, for these changes to take place not only must the contents of the cortical granules be discharged but also water must get under the zona radiata (Zotin 1961).

In sturgeon, like amphibians, the surface layer of the ooplasm matures earlier than the endoplasm and the capacity for cytokinesis also appears earlier (Dettlaff et al. 1964).

1.2.3 Chronology of Oocyte Maturation

Figures 7 and 9 depict sections of the *A. gueldenstaedti* oocytes at consecutive stages of maturation. The timing of these stages is expressed in relative units of development time, τ_0. The Greek letter τ with a zero index was used to denote the duration of one mitotic cycle during the period of synchronous cleavage divisions; in sturgeon it is equal to the interval between the appearance of the furrows of the 1st and 2nd divisions on the surface of the egg (Dettlaff and Dettlaff 1961) (see Sect. 2.4). Special investigations have shown that this value can be used to measure the duration of different periods of development, including the period of maturation, as the duration of these periods changes linearly within the range of optimal temperatures (Dettlaff et al. 1987; Dettlaff 1988, 1989, 1990). If we know the value of τ_0 at different temperatures (Fig. 42), we can determine by simple calculation the duration of a given period in hours and minutes at any temperature within the optimal range. At the same time, it should be remembered that the period from the injection of pituitary up to the GV breakdown in the oocytes may vary in different females, whereas the length of the subsequent phases of nuclear transformations hardly differs in them.

From the point of view of fish culture, the question of synchrony of oocyte maturation in different parts of the ovary is extremely important. A study of this question on oocytes that have passed from prophase I to subsequent phases of meiotic divisions, i.e., during the period when the stage of nuclear transformation can be determined with high accuracy, has shown that the differences in the stage of maturation between the oocytes from the anterior, middle and posterior regions of the ovary are not great, and do not exceed those from the same region of the ovary (Dettlaff and Zuichenko 1963; Vassetzky 1970). The following example gives us an idea of the scope of these differences. If there still is a GV in some oocytes, the others are usually at the prometaphase I.

The most advanced oocytes reach the final stage at which they can be fertilized (metaphase II) 2–3 h prior to ovulation (Kazanskii 1953a; Dettlaff and Zuichenko 1963; Vassetzky 1966). This interval is long enough for all the other oocytes to reach this stage by the onset of ovulation.

1.2.4 Changes in Oocytes During in Vitro Maturation

Follicle-enclosed oocytes undergo the same morphological changes when they are incubated in a modified Ringer solution containing 1.5–2 g $NaHCO_3$/l (Goncharov 1978) and a suspension of sturgeon pituitaries (10 μg/ml) or progesterone (5 μg/ml).

During maturation of the follicle-enclosed oocytes under the effect of gonadotropic hypophysial hormones in vitro as well as in vivo, there is a period when their contact with the hormone is required (hormone-dependent period) and a period when the hormone presence in the medium is no longer needed (period of maturation inertia). Meanwhile, a different time is required for maturation of the oocytes from the same female fish, which have been placed in Ringer solutions containing different concentrations of acetone-dried pituitaries (Goncharov, unpublished data). Thus, the duration of the hormone-dependent period depends not only on the readiness of the female for spawning, but also on the concentration (and activity) of the gonadotropic hormone in the solution.

Unlike the gonadotropic hypophysial hormone (which acts on the follicle cells), progesterone binds to the surface of the oocyte during the first minutes and, at high concentrations, even within the first seconds after exposure to it. This short interval suffices for the oocytes to mature almost simultaneously in a hormone-free Ringer solution.

The possibility of achieving maturation of strugeon oocytes in vitro enables us to study the chronology of this process with far greater ease than was possible earlier, since the oocytes are available for fixation at any intervals.

With this method, it was established for the *A. stellatus* oocytes (Ginsburg and Nikiforova, unpublished data) that approximately 2 τ_0 pass from the time of the GV breakdown in individual oocytes to the time of its breakdown in 95–100% of the oocytes in batches of good-quality oocytes in which the GV breakdown begins relatively early (8–12 τ_0 after being placed in a solution of progesterone) and is synchronous. In batches like this, the whole process of oocyte maturation from the GV breakdown to the appearance of metaphase II takes approximately 13 τ_0, metaphase I being detected after 7–8 τ_0 and the period between metaphase I and metaphase II lasting 5–6 τ_0. Less complete data obtained in vivo with an *A. stellatus* female injected with a pituitary suspension suggest that the onset of certain stages of maturation (timed from the GV breakdown) is similar in vivo after treatment with gonadotropic hypophysial hormones and in vitro after treatment with progesterone.

Similar results have been obtained in experiments carried out with the *A. gueldenstaedti colchicus* oocytes, when the female was injected with a pituitary suspension and the oocytes taken from the ovary after gaining maturation inertia were incubated in vitro (Vassetzky 1970). As in *A. stellatus*, the period between metaphase I and metaphase II lasted approximately $6\tau_0$. The first meiotic division lasted a long time, $\sim 8.6\tau_0$. Individual phases of the I meiotic division lasted as follows: prometaphase $\sim 2.2\tau_0$, metaphase $\sim 3.7\tau_0$, anaphase $\sim 1.7\tau_0$, and telophase $\sim 1.0\tau_0$. Hence, the I meiotic division in sturgeon lasts much longer than the division during the period of synchronous cleavage which lasts $1\tau_0$ (see Sect. 2.4.2).

1.2.5 Mechanisms of Oocyte Maturation

These described changes of the oocytes during their maturation are induced by the influence of progesterone (or a progesterone-like substance) secreted by the follicular epithelium cells. Its effect on the surface layer of the ooplasm initiates some changes, including protein synthesis (Dettlaff et al. 1968; Dettlaff and Skoblina 1969; Dettlaff 1970b). After some time, a substance (or substances) appearing in the ooplasm in sturgeon (Dettlaff and Skoblina 1969), as in amphibians (Dettlaff et al. 1964) induce the GV breakdown and maturation of the cytoplasm. This substance was called maturation promoting factor (MPF) (Masui and Markert 1971).

If nucleoplasm or cytoplasm is taken from the sturgeon oocyte at stages close to the GV breakdown (Nikitina and Dettlaff 1986), or at later stages, and injected in fully grown intact oocytes, such an injection induces their maturation. The active factor may be transferred from these oocytes successively through a number of oocytes and induce their maturation if protein synthesis in them is not inhibited (Dettlaff et al. 1977, 1980; Dettlaff and Felgenhauer 1980).

Unlike protein synthesis inhibitors, inhibitors of mRNA synthesis do not prevent maturation of oocytes induced by progesterone or the active cytoplasm of maturing oocytes. Hence, the hereditary apparatus of the nucleus is not involved in progesterone-induced oocyte maturation (Skoblina 1968; Dettlaff and Skoblina 1969). However, the mixing of nucleoplasm with cytoplasm occurring after the GV breakdown is essential for oocyte maturation.

Comparison of the mature intact oocytes with those in which the GV was removed prior to progesterone treatment, has shown that some features of the mature egg arise in the oocytes in the absence of GV, while the presence of GV is essential for the other features to be expressed. In the oocytes matured after the GV removal, contractility of the surface cytoplasm layer and capacity for the cortical reaction develop (Skoblina 1968, 1969), and annulate lamellae disintegrate (Aizenshtadt and Skoblina 1970). But capacity for cytokinesis does not develop in these oocytes (Dettlaff et al. 1964; Dettlaff and Skoblina 1969), chromosome condensation does not take place, and the injected sperm nuclei do not transform into pronuclei; moreover DNA synthesis is absent in the injected nuclei of spermatozoa and somatic cells (Skoblina 1976). It has been shown that the cytokinesis factor is already present in the nucleoplasm of small oocytes which have started rapid growth (see Fig. 2; Nikitina 1974, 1978). More complete data about the mechanisms of oocyte maturation in sturgeon and their comparison with those in teleosteans, amphibians and starfish are given in reviews of Dettlaff (1988) and Vassetzky (1988).

1.3 Ovulation of the Oocytes and the Formation of Coelomic Fluid

In sturgeon, just as in other species (Wright 1945; Schuetz 1972), oocyte maturation and ovulation are coordinated in time but relatively independent (Kazanskii 1962; Dettlaff 1970a). In females capable of responding by maturation to the gonadotropic hormone, oocyte maturation is normally followed by ovulation. However, under unfavorable conditions, these processes can be desynchronized or one of them can be completely inhibited. In some *A. stellatus* females after pituitary injection, oocytes ovulated at the initial stage, with a large GV, while in the others the maturated oocytes did not ovulate. Such oocytes sometimes underwent activation and even cleaved in the follicles (Dettlaff and Ginsburg 1950).

Oocyte maturation and ovulation are controlled by gonadotropic hypophysial hormones which act on the follicle wall. The duration of the hormone-dependent period for ovulation is slightly longer than for maturation (Goncharov, Skoblina, Dettlaff, unpubl. data). The follicular epithelium cells act as a target of gonadotropic hormones in oocyte maturation, but their role as the target of these hormones in the case of ovulation is yet not clear. The stimulating effect of gonadotropic hormones on both oocyte maturation and ovulation is suppressed by transcription and translation inhibitors (Dettlaff 1966; Dettlaff et al. 1968; Dettlaff and Skoblina 1969).

The follicle enclosing the fully grown oocyte is a three-layered structure, whose wall consists of theca externa, theca interna, and a layer of follicular epithelium (see Fig. 5C). According to electron microscopic studies of ovulation in *A. stellatus* (Trubnikova and Ryabova 1989), the cells of theca externa possess features characteristic of both epithelial and smooth muscle cells, whereas the structure of theca interna is typical for the connective tissue. Every follicle is supplied with a blood vessel which branches giving a network of capillaries reaching theca interna. At the follicle apex protruding into the body cavity, there is a special area with peritoneal epithelium as the surface layer. It is in this area that the follicle is ruptured.

The GV breakdown in the oocyte is followed by the preparation of the follicle for ovulation. Flattened follicle cells contract and assume a cubic shape. Long processes of these cells are contracted and, as a result, the connections between the follicle cells and the oocyte are lost. Proteolytic disintegration of theca starts simultaneously, as in amphibians, in the site of the future rupture of the follicle. In sturgeon, this process has been studied insufficiently.

The follicle cells flatten again thereafter and signs of intensive external secretion appear in them. These observations confirm the hypothesis (Gabaeva 1974) about secretory transformation of the follicle cells during which they produce the upper sticky part of the egg jelly coat.

The release of the oocyte from the follicle appears to be mediated by the contraction of the smooth muscle type cells in theca interna.

By the beginning of ovulation, all the oocytes in sturgeon are usually at metaphase II. However, their ovulation does not take place simultaneously; it usually begins in the lower regions of the ovary and gradually spreads to the upper regions.

The ovulated eggs find their way into the abdomial cavity of the female with abundant coelomic fluid accumulated by the beginning of ovulation. The origin of this fluid is still unknown. Since it appears long before the onset of ovulation and not as a result of the follicle wall rupture, there is no reason to call it ovarian fluid as some researchers do.

The coelomic fluid is clear and more or less viscous. It contains carbohydrates, but its chemical composition in sturgeon has not been studied. Coelomic fluid is a good medium for the maturation and ovulation of oocytes in vitro in the presence of gonadotropic hormones (Skoblina 1968), and for keeping mature ovulated eggs. In this fluid, the eggs retain their fertilizability at spawning temperatures much longer than in saline or fresh water (Ginsburg 1968, 1972).

However, oocytes do not have to be in coelomic fluid in order to complete maturation. Oocytes taken from the ovary at the stage of maturation inertia and brought to maturation in vitro in saline without ever coming into contact with coelomic fluid can be fertilized, and produce viable embryos (Dettlaff et al. 1968; Dettlaff 1970a).

1.4 Hormonal Stimulation of Oocyte Maturation and Ovulation

In sturgeon, oocyte maturation and ovulation can be stimulated by preparations of sturgeon pituitaries (a suspension of powdered acetone-dried pituitaries, glycerol extract), as well as by synthetic mammalian LH–RH and its analogs. Progesterone, which stimulates oocyte maturation in vitro, does not exert a stimulating effect when injected to the females (Goncharov 1981).

In vitro, when the follicle-enclosed oocytes are placed in a modified Ringer solution for cold-blooded animals with progesterone but without gonadotropic hormones, a high percentage but not all oocytes mature but do not always ovulate. For oocyte maturation and ovulation to be obtained in vitro in a maximum percentage of cases, the follicle-enclosed oocytes are incubated in a solution containing both gonadotropic hormones and progesterone.

Sturgeon culture needs information regarding the doses of gonadotropic hormones and LH–RH for injections, as well as of methods for testing the activity of these preparations. Injections of a suspension of acetone-dried pituitaries to stimulate the oocyte maturation in sturgeon have been used at sturgeon hatcheries for more than 50 years. At first, the doses were calculated by the number of pituitaries per spawner. However, since the pituitaries in different Acipenserid species and even different individuals of the same species vary in size, the dose was eventually given in milligrams of powdered acetone-dried pituitaries per spawner. At the same time, the empirical doses were established for individuals of different species and for different temperature conditions, i.e., females are given a larger dose of the powder in comparison with the male fish, *H. huso* is given more than *A. gueldenstaedti*, the latter is given more than *A. stellatus*, and every species is given slightly more at low temperatures than at higher ones. There has been a tendency to reduce the dose of pituitaries per injection. For example, during 1962–1964, the *A. gueldenstaedti* females at the Rogozhkino sturgeon hatchery were injected with 80 mg of powdered acetone-dried pituitaries, and the *A. stellatus* females with 60 mg during the whole breeding season. In accordance with the recommendations of Barannikova and Boev (1977), the following doses are now being administered on the lower Volga, depending on temperature conditions: 250–500 mg for the *H. huso* females, 40–70 mg for the *A. gueldenstaedti* females and 25–40 mg for the *A. stellatus* females (Boev and Artyukhin 1978), and from a third to a half of these doses for the males of these species.

In order to check the dosing of pituitary injections further, we have to know their gonadotropic activity, which is especially important when this activity differs significantly in different batches of pituitaries.

Three methods have been proposed to test the gonadotropic activity of pituitaries: with loach (*Misgurnus fossilis* L.) females used as the test object in the first method (Kazanskii and Nusenbaum 1947), with male frogs in the second one (Alpatov and Stroganov 1950; Barannikova and Boev 1977), and with the ovarian follicles of sturgeon in the third one.

With the last, the gonadotropic activity of pituitaries is assayed by the maturation of sturgeon oocytes in vitro in a Ringer solution containing different amounts of this preparation (Goncharov 1972). The method is described in detail in Appendix C. The method enables us to obtain comparable results only when the pituitary preparations are tested on follicles from the same female, since the follicles of different females differ in their sensitivity to gonadotropins. In our opinion, the third method has a number of advantages over the first two: (1) the activity of the gonadotropic hypophysial hormones is tested on the basis of their effect on the maturation of sturgeon oocytes, i.e., the effect for which a pituitary injection is carried out; (2) when this method is applied, a large number of follicles can be used for each test and, hence, statistically reliable results are obtained; (3) this method is far more sensitive (Goncharov 1972). The disadvantage of this method is that it can be used only during the breeding season.

However, when necessary, this method can also be used during other seasons if the ovarian follicles of amphibians are used as test objects (Goncharov 1971a, b), because they respond by maturation to the gonadotropic hormones from the sturgeon pituitaries.

A question regarding the effect of high doses of gonadotropic hormones on the quality of the eggs merits special attention. Though the literature sometimes indicates that overdosing of pituitaries spoils the quality of the eggs, there is no strong evidence corroborating this statement. On the contrary, we have every reason to believe that the doubling and even quadrupling of the effective dose does not have any adverse effect on oocyte maturation and does not spoil the quality of the eggs. During 1962–1964, we conducted a special study on the effect of a dose of a pituitary preparation on the timing of maturation and egg quality in *A. gueldenstaedti colchicus* and *A. stellatus* (Ginsburg and Dettlaff 1969). When the preparation was used in doses exceeding those currently used (80 mg for *A. gueldenstaedti* and 60 mg for *A. stellatus*), the females matured normally and produced good-quality eggs. No correlation was observed between the quality of the eggs and the size of the females. Two-fold differences in the dose of the pituitary preparation per unit of female body weight, even at the high doses injected, did not have any effect on either the time of maturation, or the quality of eggs.

This question regarding the doses of pituitary preparation also has another aspect. Due to the amplification of sturgeon breeding, as well as the deterioration of the spawners' response to the pituitary injection, there has been a significant increase in the number of injected fish and, therefore, an increase in the demand for pituitaries. In this respect, an economical use of pituitaries is needed. However, the dosing should on no account be reduced to the lower limits of the effects doses, since some females would not show a maturation response because of the differences in their reactivity and unpredictable fluctuations in environmental conditions. On the other hand, when low doses are used, it is necessary to correct the doses for the weight of the females and calculate them in units of activity per kilogram of body weight. Meanwhile, weighing, especially of large fish, is difficult and can injure the fish (Barannikova and Boev 1977).

All this applies to the dosing of powdered acetone-dried pituitaries given as a single injection to females with gonads at the completed stage IV of maturity. Repeated "gradual" injections are given in order to obtain eggs from females with gonads at the incomplete stage IV of maturity (Kazanskii 1957b, 1962; Molodtsov 1972; Kazanskii et al. 1978). A method of fractional injections has been proposed for spawners in a poor condition that do not react to a single injection (Barannikova and Burenin 1971; Barannikova 1975, 1978).

When using the method of gradual injections, very small doses of pituitaries are first injected, which speed up polarization of the oocytes and the transition of the gonads to completed stage IV of maturity, and 24 h later a single large dose is injected, which induces oocyte maturation and ovulation.

When using this fractional injection method, the overall dose of pituitaries is not increased, but divided in two: the smaller portion is injected first (5–8 mg in experiments with *A. stellatus*), and the larger portion (32–35 mg) 6–8 h later.

Until recently, preparations of acetone-dried sturgeon pituitaries were only used for stimulation of the spawners' maturation. These preparations contain, besides the gonadotropic hormones, other hormones which in principle can either enhance or, on the contrary, prevent the action of gonadotropins.

Obtaining the purified gonadotropin and study of its physico-chemical properties were important steps in the investigation of the role of hypophysial hormones in the control of reproductive function in sturgeon (Burzawa-Gerard et al. 1975a, b, 1976; Goncharov et al. 1976, 1980, 1983; Zenkevich and Latse 1979; Zenkevich et al. 1981; Goncharov and Kuznetsov 1981; Kuznetsov et al. 1983).

In addition to basic scientific interest, these studies were important for sturgeon culture. For example, it was established that the purified gonadotropic hormone of *A. stellatus* can induce in vivo maturation of gametes (Barannikova et al. 1981; Goncharov et al. 1981). This preparation did not have apparent advantages in its action over the preparation of acetone-dried pituitaries (Barannikova et al. 1981), but standardization of its activity was more reliable. However, under the present conditions of hormone purification, a considerable part of the hormone is inevitably lost, and this makes its use in sturgeon culture impractical.

Studies on the involvement of hypothalamic gonadotropin releasing factors in the control of sturgeon reproduction have shown that it is possible to use mammalian LH-RH (Barannikova et al. 1984, 1985, 1989; Barannikova and Bukovskaya 1984) and its superactive analogs for stimulation of gamete maturation (Goncharov 1984a, b, 1985, 1990; Igumnova and Goncharov 1986; Goncharov et al. 1989, 1991).

Both preparations were, at least, as effective as pituitary preparations and the quality of gametes obtained with them did not differ from that obtained with traditional method of pituitary injections.

Superactive analogs have an advantage over LH–RH in that response of both females and males can be induced by a single dose of the preparation and at a far lower concentration. Extensive hatchery testing of the synthetic analog of LH–RH on three main species of sturgeon culture in the USSR, *H. huso*, *A. gueldenstaedti*, and *A. stellatus*, have shown its high efficiency at a dose of about 1 μg/kg body weight.

In the course of experiments and hatchery testing, some peculiarities in the action of this preparation were discovered. First, even at doses exceeding the threshold by more than one order of magnitude no overdosing was observed. Second, the mean time from the injection to *A. gueldenstaedti* and *A. stellatus* females until ovulation was longer by 11 and 14%, respectively, than after the injection of pituitary preparation. No such difference was noted for *H. huso* (Goncharov et al. 1991).

Unlike the three above mentioned Acipenserid species, *A. ruthenus* spawners are much more heterogeneous in their sensitivity to the LH–RH synthetic analog. A single injection of this preparation at a dose of about 1 μg/kg body weight induced response in a much lesser number of females than the pituitary injection (Goncharov et al. 1991). The efficiency of the LH–RH analog markedly increased with the dose or when two injections were made with an interval of 12 h (Goncharov et al. 1991). The sensitivity of another representative of river-resident sturgeons, *A. baeri*, to a single injection of the LH–RH was also low. Simultaneous injection of sulpiride, an antagonist of dopamine D_2-receptors, did not increase the efficiency of the LH–RH analog action on the *A. baeri* females (Goncharov, pers. comm.), although synergism of these preparations was observed in experiments on various teleostean species (van Asselt et al. 1988; Motlokh et al. 1989).

Hence, LH–RH analogs could be effective and highly practical substitutes of sturgeon pituitaries in the culture of three main species of anadromous sturgeon.

1.5 Methods of Working with Sturgeon Spawners Migrating in Rivers at Different Seasons and Spawning During the Same or Following Year

The first advances in using various groups of Acipenserids for hatchery reproduction were already made by the beginning of the 1950s while working with the hiemal *A. stellatus* and late vernal *A. gueldenstaedti* of the summer and autumn migration in the Kura river (Gerbil'skii 1951; Kazanskii 1953b). In subsequent years, biological differentiation of Acipenserid fish, time of spawning, and possibility of obtaining good-quality eggs from females in different seasons of their reproduction have been actively studied (see Barannikova 1957; Kazanskii 1962, 1963; Popova 1973). The last problem is considered to be practically solved for all types of migrants of the Acipenserid fishes (Kazanskii and Molodtsov 1973, 1974; Kazanskii 1979). In order to introduce various groups of Acipenserids in fish culture and extend the period, during which the good-quality gametes are obtained, spawners are kept for more or less long period in holding tanks with regulated temperature (Kazanskii production line). Various conditions are made for fish from different groups to acquire (or retain) the ability to respond to pituitary injection by giving mature gametes. Different variants of the method of pituitary injections are also used to stimulate maturation of the fish from different biological groups (see below).

1.5.1 Vernal Groups of Acipenserid Fish

Vernal spawning *H. huso*, *A. gueldenstaedti*, *A. stellatus*, and *A. nudiventris* were the main objects of hatchery reproduction in the lower flow of the Volga, Kura,

Don, and Kuban' before their regulation. Spawners of these groups migrated the first in the rivers and responded well to pituitary injections. After the flow had been regulated, the hydrological regime of the rivers underwent sharp changes. In spring, due to delay or, often, untimely high water discharge via dams, the normal course of migration of the Acipenserid fishes was disturbed. Selectivity of migration of spawners with gonads at the completed stage IV of maturity toward the high water current decreased, the physiological condition of fishes migrating in rivers in spring and the structure of spawning populations changed (Barannikova 1972a, b, 1979). For example, in May 1973 at the time of the mass migration of *A. stellatus* in the Kuban', 60% of the females had gonads at incomplete and only 40% at the completed stage IV of maturity, and in June there were also females with gonads at stage III–IV of maturity (Badenko et al. 1976). In low water years, the number of fishes migrating in rivers, especially in the Kuban', Don, and Kura, is sharply reduced. In this respect, a question was raised about using, for hatchery purposes, *A. stellatus* spawners caught in pro-estuarine sea regions and from sea catches (Mailyan and Alekperov 1977).

The total decrease and extended migration of the spawners in rivers, as well as partial overlapping of migration times of representatives of different biological groups in rivers led to heterogeneity of spawners taken at fishing places for hatchery reproduction, thus decreasing the efficiency of fish culture. For example, in 1975–1976 at the sturgeon hatcheries in the lower Volga flow, 78% of *A. gueldenstaedti* females and 59% of *A. stellatus* females responded to pituitary injections (Ramanov 1979).

Only those hatcheries which are located higher up the river flow are in a better position. For example, in the region near the dam of the Volgograd hydroelectric station (in the region of *A. stellatus* spawning grounds), about 9% of females were ripe (with eggs leaking out) and the others had gonads at the completed stage IV of maturity (Veshchev 1979). Correspondingly, at the Volgograd sturgeon hatchery, practically all *A. stellatus* females responded to pituitary injections by giving good-quality eggs. Downstream of the Fedorovskii dam on the Kuban' river, *A. stellatus* females gather in May; most of them (90%) have gonads at the completed stage IV of maturity (Badenko et al. 1976). Construction of artificial spawning grounds in high-water years provides conditions for natural spawning of *A. stellatus* (Barannikova et al. 1979).

Hence, while working with spring-spawning sturgeons at hatcheries in the lower flow of the regulated rivers, it is very important to be able to select among the females, which differ in their preparedness for spawning, those which are able to respond to the pituitary injection by giving good-quality eggs. There are no convenient and safe methods which allow us to do this. The size-weight criteria of estimation of the females (Badenko 1967; Badenko et al. 1971, 1972) only allow selection of a group of fish with the mean size-weight indices in which the probability of maturation is higher than in the extreme variants.

In order to determine the condition of oocytes and the stage of maturity of gonads, an express method is applied which consists in taking follicle-enclosed oocytes from the ovary with a probe, fixing with boiling, and cutting along the animal-vegetal axis (Dettlaff et al. 1965; Kazanskii et al. 1978). When the ovaries pass from the incomplete to the completed stage IV of maturity, the GV moves towards the animal pole, in the area of fine-granular yolk. The distance between the GV and the membrane in the animal pole region becomes equal to or less than half of the GV diameter or 0.07 of the greater oocyte diameter (see Appendix A). Hence, the spring-spawning fish can be distinguished from the vernal summer-spawning and hiemal ones.

In a group of spawners with gonads at the completed stage IV of maturity, there are females which do not respond to pituitary injection or give poor-quality eggs. Some of them can be rejected using the method of estimating the degree of maturity proposed by Goncharov (1976). Oocytes taken with a probe are placed in a Ringer solution with progesterone and, after 18 τ_0, cut and examined for presence of the GV. The females in which the GV is retained later than 18 τ_0 should not be used for hatchery purposes (see Appendix B).

Besides, using the samples taken with a probe, it is possible to reject those females, whose oocytes have signs of degeneration, such as disturbance of pigment pattern (Faleeva 1979).

In order to increase the efficiency of working with vernal spring-spawning females and extend the period, during which the eggs can be obtained from them, Kazanskii (1962, 1963, 1975) proposed catching fish at the very beginning of their migration at temperatures below the spawning ones, placing them in thermo-regulated basins with running water and keeping at a temperature of 2–3 °C for different periods of time, up to three months. A case has been described when good-quality eggs were obtained from *A. stellatus* spawners caught in the river already at spawning temperatures (Molodtsov 1975).

By the time when gametes are needed, temperature in the holding tanks is gradually (by 1–2 °C/day) increased up to the spawning ones and the females and males are injected with a suspension of acetone-dried pituitaries. At hatcheries where there are fish tanks and thermo-regulated basins (Kazanskii production line), for example, at the Aleksandrov and Bertyul hatcheries, the second cycle of fish culture if provided with material obtained in this way (Kazanskii 1979).

1.5.2 Hiemal Groups of Acipenserid Fish

These groups include hiemal *A. gueldenstaedti* of summer and autumn migration in the Volga, hiemal *A. stellatus* in the Kura, Volga, Don, and Kuban', and hiemal *H. huso* in the Volga and Ural. The hiemal *A. gueldenstaedti* of the summer migration accounts for more than 50% of the whole *A. gueldenstaedti* stock in the Volga. It is quite obvious that its introduction in sturgeon.culture is very important.

There are different methods of obtaining eggs from hiemal sturgeons: ecological one combined with one pituitary injection and the method of repeated gradual injections of pituitary suspension.

When using the first method, the normal conditions of wintering are reproduced. The hiemal fish are placed in autumn in wintering ponds (Gerbil'skii

1951; Barannikova 1954), in thermo-regulated basins in which temperature is kept at 2–3 °C (Kazanskii and Molodtsov 1974), or in basins with non-regulated temperature in which the spawners winter at the same temperature as in the river (Molodtsov and Meshcheryakov 1972; Molodtsov 1979). Under these conditions, rapid growth of the oocytes takes place and, within 9 months, the gonads reach the completed stage IV of maturity. In the experiments carried out by Molodtsov, after gradual increase in the temperature in the basins from 2–3 °C to the spawning ones and 5 days of acclimation at 18–19 °C, the females responded to pituitary injection by simultaneous maturation and giving good-quality eggs which developed into viable young fish. The females, which wintered in basins with the same temperature as in the river, also matured well.

The eggs were first obtained from hiemal females of *A. stellatus* by combining long-term keeping of spawners and subsequent pituitary injection in 1950 by Kazanskii (Gerbil'skii 1951). The hiemal *A. stellatus* is used in sturgeon culture in the Kura (Mamedov 1977) and Volga (Molodtsov 1978).

Good quality eggs have also been obtained from hiemal *H. huso* in the Volga delta after long-term keeping (for 180–220 days) in wintering ponds and special basins for long-term keeping of spawners (Popova 1978).

The transition of the gonads from incomplete to the completed stage IV of maturity can be stimulated not only by means of long-term keeping of females, but also for a relatively short time by means of repeated gradual injections of pituitary suspension. For example, the females of hiemal *A. gueldenstaedti* with gonads at incomplete stage IV of maturity were injected with 15 mg of pituitary preparation two or three times and, after 24 h, with 40 mg more. Probe samples have shown that polarization of the oocytes was enhanced as a result of the first injections, while the last injection induced their normal maturation and ovulation (Kazanskii et al. 1978).

1.6 Introduction in Culture of Various Acipenserid Species

We have already mentioned that one of the main tasks of the modern sturgeon culture consists in preservation and propagation of the entire diversity of Acipenserid species, as well as enlargement of their area (Barannikova et al. 1979). Nowadays, *H. huso*, *A. gueldenstaedti*, *A. stellatus*, and, to a much lesser scale, *A. persicus* of the Kura and *A. nudiventris* are the main objects of hatchery reproduction. Good-quality eggs are regularly obtained from the *A. persicus* females in the Volga. Perspectives of its use in sturgeon culture are discussed (Barannikova 1957, 1970; Artyukhin 1979). There is experience of hatchery reproduction of *A. schrencki* and *H. dauricus* (Svirskii 1967, 1970, 1971) but their stock in the Amur is very small. Therefore, it was considered necessary to prolong closure of sturgeon fishing in this river, which has already lasted for 18 years (Krykhtin 1979).

Recently, *A. baeri* and, especially, its population in the Lena river attracted great interest (Sokolov 1965; Berdichevskii et al. 1979). The Lena sturgeon are very resistant to unfavourable conditions at all developmental stages and, in water bodies of the European part of the USSR, grow rapidly at temperatures higher than in the Lena. This fish is considered one of the most promising for commercial sturgeon culture in these waters.

Oocyte maturation in *A. baeri* was stimulated by two injections of pituitary suspension, with a high percentage of fertilization (97–99%). The technique of transportation of the embryos at stages of the 7th–8th cleavage divisions on nets in a humid medium at a temperature of 7–8 °C was developed. The mortality rate was less than 10% for two days of transportation and 30% for the whole subsequent incubation (Nikol' skaya and Sytina 1978b; Berdichevskii et al. 1979).

Despite its wide distribution, undemanding character, and great plasticity (presence of semianadromous, river-resident, and lake-river forms), *A. baeri* has not yet become the object of sturgeon culture due to the small number of sexually mature fish. The abundance of the Siberian *A. ruthenus* is still lower (Votinov and Kas'yanov 1978, 1979). In order to improve this situation, it is necessary to control the catch of *A. baeri* and *A. ruthenus* in Siberia and, especially, to prevent catching sexually immature fishes.

Special attention should be given to the preservation of and increase in numbers of *A. sturio*, since representatives of this species are characterized by a very important (in contemporary conditions) ability to live in water of high (oceanic) salinity. At present in the USSR only a small population of this species exists in the Rioni river basin where the construction of hydroelectric power station affects the conditions necessary for its natural spawning (Ninua 1976). In this respect it is proposed to use all caught spawners of *A. sturio* for artificial reproduction (Barannikova et al. 1979).

The necessity of hatchery reproduction of *A. ruthenus* has been repeatedly stressed and the expediency of its utilization for commercial fishery has been underlined (Persov 1963; Kozhin 1970; Shilov and Khazov 1971, 1979). After the Volga flow was regulated and the Cheboksary and Nizhnekamskii hydroelectric stations were built, only a small part of the spawning grounds of this species remained and the conditions for its natural spawning in the Volga were seriously affected. The mass catching of males for production of hybrids with *H. huso* seriously restricts the natural reproduction of *A. ruthenus*. The problems of construction of a hatchery for reproduction of *A. ruthenus* and its introduction into inland water bodies and water reservoirs where it will fatten are now being discussed. Its natural spawning is also possible in rivers flowing into these reservoirs (Bratsenyuk 1974; Lukin 1979).

At the same time, in the regions where *A. ruthenus* has common spawning grounds with the other Acipenserid species and its abundance is high, a suggestion of Ginzburg (1967) concerning the partial catch of these fish during their migration to the spawning grounds in spring, before the onset of spawning of *H. huso* and *A. gueldenstaedti*, is still valid. In May 1963, 900 to 9750 eggs of *H. huso* and *A. gueldenstaedti* were found in the stomachs of *A. ruthenus* individuals (Ginzburg 1967). It is evident that decrease in the population density of *A. ruthenus* in the spawning grounds will reduce the intensity of egg eating.

The eggs of *A. ruthenus* were obtained by the use of pituitary injections by Persov (1957); he organized a station for artificial propagation of this species in the region of Rybnaya Sloboda, on the Kama. Later, the eggs of *A. ruthenus* were repeatedly obtained and incubated at the Ikryanino and Volgograd sturgeon hatcheries using the biotechnique developed for the other Acipenserid species (Bykov and Shilov 1974; L'vov 1976; Shilov and Khazov 1979).

First generation hybrids between *H. huso* and *A. ruthenus*, "bester" (abbreviations of the Russian names of the respective species: "**be**luga" and "**ster**lyad'"), have recently become one of the main objects of commercial sturgeon culture; these fish rapidly gain the commerical weight (Burtsev 1967; Nikolyukin and Burtsev 1969; Romanycheva 1976, 1979). Second generation hybrids cannot be used in commercial sturgeon culture because of their high mortality (see Berdichevskii et al. 1979). Hence, besters are now obtained at the expense of reduction of the bulk of artificial reproduction of both *H. huso* and *A. ruthenus*. Construction of a special hatchery and complete isolation of water bodies used for rearing besters (to avoid their mating with pure Acipenserid species) are necessary for the production of besters not to interfere with the artificial reproduction of pure Acipenserid species (Berdichevskii et al. 1979). The importance and admissible bulk of bester production are now being re-estiamated, since the harm inflicted to the reproduction of the main Acipenserid species is not compensated by the profit from the industrial bester production. This is stressed by the decisions of some meetings of experts and the Scientific-Consulting Council affiliated to the Ichthyological Commission, USSR Ministry of Fisheries.

Studies of vitellogenesis, developmental defects in embryos and prelarvae, chromosome abnormalities in hybrids of various generations are of undoubted interest (Burtsev 1967, 1969, 1983; Krylova 1970; Burtsev and Serebryakova 1980).

Three Middle-Asian species of *Pseudoscaphirhynchus* are at the brink of disappearance. As a result of economic activity of the man, life conditions of these species in the Amu Darja and Syr Darja markedly deteriorated. These species have been introduced into the Red Book of the USSR (1984) in the first category species, i.e., those at the brink of disappearance. Now, methods of keeping spawners and reproduction of *Pseudoscaphirhynchus kauffmani* in captivity are being developed with the aim of producing populations maintained in captivity (Goncharov et al. 1986; Shubravyi et al. 1989).

1.7 Influence of Environmental Conditions on the Capacity of Females to Respond to Pituitary Injection and the Hatchery Quality of Eggs

Besides the initial condition of the females, the conditions under which they are kept before pituitary injection and during maturation markedly affect the result of injection and the hatchery quality of the eggs. These conditions include water flow, oxygen content of water, and temperature. Here, we deal mainly with the influence of temperature on the condition of females and their response to gonadotropic hormones, as well as that of unfavorable temperatures on ovulation and egg quality.

We have already mentioned that the spawners which do not respond to pituitary injection are sometimes kept at the spawning temperatures and, in this way, from some of them eggs suitable for hatchery reproduction are obtained.

The females initially capable of responding to pituitary injection by oocyte maturation react to being kept at the spawning temperature in a completely different way. Already short-term (during two-four days) keeping of such females in fish tanks at the spawning temperatures before the pituitary injection leads to the loss of their capacity to respond to the effect of gonadotropic hormones by oocyte maturation. The higher the water temperature, the more rapidly the females lose this capacity. Unfavorable temperatures exert on the females an effect similar to that of long-term keeping. Some data about the mechanism of these influences and their possible reversibility have recently been obtained. They are considered in more detail in the next Section.

1.7.1 Influence of Unfavorable Temperatures and Keeping of Females at Spawning Temperatures on the Transition of Oocytes to Maturation Under the Influence of Hypophysical Gonadotropic Hormones

Both during natural reproduction in the river and after the pituitary injection into the females at the sturgeon hatchery, the oocytes begin to mature only at spawning temperatures. The females of *A. stellatus* cease, as a rule, to respond to pituitary injection by oocyte maturation at water temperatures below 12–13 °C or above 25–26 °C. What does this depend on? The answer to this question was obtained in studies of the influence of various temperatures on the capacity of ovarian follicles to respond to hypophysial gonadotropic hormones and progesterone in vitro in saline (Davydova 1972; Dettlaff and Davydova 1974, 1979b). Follicles taken from one female were incubated in a Ringer solution with hypophysial gonadotropic hormones or progesterone and in a hormone-free solution at temperatures below the optimum (9–10 °C), optimal temperatures (17–18 °C), and temperature above the optimum (28–30 °C). Within a period of time comparable for different temperatures and measured by the same number of τ_0, the follicles incubated at the low and high temperatures for 2–10 τ_0 were transferred under the optimal temperature conditions. Those follicles which were incubated in the presence of hormones were placed in a hormone-free Ringer solution and vice versa. In all variants with the use of progesterone, the oocytes matured at all tested temperatures and some of them ovulated. The ovulated eggs were inseminated and viable prelarvae developed from them. Contrary to this, in the experiments with the use of pituitary suspension, only those oocytes matured which were incubated at the optimal temperatures. The follicles which were incubated at 8–10 °C or 28–30 °C and subjected to the influence of gonadotropic hormones both during incubation at unfavorable temperatures and after the transfer in the optimal temperature conditions, did not respond to the effect of the pituitary hormones: all oocytes remained at the initial stage and retained a large GV.

The same results were obtained when the whole female was cooled rather than the ovarian follicles isolated in saline. After rapid cooling of the female (decrease in water temperature from 20–25 °C to 4–8 °C), keeping it at the low temperature for 2–4 h, and returning to the initial temperature conditions, the follicle-enclosed oocytes did not respond to the effect of hypophysial gonadotropic hormones by maturation. This was true both for pituitary injection to the female and the addition of this suspension to the medium in in vitro experiments. At the same time, the oocytes of such females matured in a Ringer solution with progesterone.

Similar experiments were carried out with *A. stellatus* females kept in fish tanks for three to seven days at the spawning temperatures. Oocytes were taken every day from these females with a probe; some of them were incubated in a Ringer solution with pituitary suspension and the others in a solution with progesterone. In the former solution, the oocytes ceased to mature already after 3 to 4 days, while in the latter they matured even after the females were kept in the tanks for seven days. Hence, the loss by the females of the ability to mature in response to pituitary injection as a result of the effect of unfavorable temperatures or keeping at the spawning temperatures is explained by the suppression of the ability of the follicular epithelium cells to secrete progesterone (a progesterone-like substance) and, thereby, induce oocyte maturation.

It is important that the ability of the follicular epithelium cells to respond to the effect of gonadotropic hormones can be restored under certain conditions. This was achieved experimentally by injecting the thyroid gland hormone, triiodothyronine into the held-back or cooled females. These females, after they received the pituitary injection, gave ripe eggs which, in a few cases, were used for hatchery incubation and developed into viable prelarvae (Dettlaff and Davydova 1974, 1979b). However, it should be stressed that, according to our data (Dettlaff and Davydova 1979a), triiodothyronine does not affect the hatchery quality of the eggs and does not relieve the, already begun, phenomena of degeneration.

Now we can answer the question asked at the beginning of this Section: how does the maturation of females during natural spawning and after pituitary injection occur only at the spawning temperatures? The reason (during natural spawning, probably, not the only one) is that at temperatures beyond the limits of the optimal zone the ability of the follicular epithelium cells to respond to gonadotropic hormones by stimulation of oocyte maturation is reversibly suppressed. The fact that, after being suppressed under the influence of unfavorable temperatures, it can be restored is very important, since it determines the possibility of oocyte maturation after transition of the fish from the low or high temperatures again to the optimal spawning temperatures. In these experiments, we succeeded in restoring the ability of the follicular epithelium cells to respond to the effect of hypophysial gonadotropic hormones by injecting triiodothyronine into the held-back or cooled females. It is noteworthy that the stimulating effect of the thyroid gland hormone, thyroxin, on maturation of the *A. stellatus* females under the influence of hypophysial gonadotropic hormones was earlier shown by Kalashnikov and Skadovskii (1940). It cannot be excluded that thyroid gland hormones are involved in preparing the transition of fish into the spawning condition under the normal conditions also. This is corroborated by coordination of the active period of the thyroid gland with the spawning migration and spawning (Skadovskii 1949; Ivanova 1954).

1.7.2 Influence of Unfavorable Temperatures on Oocyte Ovulation and the Quality of Eggs

The influence of unfavorable temperatures in the second half of the oocyte maturation period markedly differs from that during transition of the oocytes to maturation. If, during the transition to maturation, the follicular epithelium cells are most sensitive (the oocytes do not display signs of damage), at the end of the maturation period the oocytes become very sensitive to the effect of unfavorable temperatures. Also, the normal relationship between the maturation and ovulation of the oocytes is often disturbed. Under unfavorable conditions, firstly those of temperature, the coordination between the maturation and ovulation of the oocytes is easily disturbed and, as a result, immature oocytes can ovulate and the mature eggs can be retained in the follicles, which leads to their damage. Obtaining the eggs under unfavorable temperatures leads to a sharp decrease in the fish breeding quality of the eggs and large industrial losses.

A study of the influence of temperature during the second half of the oocyte maturation period is largely facilitated by the fact that, after maturation inertia sets in, the oocytes are able to complete maturation in saline. This makes it possible to study the influence of various temperatures on maturation and ovulation of the oocytes of the same female in vitro and, then, compare the results obtained with the pattern of maturation and ovulation of the oocytes in vivo in various females injected at different temperatures.

The females were injected with a suspension of acetone-dried pituitaries and, after the maturation inertia set in, pieces of the ovaries were removed. Small pieces of the ovaries of each female were placed in a Ringer solution at different temperatures for completion of maturation and ovulation. As the ovulation proceeded, the oocytes were inseminated and their development was followed.

Figure 10 demonstrates the results of one such experiment with the oocytes of two females of the spring-spawning *A. gueldenstaedti colchicus* from the Don and one female of autumn-spawning *A. persicus* from the Kura (for details see Dettlaff 1970a). In these experiments the percentage of ovulation, fertilization, and activation of the unfertilized eggs were estimated. It is noteworthy that the

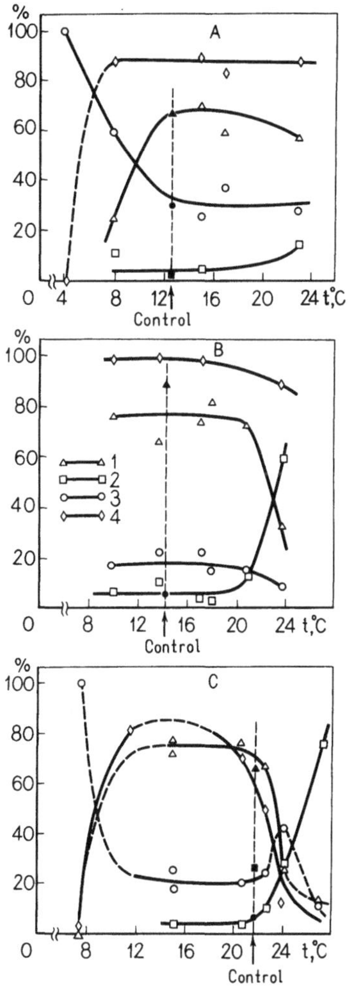

Fig. 10A–C. Maturation of the *A. gueldenstaedti* and *A. persicus* oocytes in vitro in a Ringer solution (Ginsburg and Dettlaff 1969). **A, B** *A. gueldenstaedti*; **C** *A. persicus*. *Ordinate*, percentage of fertilized (*1*), unactivated (*2*), and activated (*3*) eggs, as well as ovulated eggs (*4*); *abscissa*, temperature, °C. The same indices for the eggs matured in vivo serve as the control

behavior of the oocytes of spring- and autumn-spawning females at identical temperatures is very similar.

Temperatures below the optimal ones (4–7°C; Fig. 10A, C) reversibly suppress ovulation but do not prevent the completion of maturation of the oocytes. They also exert an activating effect on the matured oocytes and induce them to parthenogenetic development inside the follicles. If the oocytes which completed maturation at 4°C are rapidly transferred to spawning temperatures, they ovulate synchronously but only those are fertilized which were not activated during incubation at the low temperatures. A somewhat higher temperature (8 °C; Fig. 10A) does not prevent ovulation but, among the ovulated oocytes, there are still many activated eggs incapable of fertilization. Here, we encounter early activation of the eggs which is one of the main causes of the loss of their fertilizability (for details, see Dettlaff and Ginsburg 1954).

In a range of temperatures from 10 to 20–21 °C, the greatest number of oocytes normally complete maturation, ovulate and are fertilized. Such oocytes develop into normal prelarvae and feeding larvae.

At temperatures of about 24 °C and higher, the oocytes are markedly damaged: the percentage of fertilization is sharply reduced and a group of non-fertilized (largely, non-activated) eggs increases. Among the fertilized eggs many develop abnormally. At the same time, the percentage of ovulated eggs also decreases.

Similar changes are observed in the oocytes maturing in vivo. When the temperature in fish tanks is raised to 25–26 °C, the oocytes are damaged and ovulation is arrested. During natural spawning, such a rise in temperature leads to the females leaving the spawning grounds and lying in depressions in the river bed (Derzhavin 1947; Dettlaff 1970a).

The results of this experiment with the oocytes of spring- and autumn-spawning sturgeon suggest that the limits of the zone of temperatures favorable for the completion of oocyte maturation and the ovulation do not depend on the breeding season. This suggestion as if contradicts the experience of fish culturists, since the good-quality eggs are usually obtained from spring- and autumn-spawning sturgeons at different temperatures. However, there is no contradiction whatsoever. It is known that long-term keeping of females at the spawning temperatures causes the egg quality to deteriorate. Therefore, in autumn-spawning sturgeon which start spawning after the water temperature decreases to the upper limit of the zone of favorable temperatures, keeping reduces the egg quality earlier than the lower limit of the zone of spawning temperatures is achieved. In spring-spawning *A. gueldenstaedti*, and *H. huso* and *A. stellatus*, a keeping at spawning temperature results, on the contrary, in a decrease of egg quality before the upper limits of the zone of spawning temperature are achieved.

The results of these experiments on the influence of various temperatures on the ovarian follicles incubated in saline agree quite well with the data about their changes in vivo under the influence of low temperature (Dettlaff and Davydova 1978). The *A. stellatus* females with gonads at the completed stage IV of maturity were injected with a suspension of pituitaries and kept at the normal spawning temperatures (17.5–21 °C) before the oocytes acquired the maturation inertia. The females were then subjected to the influence of low temperature at different times during the period of maturation inertia: water temperature was decreased by 11–15 °C, the female was kept at 3–6 °C for 1.5–3 h and, then, the temperature was rapidly increased by 12–17 °C. When this treatment was applied shortly before the GV breakdown in the oocytes, the oocytes normally matured and ovulated. The eggs of these females were used for hatchery incubation and viable prelarvae were obtained. When cooling females with the oocytes at prometaphase and metaphase I, the fraction of ovulated oocytes decreased in some of them. Cooling at a later stage, before the onset of ovulation of the oocytes or in the beginning of their ovulation, led to more or less complete suppression of ovulation. In two latter cases the percentage of fertilization and capacity for normal development were reduced among the ovulated oocytes. Unfavorable temperature influence at the final stages of maturation resulted in changes which fish culturists call "hardening" of the female. The data obtained suggest that the decrease of temperature below the spawning ones in the end of maturation inhibits ovulation and damages the oocytes.

Comparison of the data about temperature limits of natural spawning (see Sect. 4.2), maturation of the stimulated females and maturation of the oocytes of one female in vitro at different temperatures suggests that at the end of the period of oocyte maturation, when the oocytes are especially sensitive to unfavorable conditions, it is essential that water temperature in tanks with spawners be within the limits of the zone of spawning temperatures.

1.8 Release of Eggs During Natural Spawning, Time of Oocyte Maturation and Ovulation and Time of Obtaining Eggs at Hatcheries

During natural spawning, in the presence of a male, the female does not hold eggs: as ovulation proceeds, portions of eggs together with the coelomic fluid are released in water. Active movements of the female enhance ovulation of the oocytes, their steady release through a sexual pore and their dispersion in water. In females with gonads at stage V of maturity (ripe with the eggs leaking out) caught at spawning grounds, there are, as a rule, few ovulated eggs in the body cavity.

In the experiments of O. B. Chernyshev (see Sadov 1957; Kazanskii 1957a) and P. S. Yushchenko (pers. comm.), the females matured after pituitary injection and, placed in a holding tank with a male, behaved in a similar way: they did not hold eggs in the body cavity and started spawning which lasted a few hours.

At hatcheries, females and males injected with pituitary suspension are kept separately, in different tanks. Under these conditions, the female does not display spawning behavior and the ovulated oocytes accumulate in the body cavity; only a few of them are released in water. By the time the ovulation is completed, the oocytes ovulated at the beginning of this process are already damaged.

If all the ovulated eggs are periodically stripped, as was done in the experiments on *A. ruthenus* (Persov 1957), the eggs from successive portions are fertilized equally well. If the eggs are kept in the body cavity and taken at different intervals after the onset of ovulation (experiments on *A. gueldenstaedti* and *A. stellatus*—Vernidub and Kiseleva 1953; Vernidub 1957, and unpublished data of the authors), the percentage of fertilization rapidly decreases. The fraction of non-activated eggs incapable of fertilization increases. The last portions of eggs are damaged and contain empty envelopes from which the contents leaked out ("broken eggs"). Such changes of eggs in the body cavity are sometimes observed even before the ovulation is completed; the oocytes "slipping down" at this time from the ovary (i.e., just ovulated) still give a high percentage of fertilization. Hence, inside the follicles the oocytes are better preserved than in the body cavity. However, it should be borne in mind that the broken eggs result not only when the ovulated oocytes are kept too long in the body cavity, but also when the degenerative changes began in the oocytes already at the initial stage of maturation (Faleeva 1970).

When the eggs are gradually released, the extended ovulation of the oocytes is an important adaptation due to which no long-term delay of the ovulated eggs in the body cavity takes place during spawning. At hatcheries all the mature eggs are taken at one time. In these conditions, the extended ovulation of the oocytes loses its adaptive importance and creates certain difficulties which should be overcome by fish breeders.

The time for taking the eggs is chosen correctly, when some eggs have already ovulated and, with coelomic fluid, are in the oviducts and in the body cavity and the others easily "slide off" from the ovary, i.e., ovulate when the female is hung vertically before cutting the abdomen and collecting the eggs. At this stage the rupture of follicles which hold the remaining oocytes is well prepared and the minimal effort is needed for their liberation. Experience of sturgeon hatcheries, in particular Rogozhkino and Volgograd ones, and special studies (Vernidub 1957; Kazanskii 1957a; Dettlaff and Zubova 1962; Dettlaff et al. 1965; Igumnova 1974) suggest that the eggs taken from the females in such a condition are, as a rule, highly fertilizable and develop normally.

Thus oocyte maturation in the ovary by the moment of mass ovulation and preparedness of the follicles for rupture make it possible to obtain all the eggs at one time before the ovulation is completed.

Let us consider the duration of the period from the moment of hormonal stimulation of oocyte maturation to the onset of their maturation and ovulation.

In the preceding sections it was shown that the quality of eggs depends not only on the initial condition of females and temperatures at which they are kept after pituitary injections but also on a timely obtaining of eggs. In order to determine when the females should be examined for obtaining eggs of good quality it is important to study maturation of females of various Acipenserid fish as a function of temperature (Persov 1957; Dettlaff and Zubova 1962; Dettlaff et al. 1965; Semenov 1965; Igumnova 1974, 1975b).

The period from injection of a pituitary suspension to females to obtaining ripe eggs from them is called the period of maturation (τ_m). Its duration depends mainly on the temperature at which the spawners are kept after the pituitary injection, as well as on the initial condition of the females.

Comparison of the durations of the periods of maturation and embryogenesis has shown that at the same mean temperatures the former is much shorter than the latter and, in the females of *A. gueldenstaedti*, *A. stellatus*, and *H. huso* matured at the low spawning temperatures, is 0.17–0.23 of the duration of embryogenesis (Dettlaff and Zubova 1962; Dettlaff et al. 1965).

Figure 11 demonstrates the distribution, according to the time of maturation, of 210 females of *A. gueldenstaedti colchicus* and 87 females of *A. stellatus*. These observation were performed at the Rogozhkino sturgeon hatchery in 1962–1964, i.e., before the Don flow was regulated, when the spawners migrated from the sea to the river high water. The relative duration of the period of maturation was expressed in fractions of the duration of embryogenesis, period from egg insemination to hatching of prelarvae (τ_{VII}) (Dettlaff and Dettlaff 1961; Dettlaff and Zubova 1962). For *A. gueldenstaedti* a symmetrical curve was obtained with a peak at 0.19 τ_{VII} and equally and steeply descending shoulders. Seventy percent of all females matured in a range of 0.19 $\pm$ 0.01 τ_{VII} and 88% in a range of 0.19 $\pm$ 0.02 τ_{VII}. A slightly asymmetrical curve was obtained for *A. stellatus*: 78% of females matured in a range of 0.19 $\pm$ 0.02 τ_{VII} and a group of females matured at later times (0.22–0.24 τ_{VII}) increased as compared with *A. gueldenstaedti*. But the general pattern of distribution of the females according to the time of maturation is similar in these species. This allowed us to plot graphs of probable times of maturation of the females of *A. gueldenstaedti*, *A. stellatus*, and *H. huso* based on the data about the duration of embryogenesis (Dettlaff et al. 1965).

These calculations were based on materials obtained at the Rogozhkino and Volgograd sturgeon hatcheries on spring-spawning fish which responded well to

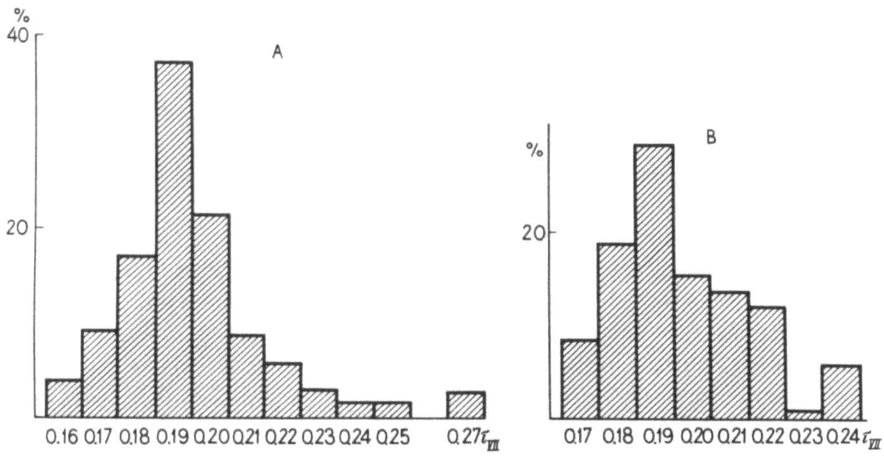

Fig. 11A, B. Distribution of females of *A. gueldenstaedti* (**A**) and *A. stellatus* (**B**) according to relative duration of their maturation (Ginsburg and Dettlaff 1969). *Ordinate*, percentage of females; *abscissa*, duration of the period of maturation (τ_m) in fractions of the total duration of embryogenesis (τ_{VII})

pituitary injections (Dettlaff et al. 1965; Ginsburg and Dettlaff 1969). However, it was soon discovered (Semenov 1965) that at higher temperatures some females of *A. stellatus* and *A. gueldenstaedti* mature at later times than predicted by our graphs. The number of females maturing at later times has recently increased, in accordance with the increase in the number of spawners used for sturgeon culture. In addition, at the hatcheries in the lower Volga flow *A. stellatus* females mature for a markedly longer time than predicted by the graphs in the whole range of spawning temperatures. As a result the problem of duration of maturation of the *A. stellatus, A. gueldenstaedti,* and *H. huso* females was reinvestigated.

In this book we provide more precise graphs for determining the time of maturation of the females of *A. gueldenstaedti, A. stellatus, H. huso,* and *A. ruthenus* after pituitary injection at different temperatures (see Fig. 78). When plotting these graphs we took into account additional data obtained at the Volgograd sturgeon hatchery concerning the time of maturation of the females of *A. stellatus* and vernal *A. gueldenstaedti* provided by L. V. Igumnova and the data about the time of maturation of females of *A. stellatus* and hiemal *A. gueldenstaedti* at the Temryuk sturgeon hatchery in the Kuban' delta in 1978–1980 and at the Sergiev sturgeon hatchery in the Volga delta in 1979 and 1980.

The graphs have been plotted on the basis of recalculation of all available data in the number of τ_0 (see Sect. 1.2.3 and Fig. 42), rather than in fractions of τ_{VII}, as was done before. This method of calculation made it possible to extend the curves in the zone of lower spawning temperatures for *H. huso*, for which there are no data about the value of τ_{VII}, and make more precise the position of the curve in the zone of higher temperatures for *A. gueldenstaedti*. New graphs have been plotted for *A. stellatus* which take into account a wider range of times of their maturation (Igumnova and Dettlaff 1983).

On the basis of the data obtained at the Volgograd hatchery for a number of years, including 1979, L.V. Igumnova has established that most of the females giving good-quality eggs mature in the range of 25 to 38 τ_0 and single females (about 2%) in the range of 38 to 40 τ_0.

One of the authors (T. A. Dettlaff) has recalculated the data about the time of maturation of the *A. stellatus* females at the Temryuk and Sergiev sturgeon hatcheries in the number of τ_0. The calculations have shown that at the Temryuk hatchery all females matured in the range of 28 to 42 τ_0 after pituitary injection. Hence, it follows that at the Temryuk hatchery the first females matured by 3 τ_0 later than at the Volgograd hatchery and, during the past years, at the Rogozhkino hatchery in the Don delta. Twenty per cent of females matured at the Temryuk hatchery in the range of 38 to 42 τ_0, and half of them gave eggs suitable for fish breeding purposes.

Hence, at the Volgograd hatchery, the maturation of the *A. stellatus* females caught at the level of spawning grounds starts and terminates somewhat earlier than that of females caught in the Kuban' river mouth or in the sea, off the

Perekop or Golubino shore. But these differences are not as great, as could be expected from different quality of females taken for fish culture at the hatcheries in the lower flow. At the same time, at the Sergiev hatchery, although it is located in the Volga delta, most of the females, which responded to pituitary injection by oocyte maturation, matured in the range of 25 to 38 τ_0 (about 60%) and only single females matured in the range of 38 to 42 τ_0, i.e., at the same time as at the Volgograd hatchery. These data allow us to recommend examining *A. stellatus* females (in order to determine their state of maturity) at all sturgeon hatcheries between 25 to 42 τ_0 after pituitary injection.

Recalculation in the number of τ_0 of the data about the duration of maturation of the females of vernal *A. gueldenstaedti*, which we obtained earlier at the Rogozhkino hatchery, as well as of the data obtained at the Bertyul hatchery (Dettlaff et al. 1965) has shown that most females matured in a range from 24 to 32 τ_0 after pituitary injections and only a small number of females at 23–24 and 32–35 τ_0. All females of *A. gueldenstaedti* matured at the same time at the Temryuk and Sergiev hatcheries. Hence, females of *A. gueldenstaedti* mature within shorter periods of time than those of *A. stellatus*. Figure 78A presents graphs which allow one to determine the time of examination of *A. gueldenstaedti* females at various temperatures in the range of 23 to 35 τ_0 (in hours).

Recalculation in the number of τ_0 of the data about time of maturation of the Volga *H. huso* at the Volgograd hatchery for a number of years has allowed us to plot graphs for the whole range of temperatures at which the eggs are obtained at the hatcheries (Igumnova 1975b, 1979; see Fig. 78C). According to these data, the first females mature within 19 τ_0 and the last ones within 34 τ_0 after pituitary injection. The females of the Don *H. huso* matured in the same time range at the Rogozhkino hatchery (Dettlaff et al. 1965). The *H. huso* females start to mature within a lesser number of τ_0 than those of *A. stellatus* and *A. gueldenstaedti*, whereas the duration of the period (in the number of τ_0) during which all females giving good-quality eggs mature, is longer in *H. huso* than in *A. gueldenstaedti* and does not differ practically from that in *A. stellatus*. These differences between females of various sturgeon species appear to characterize differences in the initial condition of females at the moment of pituitary injection.

It should be mentioned specially that spawners from different biological groups of the same species in different rivers and even from different (closely related) species mature within the same time. For example, the graphs of maturation time for the Don *A. gueldenstaedti colchicus* satisfactorily predict the time of maturation of the vernal and hiemal *A. gueldenstaedti* in the Volga. The hiemal females mature during this period both after keeping at the hatchery conditions in the Kazanskii' production lines in the Volga delta (Molodtsov 1971) and after wintering in the river at the Volgograd hatchery (the data obtained in 1979 and communicated by L. V. Igumnova). As to the late vernal *A. persicus*, some females at the Volgograd hatchery matured in 1979 at the same time that *A. gueldenstaedti*, while the others (about 50%) much later. Females of the Lena *A. baeri*, according to the data of I. I. Smol'yanov, mature within the same time range as the *A. gueldenstaedti* females (see Berdichevskii et al. 1979) and only at lower temperatures (8–10 °C) do they mature more rapidly (Khakimullin 1979). Scarce data about the duration of maturation of the Ob' *A. baeri* (Votinov 1963) and Amur *A. schrencki* (Svirskii 1967) also conform with this time range. Females of *H. huso* from the Don, Kura, and Volga mature at the same temperature within the same time range. Rare data about the duration of Amur *H. dauricus* maturation (Svirskii 1967) also coincide with the time range established for *H. huso*.

The causes of variation in the time of maturation of different females within the time range limited by curves I and II in Fig. 78 require special studies.

In order to obtain comparable data about maturation and ovulation of the oocytes from different females at the moment of obtaining eggs, we proposed to distinguish groups of females by the following features (Ginsburg and Dettlaff 1969):

1. All oocytes are in the ovary (oocytes without GV); oocytes cannot be manually separated from the follicles.
2. Oocytes are in the ovary (oocytes without GV); a small part of oocytes "slip down" or are easily separated from the ovary.
3. Some oocytes ovulated and leak out; a greater part "slip down" or are easily separated from the ovary; there are few residual oocytes in the ovary.
4. A greater part of oocytes ovulated and leak, the rest "slip down" or are easily separated from the ovary; there are no residual oocytes in the ovary.
5. All oocytes ovulated; there are few or no damaged eggs.
6. All oocytes ovulated; many damaged (broken) eggs.
7. A greater or lesser part of oocytes are in the ovary with signs of damage, sometimes degenerative changes in the pigment pattern, or with GV; there are damaged eggs among the ovulated ones.
8. All oocytes (with GV) are in the ovary; some oocytes can display signs of degeneration.
9. All oocytes are in the ovary; connective tissue of the ovary contains much fat (fatty females).

Group 1 resembles, in external appearance, group 8. In order to distinguish between them, 20 to 25 oocytes should be taken, boiled (see Sect. 1.2.2), and examined for the presence of GV. Group 1 includes the females cut open too early, before the ovulation of the oocytes started. Their eggs cannot be used for hatchery purposes. Group 2 also includes females killed too early which have advanced somewhat farther in their maturation. A small part only of the eggs from these females, which "slip down" or can be manually separated from the ovary, are fertilizable. The best results are given by females from groups 3 and 4. The females of group 5, depending on the water temperature during the period of maturation and the duration of delay of the ovulated eggs in the body cavity, can give eggs more or less suitable for hatchery purposes. Eggs from females of group 6 are not usually incubated, since they yield very few prelarvae.

Groups 7 and 8 are usually small. They include females which either do not respond to pituitary injection altogether (group 8) or respond to it less than the full extent (group 7). Group 9 consists of females of the hiemal groups which should not have been injected.

Figure 12 demonstrates the distribution of females according to the condition of their ovaries at different times after pituitary injection: in the range of 0.16 to 0.22 τ_{VII}. It can be seen that most females killed at times 0.16–0.20 τ_{VII} belong to groups 3–5. Hence, actual differences in the time of female maturation do exist.

It could be expected that in females killed at later times, ovulation would be more complete. But this was not the case. The number of females whose oocytes ovulated incompletely or did not ovulate at all progressively increases.

Single females which do not display signs of maturation by the latest time (see Fig. 78) do not, as a rule, mature or give low-quality eggs. These females belong to groups 7–9.

The causes of differences in the time of maturation of the females were specially studied. This time was shown not to depend on the pituitary dose. Within the limits of doses accepted at hatcheries during 1962–1964 (80 mg of acetone-dried pituitaries per female of *A. gueldenstaedti* and 60 mg per female of

Fig. 12. Maturation and ovulation of the oocytes taken from females of *A. gueldenstaedti* after different relative duration of maturation (Ginsburg and Dettlaff 1969). *1* Composite group including females from group 1 and, mainly groups 8 and 9; *2* composite group including females from group 2 and, mainly group 7; *3* group 3; *4* group 4; *5* group 5; *6* group 6. ***Ordinate***, percentage of females; *abscissa*, duration of the period of maturation (τ_m) in fractions of the total duration of embryogenesis (τ_{VII})

A. stellatus irrespective of their weight), both large and small females matured at the same time and the large females often matured even earlier. Hence, two-fold differences in the dose of pituitary suspension per weight unit of female do not affect the time of female maturation and the quality of eggs. At the Volgograd hatchery, when the *A. stellatus* females were injected with 35–40 mg of acetone-dried pituitaries, the time of maturation and egg quality did not depend on the dose of pituitary suspension. Both large and small females matured at the same time and good-quality eggs were, as a rule, obtained in both groups of females.

Let us return again to the problem of possible causes underlying the extended period of maturation of various females at the same temperatures. Many special studies (see, for example, Semenov 1965), as the practical experience accumulated at the hatcheries suggest that in females with gonads at the completed stage IV of maturity there is a clear correlation between the time of keeping them at spawning temperatures under unfavorable conditions and the duration of maturation after pituitary injection. Females just caught or females kept in fish tanks for less than 2 days mature the first. The longer the females are kept in fish tanks, the slower they mature and the lower the quality of their eggs (Fig. 13). Hence, distribution of females according to the time of maturation allows us to judge, to a certain extent, the conditions of maturation and the initial condition of females. It is possible that there are some differences also between the females with gonads at the completed stage IV of maturity. They can be expressed, in particular, in terms of different duration of hormone-dependent period in oocytes.

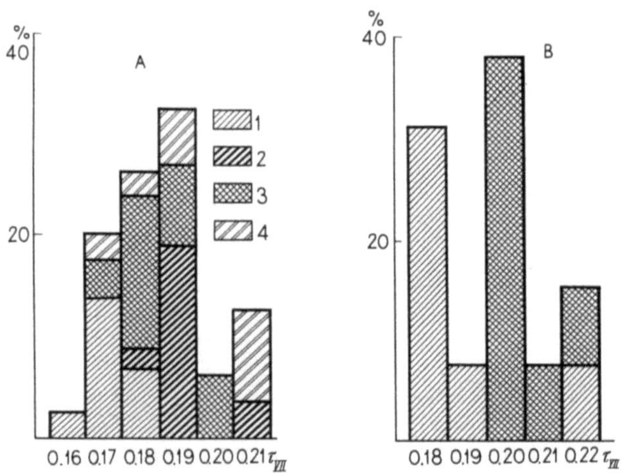

Fig. 13A, B. Distribution of females of *A. gueldenstaedti* (**A**) and *A. stellatus* (**B**) according to the relative duration of their maturation after keeping them in holding tanks for varying periods of time (Ginsburg and Dettlaff 1969). Time of keeping: *1* less than 1 day; *2* 3–6 days; *3* 6–23 days; *4* 16–26 days. *Ordinate*, percentage of females; *abscissa*, duration of the period of maturation (τ_m) in fractions of the total duration of embryogenesis (τ_{VII})

In order to determine when the eggs should be taken from the females within the limits of indicated range, between curves I and II (Fig. 78A–D), one should follow criteria accepted at sturgeon hatcheries (Kazanskii 1951). The females should be cut open when the following signs are observed: abdomen is soft, eggs are released as a strong spurt, the abdominal wall markedly sinks when the female is lifted by its tail.

As to females which did not mature in times corresponding to the graphs, it is expedient to study first the structure of the oocytes in the ovaries by taking a sample with a probe before they are killed. If the oocytes contain the GV, the females should be killed. If the GV is absent, it is possible to wait a bit longer, although the probability of obtaining good-quality eggs from such females is very low.

The curves plotted in Fig. 78A–D (Appendix E) allow us to calculate the time of pituitary injection to the females in such a way that the moment of their maturation is timed to the working hours. These curves facilitate handling of large fish, such as *H. huso*, since they allow reduction of the number of examinations of the females (they may not be examined earlier than the time corresponding to the lower curve) and help to determine more precisely the moment of taking eggs. When the eggs are taken in time, their losses due to overripening or underripening of the oocytes in the female body are markedly reduced and the eggs have better breeding quality.

Long-term experience of using these recommendations at the sturgeon hatcheries has shown their validity but changing conditions of migration of the spawners in rivers and changes in their condition required the precision of the recommendations that has been done in this book (see Appendix E).

Chapter 2 Embryonic Development

Development of the embryo starts from fertilization, i.e., the fusion of gametes: a female gamete, egg, and a male gamete, spermatozoon or sperm. The fertilized egg is already a new organism, the embryo, which transforms, under favourable conditions, through a series of sequential changes, into a prelarva (yolk-sac larva), larva, juvenile, and, finally, adult fish.

The fertilized egg is a single cell. It divides many times (period of cleavage). A cavity gradually forms among the arising cells (blastomeres), and the fertilized eggs transforms in a multicellular ball with a cavity inside (blastula). As a result of complex movements of the blastula wall cells, archenteron forms, and the embryo becomes three-layered (period of gastrulation). Later, rudiments of the most important systems of organs arise, nervous, excretory, muscle, circulatory (period of organogenesis). Soon the heart starts to beat, muscles contract, and the embryo acquires motility. The tail region rapidly grows, the embryo becomes able to perform swimming movements, which are at first restrained by the egg membranes. Finally, the embryo ruptures the egg membranes and leaves them, becoming a prelarva.

The prelarva conducts a free way of life and uses reserves of nutrients which were contained in the egg. Later, it starts to actively capture food and becomes a larva which grows and gradually transforms in a juvenile, already similar in the main structural features with the adult fish.

Embryonic development of sturgeon can be subdivided into five successive periods: fertilization, cleavage and blastulation, gastrulation, development from the end of gastrulation to the onset of heart beating, and from the onset of heart beating to hatching. For convenience of description, we distinguish 36 stages in the embryonic development of Acipenserid fish (Dettlaff and Ginsburg 1954; Ginsburg and Dettlaff 1955, 1969, 1975; Dettlaff et al. 1981).

Development of the embryos of the studied Acipenserid species — *H. huso*, *A. gueldenstaedti*, *A. stellatus, and A. ruthenus* — is very similar. Therefore, it will be described below in a general form. We will mention individual species only when there are any significant differences between them.

It is interesting that the structure of the embryos of *Polyodon spathula*, a species populating rivers and lakes of North America and closely related to the *Acipenseridae* but referred to another family (*Polyodontidae*) of the order *Acipenseriformes*, coincides, in almost all details, with that of the Acipenserid embryos (Ballard and Needham 1964; Vassetzky 1971).

The drawings in this book have been made from the embryos of *A. gueldenstaedti colchicus* obtained in the lower Don flow (except Plate VIII, stage 35A, C depicting the embryos of *A. stellatus* and *A. ruthenus*). All drawings have been made with the same magnification (scale is given in Plate I, see page 239).

Before we describe embryonic development, we will consider the structure of the mature egg and spermatozoon.

2.1 Gametes

2.1.1 The Egg

The number of eggs deposited by the females of Acipenserid fish is very high, i.e., 8000–199 000 in the relatively small *A. ruthenus* (Berg 1948), 35 000–363 000 in *A. stellatus*, and 84 000–837 000 in *A. gueldenstaedti*. It amounts up to several million eggs in the large sturgeon (up to 2 800 000 in *H. huso* and 4 100 000 in *H. dauricus*) (Derzhavin 1922). In rare cases, the fecundity of sturgeon can be even greater: a female of *H. huso* weighing 1228 kg, with 245 kg of eggs numbering 7 700 000, was caught in the northern Caspian in 1924 (Babushkin 1947).

External Appearance and Structure of the Egg

The eggs of sturgeon are globular or, mostly, somewhat elongated (Fig. 14). The size of unfertilized eggs in the various species is given in Table 1.

The eggs are brownish-gray. The color of the different regions of the egg varies, reflecting the polarity of its internal structure. The part of the egg facing upward after fertilization (animal region) is usually lighter than the lower uniformly colored vegetal part. In the center of the animal area, there is a light

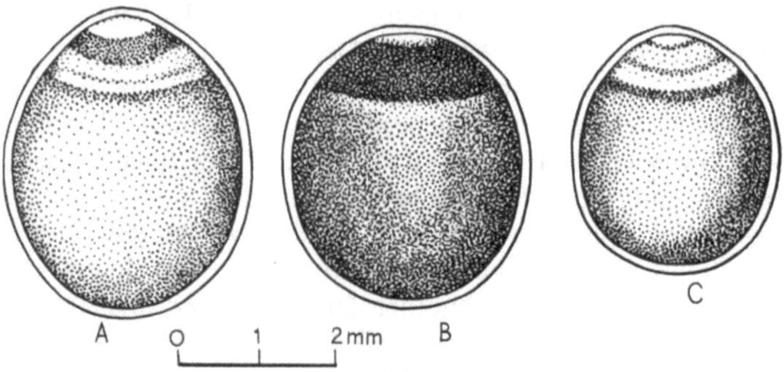

Fig. 14. **A** Eggs of *H. huso*, **B** *A. gueldenstaedti colchicus*, and **C** *A. stellatus* (Dettlaff and Ginsburg 1954). Lateral view

Table 1. Diameter of the unfertilized egg together with the membranes in Acipenserid fish

Species	Diameter, mm	Reference
Huso huso (L.)	3.6–4.0	Dettlaff and Ginsburg (1954)
H. dauricus (Georgi)	3.6–4.0	Berg et al. (1949)
Acipenser transmontanus	3.5–4.0	Cherr and Clark (1982)
A. gueldenstaedti colchicus V. Marti	3.2–3.8	Dettlaff and Ginsburg (1954)
A. gueldenstaedti Brandt	3.0–3.5	Berg et al. (1949)
A. brevirostrum Lesueur	3.0	Dodswell (1976)
A. stellatus Pallas	2.7–3.2	Dettlaff and Ginsburg (1954)
A. sturio L.	2.6–3.0	Berg et al. (1949)
A. baeri Brandt	2.5–2.7	Sokolov (1965)
A. oxyrhynchus Mitchill	2.5	Vladykov and Greeley (1963)
A. ruthenus (L.)	2.0–2.5	Sawadski (1926a)
	1.9–2.0	Berg et al. (1949)

polar spot surrounded by dark concentric rings. The typical pattern of the animal region is given in Fig. 14 and Plate I: the polar spot is encircled by a dark pigmented ring, and another ring is found at the boundary with the vegetal region. An intermediate pigmented ring is found between these two rings in some eggs (less frequently in *A. gueldenstaedti* and more frequently in *H. huso* and *A. stellatus*). The diameter of the polar spot, and the width and intensity of the pigmentation of the concentric rings and the lighter regions between them vary extensively not only in the eggs of different females, but also in those of the same fish. The animal region in the eggs of *A. gueldenstaedti colchicus* is sometimes so intensely pigmented that it becomes darker than the vegetal region. In these cases, the individual rings may be indistinguishable, with a small lighter spot seen only in the center. Sometimes, but very rarely, the eggs have no dark pigment whatsoever, and are pale yellow (eggs of this type have been found in albino fish). The *A. transmontanus* eggs are black, apart from having two lightly pigmented rings surrounding the micropylar region (Cherr and Clark 1982). Our experiments on raising individual eggs with different patterns of the animal region have shown that all the eggs with one, two, and three pigmented rings, as well as those with a uniform dark colour of the animal region, are fertilizable and can develop into normal larvae.

According to the distribution of yolk and lipid inclusions and the structure of cortical granules (Figs. 15, 16A), the mature egg does not differ from the oocytes which start to mature. The cortical layer structure has changed, the length of microvilli has been markedly reduced, and they no longer enter the zona radiata tubules (Fig. 16A). But the main differences between a mature egg and an oocyte lies in the fact that there is no GV in the former and, after the GV breaks down during maturation, the karyoplasm spreads over the animal cytoplasm without mixing with it completely, i.e., it forms a branching network of lacunae containing the remains of the GV contents (see Fig. 7E). With certain methods

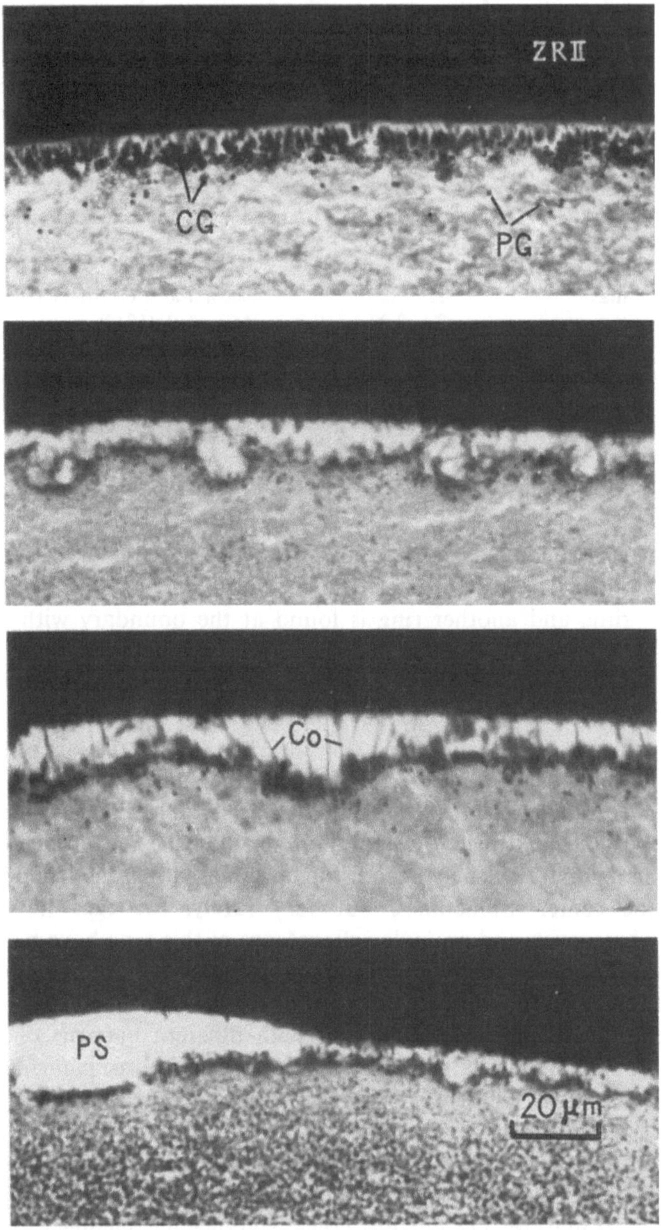

Fig. 15. Structure of the cortical layer in a mature unfertilized egg of *A. stellatus* (*upper photo*) and its changes after fertilization (cortical reaction) as seen under an optical microscope. *CG* Cortical granules; *Co* columns forming at the sites of extrusion of the cortical granules contents as a result of fixation; *PG* pigment granules; *PS* perivitelline space; *ZR II* zona radiata interna

applied, globules of hydrophilic colloid can be seen in the lacunae (Fig. 8, 3; Dettlaff 1962).

The nucleus of a mature egg is at metaphase II. The spindle of the II meiotic division lies at the surface of the egg near the animal pole, adjacent to the I polar body (Fig. 9E). As with the oocyte in the ovary, the egg is enveloped with two egg membranes, i.e., bilayered zona radiata and a superficial jelly coat (Fig. 5), which have a distinctive structure (Markov 1975; Cherr and Clark 1982; Vorob'eva et al. 1986).

Besides these membranes, a distinctive layer, less than 1 μm thick, has been found on the outer surface of the jelly coat when studying the egg membranes in *A. transmontanus* under the transmission electron microscope (Cherr and Clark 1982, 1984a). Its main component is a 110 kDa glycoprotein containing sialic acid. This layer hydrates and becomes adhesive when the eggs (fertilized as well as unfertilized ones) are in contact with fresh water. The authors believe that the hydration of the fine outer layer under the influence of water is the single cause that makes the eggs sticky (cf. Sect. 2.3.2).

The total thickness of membranes of an unfertilized egg attains, under vital estimation, 70 μm in *A. stellatus* and 110–120 μm in *A. gueldenstaedti* (Dettlaff and Ginsburg 1954). Before fertilization the egg membranes are not strong enough and the egg can easily be damaged. This is to be remembered when working with unfertilized eggs and, in particular, obtaining eggs from a female.

In the animal pole region, the thickness of the membranes is somewhat less than over the rest of the egg surface, about 50 μm in *A. stellatus* and 80–110 μm in *A. gueldenstaedti* (Dettlaff and Ginsburg 1954). Here, there are micropylar canals (Fig. 4) through which a spermatozoon can penetrate in the egg cytoplasm at fertilization. The egg membranes per se are so strong that the spermatozoon cannot penetrate them, as is the case, for example, in minnows and frogs. The number of micropyles markedly varies in various Acipenserid species, within the limits of the species in different females, and also in the eggs of one female. For example, it varies from 1 up to 13 in *A. stellatus*, to 33 in *H. huso*, and to 52 in *A. gueldenstaedti colchicus*. In *A. stellatus*, eggs with the mean number of micropyles less than five predominate, while in *H. huso* and *A. gueldenstaedti* the eggs with the mean number of micropyles can exceed 25 (Ginsburg 1968). In *A. ruthenus*, 5 to 13 micropyles were observed (Kowalewsky et al. 1870; Salensky 1878; Persov 1957), in *A. sturio*, 3 to 9 (Dean 1895, 1896), and in *A. transmontanus*, 3 to 15 (Cherr and Clark 1982).

The micropylar canals are restricted to a small area near the animal pole without any strict order (Fig. 17A, B). In *A. stellatus*, *H. huso*, and *A. gueldenstaedti*, the distance between the neighboring outer openings of micropyles is usually between 40 to 80, less frequently up to 100 μm. When there are five to ten micropyles, they are located in a region of 120–130 μm in diameter. In *A. gueldenstaedti* with 30 and more micropyles, the diameter of this region is upto 1000–1100 μm.

An individual micropyle is a funnel-shaped deepening of the jelly coat transforming into a canal (Fig. 17C). The micropylar canal forms an extension,

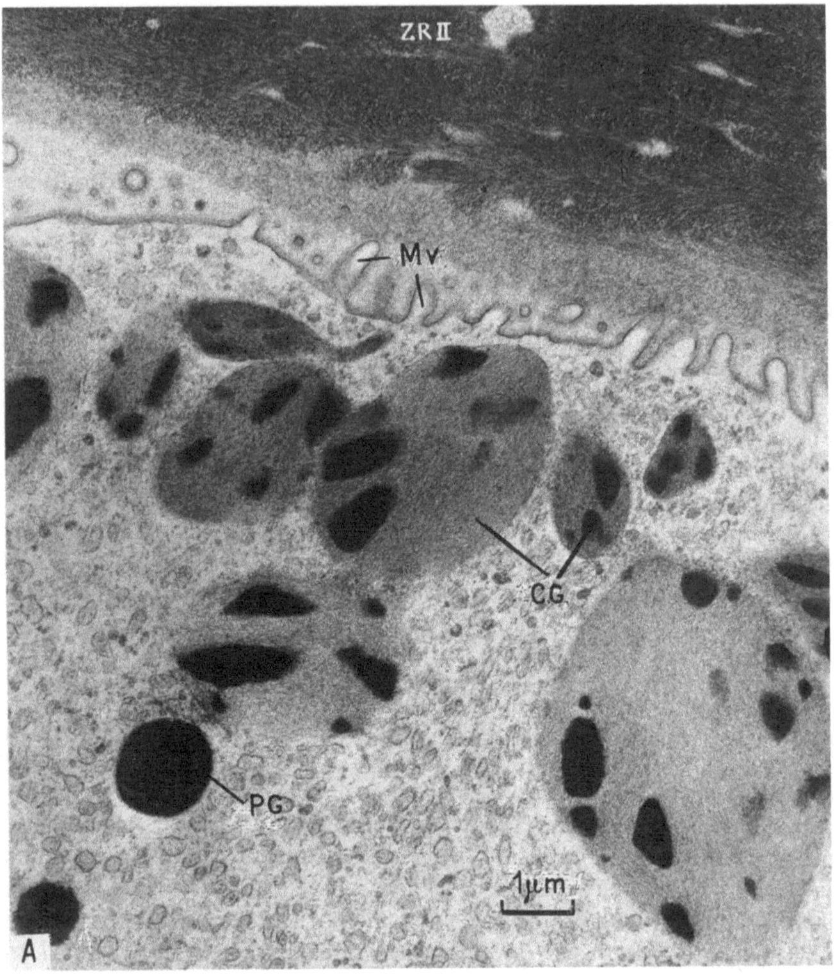

Fig. 16A–C. Cortical reaction in the egg of *A. stellatus* as seen under the electron microscope (22.1°C). **A** Cortical layer of an unfertilized egg; **B** beginning of exocytosis of the cortical granules (3 s after insemination); **C** the contents of the cortical granules have been discharged from the cytoplasm and merged in a single layer (180 s after insemination) (**A, B** from Ginsburg and Nikiforova 1978). *CG* Cortical granules; *Mv* microvilli; *PG* pigment granules; *SCG* secreted contents of the cortical granules; *ZR II* zona radiata interna

ampulla, from which a narrow terminal tubule extends. The diameter of this tubule does not markedly exceed the width of the rod-like spermatozoon head.

The terminal tubule of each micropyle is filled with a cytoplasmic process (Ginsburg 1959). When the eggs were placed in 1 M sucrose, urea, or sodium chloride, the cytoplasm was separated from the membranes and cytoplasmic processes of distinctive structure were sometimes found on the egg surface, cylinder-like columns with a large light vacuole at the end (Fig. 18). These

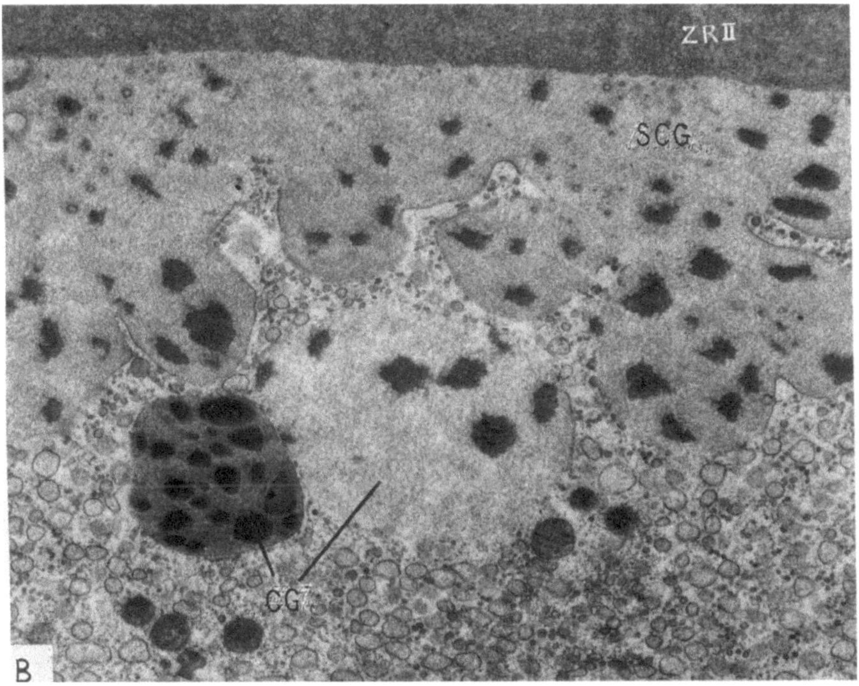

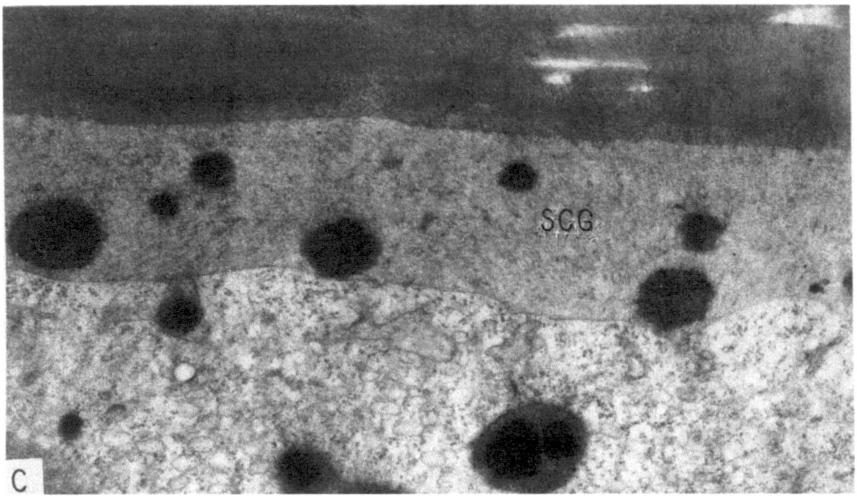

columns appear to correspond to cytoplasmic processes filling the terminal tubules of micropyles.

The sizes of mature eggs from various females of the same species can vary quite substantially. These differences do not exert any significant effect on the viability of the prelarvae.

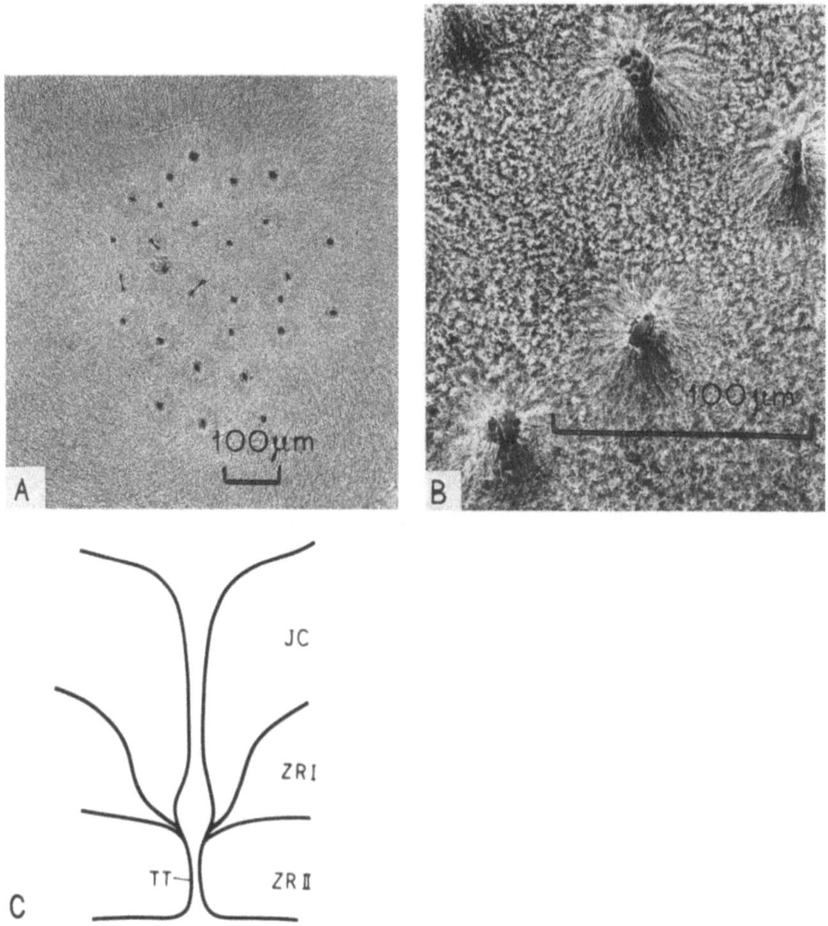

Fig. 17A–C. Openings of micropyles on the outer egg surface in *A. gueldenstaedti colchicus* (**A** a group of 31 micropyles); **B** *H. huso*, and **C** the structure of the micropyle in *A. gueldenstaedti*, diagrammatic (Ginsburg 1968). *JC* Jelly coat; *TT* terminal tubule of the micropyle; *ZR I* zona radiata externa; *ZR II* zona radiata interna

Fertilizability of the Eggs

The fertilizability of the mature ovulated eggs gradually diminishes and then disappears altogether when they are retained in the body cavity of the female or kept in the coelomic fluid, or in water in vitro. In water, this process takes place very rapidly in many teleosts. For example, the eggs of *Esox lucius* or *Vimba vimba* lose their fertilizability completely within 1 min after being place in water, and the eggs of *Oncorhynchus keta, O. tschawytscha, Salmo salar, Coregonus albula* — within a few minutes. Unlike the eggs of these species, some eggs of *A. gueldenstaedti* and *A. stellatus* can be fertilized after many hours in water. However,

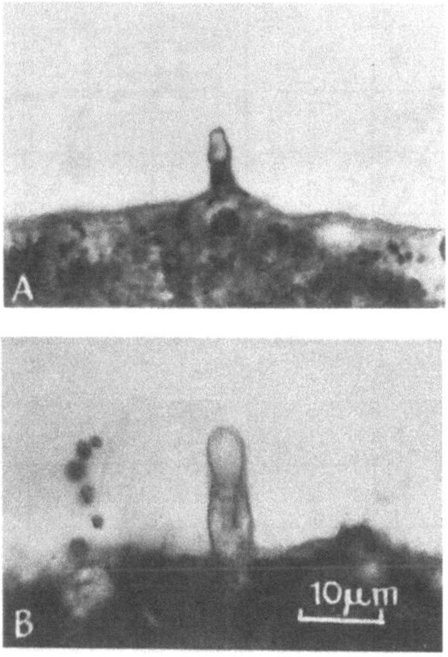

Fig. 18A, B. Cytoplasmic processes on the surface of unfertilized eggs of *A. gueldenstaedti* after they were kept in 1 M sucrose (**A** 7 min) or 1 M urea (**B** 20 min) (Ginsburg 1959). Sanfelice, iron hematoxylin

the percentage of fertilization begins to decrease much earlier, i.e., often after 1 h in *A. gueldenstaedti* and even earlier in *A. stellatus* (Fig. 19A, B).

The fertilizability of the eggs is retained much longer in saline and, especially, in coelomic fluid. The eggs of sturgeon can retain a high fertilizability for 4–6 h in the coelomic fluid at the spawning temperatures (Fig. 19A′, B′).

2.1.2 The Spermatozoon

If the female spawns a large number of eggs, the number of spermatozoa released by the male is many times greater. It is enormous: males of Acipenserid fish release several tens or even hundreds of cubic centimeters of sperm at one time (up to 200 cm^3 in *A. stellatus*, 500 cm^3 in *A. gueldenstaedti* and 1000 cm^3 in *H. dauricus*). One cubic centimeter of sperm contains up to 1×10^{10} spermatozoa, usually from 1×10^9 to 4×10^9 (Fig. 20).

A spermatozoon has a narrow rod-shaped head, a short midpiece, and a long tail, or flagellum. In comparison with the egg, which is loaded with a vast reserve of nutrients, the size of spermatozoon is negligible (Table 2).

The rod-shaped head of the spermatozoon tapers somewhat towards the anterior end (Fig. 21) where the acrosome is located. The rest of the head is

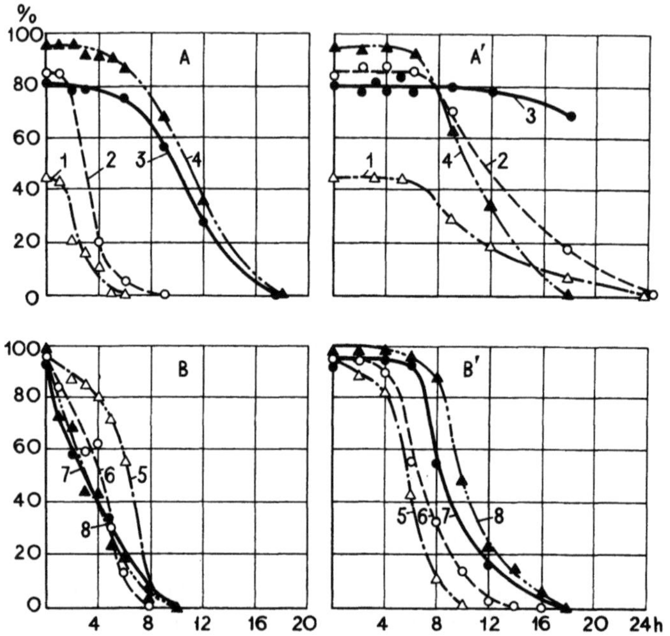

Fig. 19A, B. Retention by the eggs of *A. gueldenstaedti* (**A**, **A'**) and *A. stellatus* (**B**, **B'**) of fertilizability upon their keeping in water (**A**, **B**) and coelomic fluid (**A'**, **B'**) at the spawning temperatures (Ginsburg 1968). The mean temperature was: 14.8 °C (**A**, **A'**) and 20.1 °C (**B**, **B'**). *Numerals* on the curves designate numbers of females. *Ordinate*, percentage of fertilization; *abscissa*, time of keeping eggs in water (**A**, **B**) and coelomic fluid (**A'**, **B'**), h

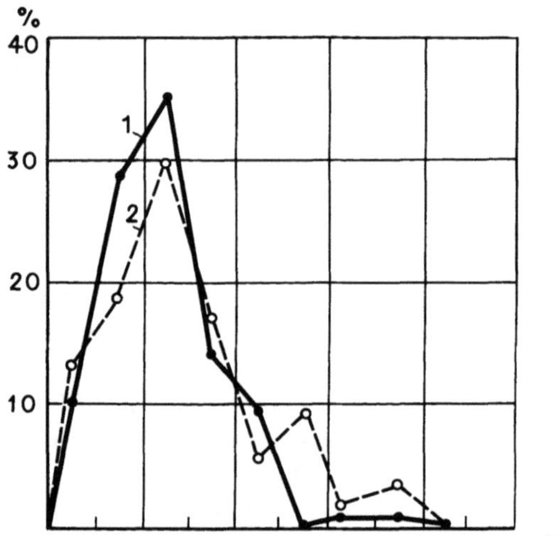

Fig. 20. Distribution of *A. gueldenstaedti* (*1*) and *A. stellatus* (*2*) males according to the concentration of spermatozoa in their ejaculates (Ginsburg 1968). Sperm samples were taken from 105 males of *A. gueldenstaedti* and 53 males of *A. stellatus*. *Ordinate*, relative frequency of males with a given concentration of the spermatozoa in the ejaculate

Table 2. Mean size of spermatozoa in sturgeon (μm)

Species	Total length	Head		Reference
		Length	Width	
Huso huso	55	7.4	1.1	Dettlaff and Ginsburg (1954)
Acipenser gueldenstaedti colchicus	58	8.9	1.9	Dettlaff and Ginsburg (1954)
A. stellatus	47	6.3	1.8	Dettlaff and Ginsburg (1954)
A. transmontanus	42	7.0	1.5	Cherr and Clark (1984b)

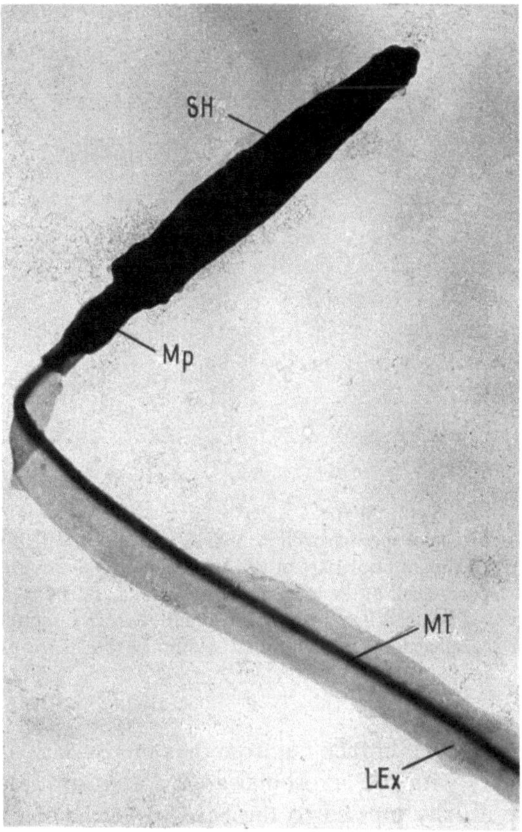

Fig. 21. The spermatozoon of *A. gueldenstaedti*. *LEx* Lateral extension of the cell membrane; *Mp* midpiece of the spermatozoon; *MT* main part of the tail; *SH* head of the spermatozoon

occupied by the nucleus. Three canals, 50–60 nm wide, with an electron-dense contents, arranged in a helix and limited by a membrane run all along the nucleus in its center (Fig. 22C, D). These canals begin in the subacrosomal area and penetrate the whole nucleus.

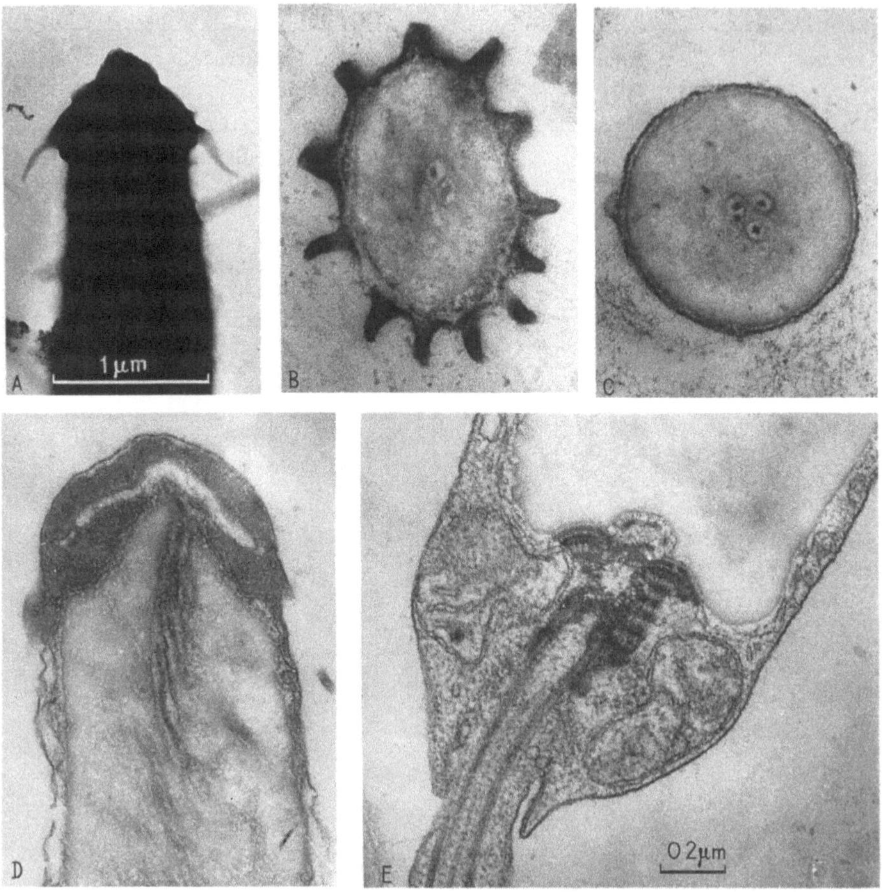

Fig. 22A–E. Ultrastructure of spermatozoa in **A–C** *A. stellatus* and **D, E** *A gueldenstaedti*. **A** Electron micrograph of the anterior part of the head of a whole-mounted spermatozoon. **B** Cross-section of the sperm head at the level of the finger-like appendages of the acrosome. **C** Cross-section of the sperm head behind the acrosomal area. **D** Longitudinal section of the anterior end of the sperm head. **E** Longitudinal section of the sperm midpiece (**A–C**, from Ginsburg 1977)

Finger-like appendages branch out from the outer surface of the sperm head, along the boundary with the acrosomal area. These appendages in the fixed spermatozoa are usually pressed to the head surface but may be sometimes separated from it (Fig. 22A, B). Their number in *A. gueldenstaedti* and *A. stellatus* varies from 10 to 13. The acrosomal area is surrounded by the sperm plasma membrane to which a flattened acrosomal bell-shaped vesicle is closely apposed (Fig. 22D; Cherr and Clark 1984b). Subacrosomal material is located behind the acrosomal vesicle.

There is a nuclear fossa at the head base (Fig. 22E). The midpiece, about 1 μm long, contains a centriolar system surrounded by one or two C-shaped mitochondrial bodies. The proximal centriole has a typical structure and is oriented, as in the spermatozoa of most other animals, with its long axis perpendicular to the longitudinal axis of the head (Fig. 22E). The distal centriole located perpendicular to the proximal one is the basal body of the flagellum. A fibrous body consisting of osmiophilic disks alternating with lighter material adheres to the centrioles laterally. This body resembles the centriolar complex appendage in spermatids of some *Elasmobranchii* and *Teleostei* and appears to fasten the centrioles to the nucleus. In the intact spermatozoon, the midpiece is cylinder-shaped and forms, in its posterior part, a cytoplasmic collar surrounding the flagellum base.

The sperm tail has an axoneme with the conventional 9 + 2 microtubular pattern: the central pair of microtubules is surrounded by nine peripheral doublets. This structure is characteristic of the microtubular complex of the tail in spermatozoa of the vast majority of animals, as well as flagella and cilia of unicellular *Protozoa* and somatic cells of multicellular organisms.

The cell membrane around the axoneme along the main part of the sperm tail has two lateral extensions or sidefins in line with two central microtubules. These extensions start asymmetrically: on one side, from the lower end of the midpiece and on the opposite side, a few microns more caudally (Fig. 21). They terminate at the site where the main part of the tail passes into the endpiece. The membranous extensions are filled with fine-granular material and do not contain any skeletal structures. The sperm nucleus contains a haploid set of chromosomes.

Motility of Spermatozoa and Their Fertilizing Capacity

The spermatozoa are immobile in the ejaculate and activated the moment water is added. At first, they perform quick forward motion; during this period, a spermatozoon of *H. huso* moves at a rate of 0.1 mm/s. Then, the forward motion of the spermatozoa gradually slows down, turns into uncoordinated oscillating movement and, finally, ceases altogether. A spermatozoon is capable of fertilizing an egg only while it is able to move forward to approach a micropyle and penetrate it. Its oscillations on the spot do not ensure fertilization.

The spermatozoa of sturgeon can retain their motility in water for different periods of time, from several minutes to many hours. At the same time, very significant differences are observed not only between the spermatozoa from the ejaculates of different males of each species, but also within a single drop of sperm. Observations carried out with the sperm of 20 males of *A. gueldenstaedti* have shown that the forward motion of the majority of spermatozoa in a suspension usually ceases 5 to 10 min after water was added. However, a small part of the spermatozoa swim for a long time, up to 20–60 min and even several hours (Ginsburg 1968, 1972). In accordance with this, the fertilizing capacity of the *A. gueldenstaedti* and *A. stellatus* spermatozoa sharply diminishes in many cases during the first 10 min after water is added, but it is not lost entirely. In

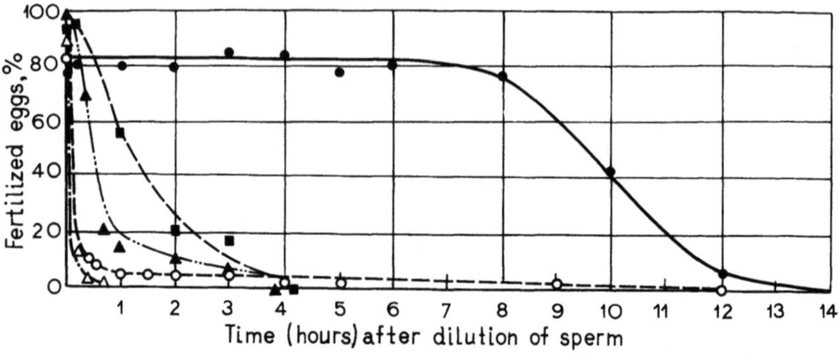

Fig. 23. Retention by *A. gueldenstaedti* spermatozoa of fertilizing capacity in water (Ginsburg 1968). Experiments with sperm of five males. Ejaculate was diluted with water at a ratio of 1:9. Water temperature was 14.8–17.4 °C

exceptional cases, the fertilizing capacity of diluted sperm is retained at a high level for many hours (Fig. 23).

Hence, the spermatozoa of sturgeons retain their motility in water for a considerably longer time than those of teleosts that spawn in fresh water: for example, the spermatozoa of *Oncorhynchus gorbuscha, Salmo salar, Salmo trutta, Coregonus lavaretus, Abramis brama, Carassius carassius, Cyprinus carpio*, and many other species swim in water only for 1–3 min.

The properties of sturgeon gametes (the ability of the eggs and spermatozoa to remain in water for a considerable length of time without losing their fertilizability and fertilizing capacity) increase the effectiveness of natural spawning which takes place in the parts of a river with a strong current that quickly disperses the eggs and sperm.

Without water, the immobile spermatozoa retain their viability for a very long time, especially at low temperatures (1–4 °C). This is why sperm designated for storage is collected in a dry container and placed on ice. When eggs are fertilized with the sperm kept under these conditions for 5–6 days, the percentage of fertilization and yield of prelarvae was in some cases the same as when the eggs were fertilized with fresh sperm. In exceptional cases, spermatozoa, stored in this way have been known to retain their viability for up to 21 days (Shmidtov 1936; Persov 1941a; Ginsburg 1968, 1972).

Storage of sperm obtained from several males in the same container reduces the viability period of the spermatozoa (Shmidtov 1936). The attempts to preserve pieces of testes in cold storage proved unsuccessful, all the spermatozoa dying within the first day (Persov 1947).

Excellent results have been obtained in the prolonged storage of deep-frozen spermatozoa for a number of animals (in dry ice at − 79 °C and in liquid nitrogen at − 196 °C). This method is actively used to store sperm of farm animals throughout the world.

The attempts to use deep-freezing with the sperm of teleosts showed promising results in some species. In particular, high percentages of fertilization were obtained when eggs were inseminated with deep-frozen sperm in *Oncorhynchus gorbuscha*, *O. kisutch*, *O. nerka*, *Salmo gairdneri*, *Thymallus thymallus* (see Horton and Ott 1976).

In the first investigation on the freezing of sturgeon sperm in liquid nitrogen (Burtsev and Serebryakova 1969) with the use of a diluent containing sucrose or lactose, egg yolk and glycerol, up to 80% of *H. huso* spermatozoa and up to 100% of *H. dauricus* spermatozoa retained, in some experiments, their capacity to be activated in water (after being stored for 34 and 16 days, respectively). However, the results of sperm freezing varied. When eggs were inseminated with sperm which, after thawing, contained 20–40% of motile spermatozoa, the percentage of fertilization proved to be very low (approximately 1%).

A higher percentage of fertilization was obtained when the *A. gueldenstaedti* and *A. stellatus* sperm was frozen at − 55 °C and the diluent contained glycerol, and egg yolk extract, chloral hydrate, and urea. About 35% of the eggs were fertilized after the sperm had been kept at this temperature for 2 h (Kasimov et al. 1974).

Pushkar' et al. (1978) have developed a method of deep-freezing sturgeon sperm, and it has already produced satisfactory and fairly stable results. The sperm is frozen in vapors of liquid nitrogen by decreasing the temperature up to − 14 °C at a rate of 0.5–5 °C/min and then to − 196 °C at a rate of 20–30 °C/min. Tris-HCl buffer with dimethyl oxide and egg yolk was used as a diluent.

In the *A. gueldenstaedti* and *A. stellatus* sperm frozen by this method and stored at − 196 °C for several days, 50–60% of the spermatozoa usually retained their ability of forward motion. When *A. stellatus* eggs were inseminated with the sperm of four males after 7–23 days of storage at − 196 °C, 63% of the eggs, on average, were fertilized. In the control, when the eggs of the same females were fertilized with fresh sperm, the percentage of fertilization was 77%. A significant percentage of fertilization was also obtained with long-term storage of sperm deep-frozen by this method. The eggs of three *A. stellatus* females were inseminated with sperm stored at − 196 °C for 1 year, and 49, 58, and 70% of the eggs were fertilized, compared with 88, 83, and 98% fertilization in the control (Pushkar' et al. 1979; Kopeika pers. comm.).

Hence, we now have a method of low-temperature conservation of sturgeon sperm, which can be used for hatchery reproduction.

Quality of Sperm

The sperm of the various males and even portions of sperm from the same male obtained at different times vary markedly in their properties. First of all, there is a great difference in the concentration of spermatozoa per unit volume of the ejaculate, which can be established already by the external appearance of the sperm. A watery sperm the color of whey contains less than 1×10^9 spermatozoa in 1 cm^3, sperm the color of skimmed milk contains 1–2×10^9 spermatozoa, and sperm the color of whole milk with yellowish tinge contains more than 2×10^9 spermatozoa in 1 cm^3. A low concentration of spermatozoa in the ejaculate does not yet suggest that it is unsuitable for fertilizing eggs. Spermatozoa from such an ejaculate are often activated quite well in water. We have obtained a high percentage of fertilization through the use of sufficient amounts of watery sperm for artificial insemination of eggs in the laboratory. However, it is not advisable to use watery sperm at the hatcheries unless it is unavoidable. When this type of sperm is used together with the more concentrated sperm of other males, its addition will lead to an unpredictable drop in the concentration of spermatozoa in a suspension of sperm from several males, thus making it difficult to dose the sperm properly, which could produce a negative effect.

Sperm differs in its ability to be activated by water. When water is added, the movement of spermatozoa may be intensive and prolonged, or weak and short-lived. Frequently, a greater or smaller part of the spermatozoa are not activated at all, and, in some cases, all the spermatozoa remain immobile. Such variations were observed not only with watery sperm, but with sound-looking concentrated sperm as well.

The following 5-point scale was developed for determining the quality of sturgeon sperm by its ability to be activated in water (Persov 1941b):

5 Points. Rapid forward motion of all spermatozoa;
4 Points. Rapid forward motion of most spermatozoa; slow-moving, zigzagging, and oscillating spermatozoa can also be seen in the field of vision of a microscope.
3 Points. Rapid forward motion of some spermatozoa, with zigzaggin and oscillating movement prevailing. Immobile spermatozoa can also be seen.
2 Points. Forward motion is rare. Some oscillating spermatozoa are seen. Approximately 75% of the spermatozoa are immobile.
1 Point. All the spermatozoa are immobile.

This scale can be useful in a comparative estimation of sperm from different males. For practical purposes (artificial insemination of sturgeon eggs), it is recommended that a mixture of sperm from several males be used in view of the high variability of sperm quality. In this case, microscopical analysis of sperm quality is not necessary (see Appendix F).

The viability periods of the spermatozoa during storage of the sperm vary significantly, depending on the quality of the sperm. For instance, as was already mentioned, the fertilizing capacity remains in some cases at a high level for 5–6 days in "dry" sperm (ejaculate stored without addition of water) at low temperatures, while in other cases this capacity is lost completely within 1–2 days.

2.2 Artificial Insemination of Eggs

In the case of artificial breeding of sturgeon, the encounter of eggs with spermatozoa is provided by the fish breeder. To make it effective, methods of taking eggs and sperm and methods of artificial insemination have been developed.

A matured female is struck on the head and bled by cutting the gill vessels. Then the female is suspended and a midventral incision made from the sexual pore to the level of pectoral fins, with a basin put under the fish. All ovulated eggs together with coelomic fluid flow into it. If some eggs are still follicle-enclosed but prepared for ovulation and easily "slip down" from the ovary, the fish breeder separates them manually by cautiously combining eggs from the ovarian folds with his fingers. With compete ovulation of the eggs, aggregates of eggs sometimes appear in its first portion; these aggregates should be removed before insemination.

The best results are obtained with insemination performed just after the eggs are taken. Before sperm is added, an excess of coelomic fluid should be poured from the eggs, since it interferes with fertilization. If it is necessary to wait for some time, e.g., if there is a shortage of males or when the eggs are obtained simultaneously from many females, the eggs should be stored under a layer of coelomic fluid, the medium most favorable for retaining their fertilizability, and excess of coelomic fluid should be poured off only shortly before insemination. When in contact with air and drying out, the eggs rapidly lose their fertilizability. The same effect is exerted by direct sun light and high temperature (above 20 °C). Therefore, the eggs should be stored before insemination and insemination performed in the shade, under a roof, or in a room. In hot weather insemination should be performed without any delay.

The sperm is collected by manually stripping of the males in dry containers, separately for each male. Eggs should be inseminated with a mixture of sperm from several, preferably three to five males. In this way, insemination failure due to poor quality of the sperm can be avoided.

When the eggs and sperm are obtained, insemination is performed. In sturgeon culture, different methods of insemination have long been used: dry, semi-dry, and wet methods. With the semi-dry method of insemination developed by Vrasskii on salmon in 1856, the eggs and sperm are collected in dry containers but shortly before addition to the eggs the sperm is diluted with water. Using the dry method (modified Vrasskii method), sperm is added directly to the eggs with coelomic fluid, carefully mixed, and, only after this, is water added. For the wet method, which was first used for sturgeon at the beginning of the century by Derzhavin (see also Maslov 1919; Kasimov et al. 1964), the eggs are carefully washed with water and, in this way, all of the coelomic fluid is removed before insemination. Thereafter the sperm is added.

Comparison of the effectiveness of these three methods of insemination carried out on the eggs of *H. huso*, *A. gueldenstaedti*, and *A. stellatus* at the

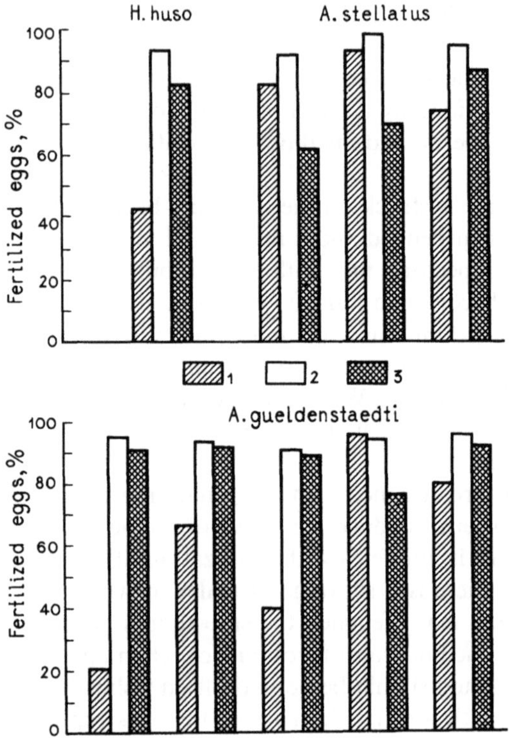

Fig. 24. The results of insemination of the eggs of *H. huso*, *A. stellatus*, and *A. gueldenstaedti* by different methods (Ginsburg 1963b). *1* Dry method; *2* semi-dry method; *3* wet method

hatchery has shown that the best and most consistent results are obtained with the semi-dry method (Fig. 24; see Ginsburg et al. 1963; Ginsburg 1968, 1972). In 1963 this method was recommended by the USSR Ministry of Fisheries (see Ginsburg 1963b) and since that time has been successfully used at the sturgeon hatcheries (Romanov 1979).

The lesser suitability of the dry method of insemination for sturgeon is due to the fact that sperm is mixed first with the coelomic fluid and only after this with water. Meanwhile, spermatozoa of sturgeon, unlike those of many teleosts, are not activated by the coelomic fluid but, in a mixture containing this fluid, are more poorly activated by water. This unfavorable influence of coelomic fluid is especially strong during insemination of the *A. gueldenstaedti* eggs with the dry method: in this species the coelomic fluid is more viscous. It is also true for all other Acipenserid species given a small amount or insufficient quality of sperm.

The semi-dry method has advantages not only over the dry method but also over the wet method. The main disadvantage of the wet method is that some eggs are frequently activated even before insemination, under the influence of water and mechanical mixing. Such eggs lose their fertilizability. A specially significant decrease in the percentage of fertilization was noted when inseminating *A. stellatus* eggs with the wet method. In addition to this disadvantage, there is another: when the unfertilized eggs are placed in water, the eggs which were activated whilst still in the female body become sticky. If a batch of eggs with a high percentage of such eggs is inseminated with the wet method, many clusters can form and this will make the eggs unsuitable for hatchery purposes. When the semi-dry method is used, the activated eggs also become sticky when in water but no clusters form, since these eggs lose their stickness simultaneously with the fertilized ones.

During insemination, it is essential that the sperm be diluted with water shortly before its addition to the eggs, since even short-term stay of the spermatozoa in water can reduce their mobility and, hence, fertilizing capacity.

The dosage of sperm is very important for the results of insemination. If an insufficient amount of sperm is taken, some eggs remain unfertilized (see, for example, Persov 1953). Apparently, in a low-concentrated suspension of spermatozoa only few of them get into the immediate vicinity of the micropylar canals of eggs, while the rest have no time to cover the distance separating them in the few minutes during which insemination takes place.

If too much sperm is used, the percentage of fertilization will be high, but some eggs (those in which several spermatozoa penetrate, polyspermic eggs) will prove unviable (see Sect. 2.4.4). Losses due to polyspermy can attain 15–20%.

Concrete recommendations on the technique of insemination are given in Appendix F.

2.3 Fertilization

In sturgeon, the spermatozoon reaches the egg surface by passing through the micropylar canal. It enters the cytoplasm and, after complex transformations, the paternal haploid set of chromosomes, contained in the sperm head, unites with the maternal one. As a result, the fertilized egg, zygote, acquires the normal, diploid chromosome number. In Acipenserids, this number is very large: about 120 in some species and about 240 in the others. The karyotype of the Acipenserids contains many dot-like microchromosomes. This complicated accurate counting of the chromosomes and led, in some cases, to notable errors. The group of species with about 120 chromosomes includes *H. huso*, *A. sturio*, and *A. ruthenus*, as well as, apparently, *A. stellatus* and *A. nudiventris*. The group of species with about 240 chromosomes includes *A. naccarii* and, apparently, *A. gueldenstaedti* (Fontana and Colombo 1974; Vasil'ev 1980). The two latter species appear to be polyploid.

The initial period of development, period of fertilization, starts with the encounter of the spermatozoon and the egg and lasts until the male and female nuclear material unite to form the zygote nucleus. During this period, deep changes of both spermatozoon and egg take place.

2.3.1 Transformations of the Spermatozoon upon Its Encounter with the Egg (Acrosomal Reaction)

Transformations of the sturgeon spermatozoon which take place as a result of its contact with the egg can be induced by treating spermatozoa with egg water (egg infusion in fresh water), excess of Ca ions in the medium, Ca ionophore A 23187, alkalinization of water to pH 8 or 9 (Dettlaff and Ginsburg 1963; Ginsburg 1977; Cherr and Clark 1984b). Upon acrosomal reaction the material contained in three intranuclear canals of the sperm head is ejected forwards and a long fine acrosomal process forms. Three elastic filaments encased in a transparent membrane serve as a stem of this process (Fig. 25). The material of the head subacrosomal region appears to be also involved in the process formation, since its amount decreases after the acrosomal reaction is completed. The membrane is frequently destroyed over its greater or lesser length, leaving the process-enclosed filaments exposed and separating from each other.

The length of the acrosomal process is usually close to that of the rod-shaped head of the spermatozoon: 4–6 μm in *A. stellatus* and 5–8 μm in *A. gueldenstaedti*. Sometimes, it is much longer, up to 14 μm in *A. gueldenstaedti*. In *A. transmontanus*, the acrosomal process length is 10 μm (Cherr and Clark 1984b).

Simultaneously with the acrosomal process formation, the mitochondrial body of the midpiece loses its intimate connection with the axoneme of the flagellum and is either displaced laterally assuming a spherical shape (Fig. 25A), or is fully separated from it and lost. These changes are regular: apparently, after the spermatozoon becomes attached to the cytoplasm surface and loses motility there is no need for the close connection between the mitochondrial body, a generator of energy for the flagellum movements, and the axial complex of the sperm tail.

A water-soluble 66 kDa glycoprotein is responsible, according to Cherr and Clark (1985a), for induction of the acrosomal reaction. This glycoprotein is present in the jelly coat (layer 3, according to the terminology of the authors) only after exposure of the eggs to fresh water. The authors supposed that it may be proteolytically derived from a high molecular weight complex in layer 3.

For acrosomal reaction to be realized, the presence of Ca ions in the medium is necessary. Under experimental conditions, they can be substituted by other divalent cations: if spermatozoa are placed in egg water prepared on Ca-free Holtfreter solution complemented with Sr or Mg ions, the acrosomal reaction proceeds, as in the conventional egg water (Dettlaff and Ginsburg 1963). No acrosomal reaction took place in Ca-free egg water complemented with Ba or Mn ions.

The presence of Ca^{2+} or Sr^{2+} is necessary for the acrosomal reaction and, hence, for fertilization. However, their role in gamete fusion is not limited to this. This follows from the observation that in the presence of Mg the acrosomal process forms but no fertilization takes place: the spermatozoa reach the egg surface through the micropylar canals, come into contact with its plasma membrane, but do not enter the cytoplasm. The presence of Ca^{2+} appears to be

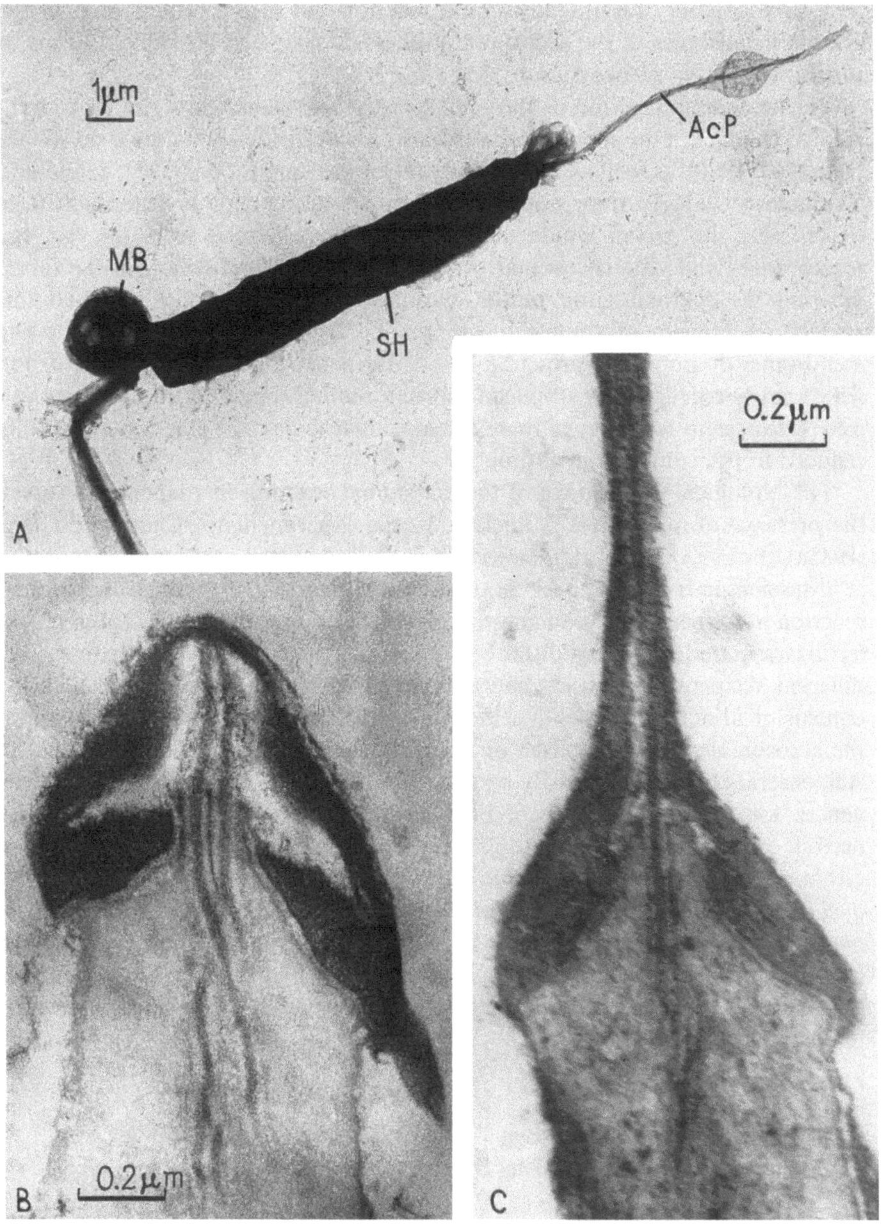

Fig. 25.A Acrosomal reaction in *A. stellatus* and **B, C** *A. gueldenstaedti*. **A** Acrosomal process formed, the mitochondrial body of the sperm midpiece displaced. Whole-mounted preparation. **B, C** Sections of the anterior end of the sperm head: **B** beginning of ejection of intranuclear canals contents; **C** base of the formed acrosomal process. (**A**, from Dettlaff et al. 1981; **B** and **C**, from Ginsburg 1977)

necessary not only for the acrosomal reaction but also for the fusion of the plasma membranes of the acrosomal process and the egg. For this, Ca can be substituted by Sr, rather than by Mg.

The acrosomal reaction of sturgeon spermatozoa is basically similar to well-studied changes in the spermatozoa of many invertebrate animals (for review see Dan 1967, 1970), as well as the lamprey (Nicander and Sjöden 1971; Jaana and Yamamoto 1981). In these animals, the acrosomal reaction is a mechanism of overcoming the barrier which separates the spermatozoon from the egg, egg membranes, while the acrosomal process is a site of application of the forces drawing the spermatozoon in the cytoplasm. Acipenserid fish retained this ancient mechanism of gamete fusion despite the fact that, in them, the egg membranes do not really present a barrier for penetration of the spermatozoon which easily reaches the cytoplasm through the micropyle. With more specialized *Teleostei*, in whose eggs there are also micropyles, the acrosome was fully reduced in the course of evolution.

The biological importance of the acrosomal reaction in Acipenserids given the presence of micropyles is unclear. In the experiments of Cherr and Clark (1985a), the egg water of *A. fulvescens* did not induce the acrosomal reaction in *A. transmontanus*. On the basis of these results they suggest that the acrosomal reaction in Acipenserids is an adaptation which ensures the species specificity of fertilization under the conditions of spawning in a river where gametes of different Acipenserid species can encounter each other. We think that the conclusion about the species-specific response of spermatozoa to the inductor of the acrosomal reaction is based on an insufficient material. It is known that in Acipenserids hybrids can easily be obtained between different species and even genera, for example "bester", a hybrid between *H. huso* and *A. ruthenus*, widely used in commercial sturgeon culture. Interspecific hybrids occur also in nature, although they are not frequent. Hybrids have been described between *A. gueldenstaedti* and *A. ruthenus*, *A. gueldenstaedti* and *A. stellatus*, *H. huso* and *A. gueldenstaedti*, *H. dauricus* and *A. schrencki*, etc. (Nikolyukin 1952). Such hybrids appear because different Acipenserid species can use the same spawning grounds and for many of them there is a narrow zone of coinciding spawning temperatures.

2.3.2 Transformations of the Egg After Contact with the Fertilizing Spermatozoon

After the fertilizing spermatozoon enters the micropyle and the acrosomal process comes in contact with the cytoplasm surface, the plasma membranes of the gametes fuse, and a number of changes are initiated in the egg.

The effective contact between the gametes in Acipenserids is established within 2–10 s after insemination (Ginsburg 1957, 1968, 1972). This was shown in experiments on *A. stellatus* and *A. gueldenstaedti* in which the spermatozoa were inactivated by a detergent within different periods of time after insemination.

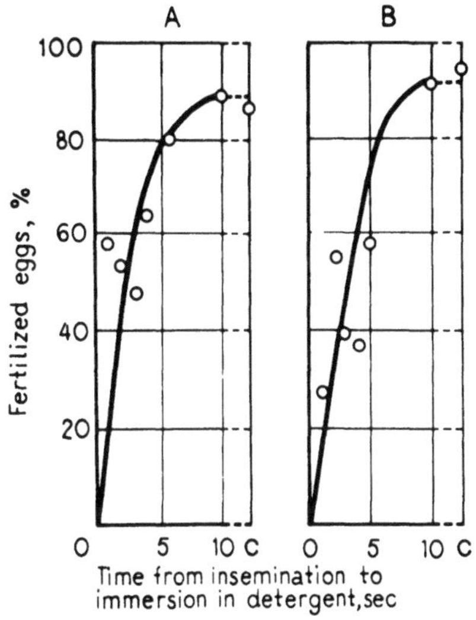

Fig. 26A, B. The time of successful sperm-egg encounter in *A. stellatus* (Ginsburg 1968). Sperm concentration in the fertilizing suspensions: **A** 5.9 $\times 10^7$; **B** 8.4 $\times 10^7$ spermatozoa/ml; the detergent was added 1–10 s after insemination. *C* Control

The eggs which have established effective contact with the spermatozoon before the treatment with the detergent divided during subsequent incubation, while all others remained unchanged. It can be seen in Fig. 26 that 40 to 60% of the eggs are fertilized within 2–3 s and all eggs capable of fertilization within 10 s.

The establishment of the effective contact with the fertilizing spermatozoon is followed by the appearance and spreading of the activation impulse, which relieves the meiotic arrest; cortical reaction, which leads to the formation of a narrow perivitelline space and liberation of the egg from its connection with the egg membranes; secretion of the hydrophilic colloid and extension of the perivitelline space; swelling of the egg membranes and acquisition of stickiness by them. All these changes can take place also in a mature unfertilized egg after its artificial activation, for example, by needle pricking.

Activation Impulse and Breaking of Meiotic Arrest

In order to determine the rate of spreading of the impulse of activation, which breaks the meiotic arrest and leads to the completion of the II meiotic division and to compare it with the rate of spreading of the cortical reaction, an experiment was carried out with activation of the egg by needle pricking at various distances from the site where the nucleus is located (Dettlaff 1962). Mature unfertilized eggs of *H. huso* and *A. gueldenstaedti* were activated by pricking with a fine glass needle at the boundary between the animal and vegetal parts, i.e., not far from the nucleus located near the animal pole, or in the vegetal

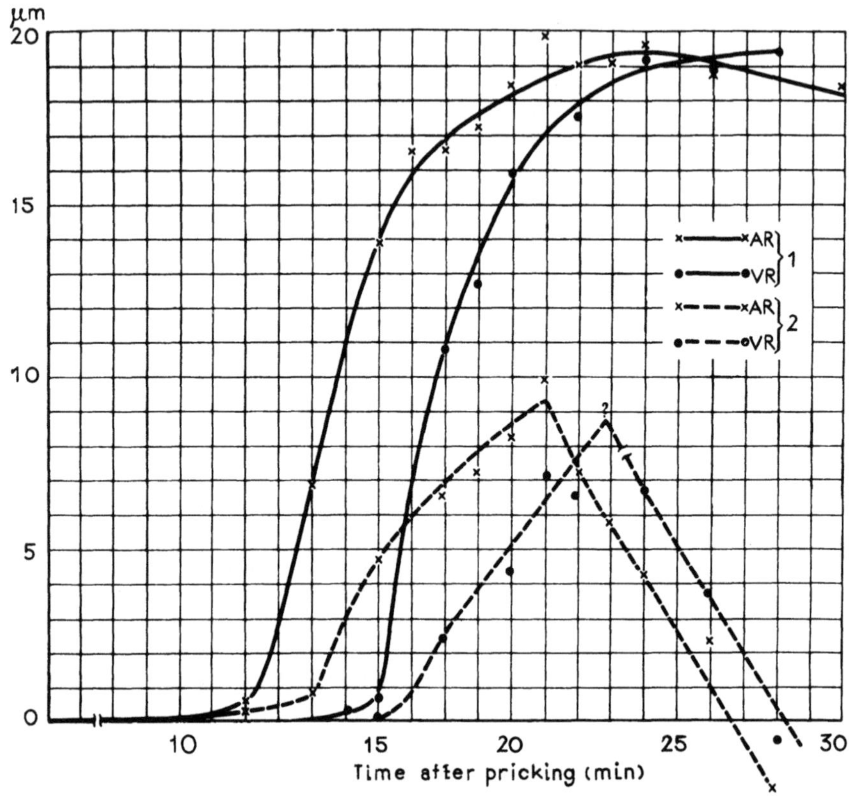

Fig. 27. Changes of distance between sister chromosomes (*curves 1*) and increase in the spindle length (*curves 2*) during the II meiotic division in *A. gueldenstaedti* eggs activated by pricking at the boundary of the animal region (*AR*) and at the vegetal pole region (*VR*) at 14.9–15.2 °C (Dettlaff 1962)

pole region. The eggs were fixed with an interval of 1 min and serial sections were prepared. The distance between homologous chromosomes and the length of the spindle were measured on a drawing of the mitotic figure made with a drawing device (Fig. 27).

When the egg was pricked at the boundary between the animal and vegetal areas, separation and very slow divergence of homologous chromosomes were observed suggesting the re-initiation of meiosis. These processes occupied a significant period of time, 12 min in the described experiment. They were followed by a rapid chromosome divergence together with spindle elongation (characterizing the stage of anaphase—Dettlaff 1963) and, then, extrusion of the II polar body. Hence, the time required for the transition of the blocked metaphase II to rapid chromosome divergence (anaphase) amounts to 0.2 τ_0 and corresponds to the duration of metaphase during the period of synchronous cleavage divisions (see Sect. 2.4).

When the egg was activated by pricking in the vegetal part, these processes were retarded. This difference was especially distinct during the period of rapid chromosome divergence. The difference in timing of this period was 3 min. At the same time, the cortical reaction in an *A. gueldenstaedti* egg inseminated at a somewhat higher temperature (18 °C), covered the whole egg surface, from the animal pole to the vegetal, for a period slightly exceeding 3 min. Hence, the rate of spreading of the activation impulse, which induces re-initiation of the II meiotic division, is close to that of cortical reaction.

Exocytosis of the Cortical Granules (Cortical Reaction) and Protection of the Egg from Penetration of Supernumerary Spermatozoa

The cortical reaction starts in the center of the animal area, where the fertilizing spermatozoon penetrates through one of micropyles. From there the cortical reaction spreads in all directions in a wave-like manner and covers, in succession, the animal area, the equatorial zone, and is completed at the vegetal pole.

Exocytosis starts after a very short lag period, no more than 2 s long. This is corroborated both by studies of the eggs fixed at short intervals after insemination and vital observations of the eggs after their artificial activation (Ginsburg 1961, 1987). The unfertilized eggs of *A. gueldenstaedti* were placed in water and activated by needle pricking at a site close to the animal pole. Within 1 or 2 s (at 21–22 °C) a light spot appeared at the site of pricking which was then spreading in all directions.

Electron microscope studies of fertilized *A. stellatus* eggs (Ginsburg and Nikiforova 1978) show significant changes of the cortical layer in the region of micropyles within 3–6 s after insemination (Fig. 16B). The cortical granules located near the plasma membrane opened up and discharged their contents under the zona radiata interna. Within 10 to 30 s, the cortical granules located at some distance from the plasma membrane were fusing with the above lying granules, and their contents joined the previously released material. Within 2 or 3 min, the discharged contents of the cortical granules forms a single continuous layer closely adhering to the egg membrane (Fig. 16C).

Under an optical microscope, the first changes in the cortical layer are also found within a few seconds after insemination (Dettlaff 1957, 1962; Ginsburg 1968, 1972, 1987). Vacuoles arise at the level of cortical granules and the regions of cytoplasm in between assume the shape of columns. Soon after this, the columns are elongated and, together with the cortical granules, separated from the egg (Fig. 15). Such an early separation of the cortical granule material is an artifact caused by the treatment of the eggs for cytological study. However, the early changes in the cortical layer, as seen under the light microscope, coincide in time with the onset of the cortical reaction according to the electron microscopical data and can be used as a criterion of the cortical reaction spreading over the egg surface (Fig. 28).

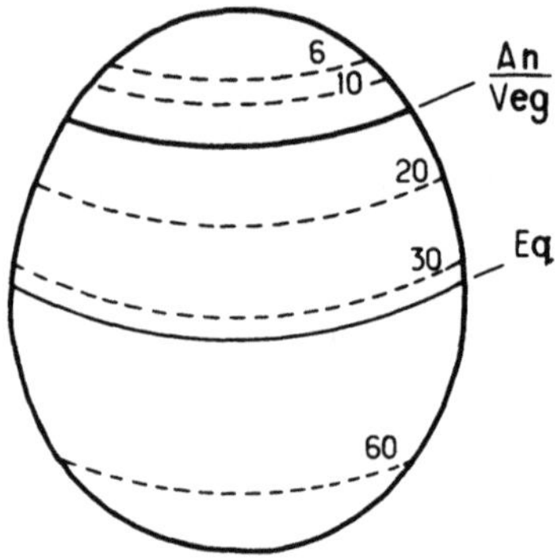

Fig. 28. Spreading of the cortical reaction over the surface of a fertilized *A. stellatus* egg at 21.9 °C (Ginsburg and Nikiforova 1978). *An/Veg* Boundary between the animal and vegetal areas; *Eq* equator. *Numerals* designate seconds after insemination

The rate of spreading of the cortical reaction is especially high during the first 5–6 s after insemination, when the reaction embraces the apex of the animal region: it averages 360 μm/s in the eggs of *A. stellatus* (Ginsburg and Nikiforova 1978). As the cortical reaction progresses in the vegetal direction, its rate decreases. It covers the whole egg surface within 2–5 min or more (depending on temperature).

Exocytosis of the cortical granule contents and its slight swelling in the first minutes after the cortical reaction initiation lead to the formation of a narrow perivitelline space. This space is still not discernible upon vital observations of the fertilized eggs in a vertical chamber with a low magnification of the microscope (Dettlaff and Ginsburg 1954). This is the first phase of the perivitelline space formation. It leads to liberation of the egg from its connection with zona radiata interna.

At the same time, the cortical reaction fulfills an important function of protecting the egg from penetration of supernumerary spermatozoa.

As we have already mentioned, the diameter of the terminal tubule of the micropyle does not exceed markedly the width of the spermatozoon head, therefore preventing simultaneous penetration of more than one spermatozoon through the micropyle. But, since the sturgeon eggs have several, sometimes many, micropyles, several spermatozoa could have penetrated through them, thus leading, as we will see later, to abnormal development of the embryos and their subsequent death.

Polyspermic fertilization is prevented by a special mechanism, the block to polyspermy (see Ginsburg 1987). The cortical reaction plays the central part in this mechanism in sturgeon, as suggested by the results of the following experiment.

The eggs of *A. gueldenstaedti* and *H. huso* were activated by pricking with a glass needle, as when studying the spreading of the activation impulse, near the animal pole (and, hence, near the site of the micropyle location) or in the vegetal

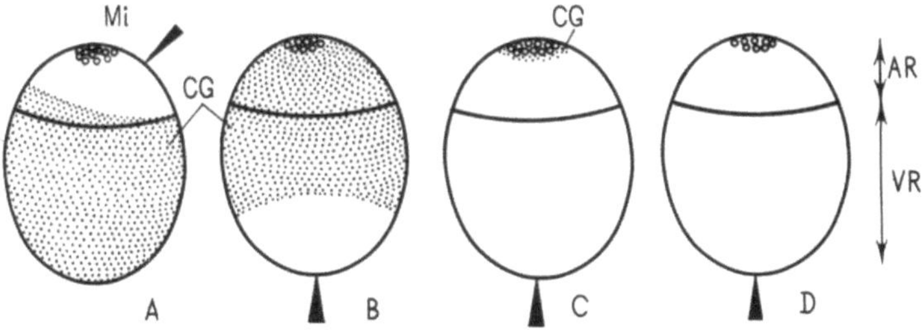

Fig. 29A–D. Diagram of the cortical reaction spreading at various times after pricking the egg of *A. gueldenstaedti* at 13.8 °C (Ginsburg 1961). **A** 5 s after pricking near the animal pole; **B** 2 min; **C** 3 min; **D** 4 min after pricking in the vegetal pole region. *AR* Animal region; *CG* cortical granules; *Mi* micropyles; *VR* vegetal region. The part of the egg surface where the cortical granule contents have already discharged is *white*

Table 3. Activation of the *A. gueldenstaedti* eggs by needle pricking near the animal pole or in the vegetal area with subsequent insemination (at 13.8 °C)

Interval between pricking and inseminantion	Pricking near the animal pole			Pricking in the vegetal area		
	n	Fertilized eggs		n	Fertilized eggs	
		n	%		n	%
2 s	60	2	3.3			
5 s	40	4	10.0			
10 s	20	0	0			
1 min				20	16	80.0
2 min				20	19	95.0
3 min				30	13	43.3
4 min				20	0	0
5 min				20	0	0

pole region (Fig. 29). The eggs were then inseminated at different time intervals, from 1–2 s to several minutes. In a typical experiment of this kind (Table 3), in the first variant the eggs lost fertilizability within 1–5 s, while in the second one they retained this ability for a long time, up to 3 min. The cytological control has shown that the loss of fertilizability in both variants of the experiment coincided precisely with the moment of exocytosis of the cortical granule contents in the region of micropyles. In the second variant, the cortical reaction in the micropylar region was completed within 3 min only in some eggs and, correspondingly, more than 40% of the eggs could be fertilized. Within 4 min, exocytosis of the cortical granule contents in the micropylar region took place in all eggs and they all lost fertilizability. It is essential that no decrease in the egg fertilizability was observed before the spreading cortical reaction reached the micropylar region.

Apparently, in sturgeon, as in *Oryzias latipes* (Nuccitelli 1980), the early electric phase of polyspermy block is absent and the block to polyspermy is a single-step process.

After the cortical reaction is completed in the micropylar region, the discharged contents of the cortical granules prevents the spermatozoa, which enter any micropyle, from penetrating the cytoplasm (Fig. 16C). In experiments on salmon (Ginsburg 1961, 1963a), when spermatozoa come in contact with the secreted contents of the cortical alveoli, they agglutinate. The surface of the spermatozoa becomes sticky, their tails are attached to each other and form loops, and they lose the ability of forward motion.

Due to polyspermy block, fertilization of the good-quality eggs can be polyspermic only rarely, when several spermatozoa penetrate through different micropylar canals strictly simultaneously. The percentage of polyspermic eggs increases when the eggs are inseminated by a highly concentrated sperm suspension. It can also be significant if the cortical reaction spreading is delayed, as is the case given the poor physiological condition of the eggs.

In natural conditions, the mechanism of polyspermy block in Acipenserids appears to be effective. The contact of the fertilizing spermatozoon with the egg initiates the cortical reaction within a short period of time (1–2 s); this reaction covers the whole area of micropyle, which usually has a diameter of less than 200 μm, within fractions of a second. Under conditions of spawning in a rapid current when the eggs and spermatozoa are disseminated by the current, this mechanism is able to ensure monospermic fertilization.

Widening of the Perivitelline Space

In Acipenserids, unlike teleostean fish (Kusa 1956), the perivitelline space formation is a biphasic process (Dettlaff and Ginsburg 1954). After the first phase related to exocytosis of the cortical granule contents is completed, the second phase begins, widening of the perivitelline space due to secretion of hydrophilic colloid (Dettlaff 1962). Globules of hydrophilic colloid which appear during oocyte maturation upon transition of karyoplasm into the cytoplasm (see Sect. 1.2.2) slowly move after egg activation into the superficial layer of animal cytoplasm. The lift of the hydrophilic colloid globules to the surface starts soon after fertilization and lasts 30–40 min (at 18 °C). At this time, the globules lose their spherical shape, swell, and vacuoles appear in them. At first, clusters of globules form islands at the egg surface, which then merge in a continuous layer. In this layer the colloid vacuolizes, hydrates, and comes into perivitelline space. During the period of intensive colloid secretion, the superficial layer of the egg loses distinct contours and some pigment granules and glycogen globules migrate together with the colloid in the perivitelline space.

During vital observations of changes in the fertilized eggs in a vertical chamber, the perivitelline space becomes discernible under low magnification of

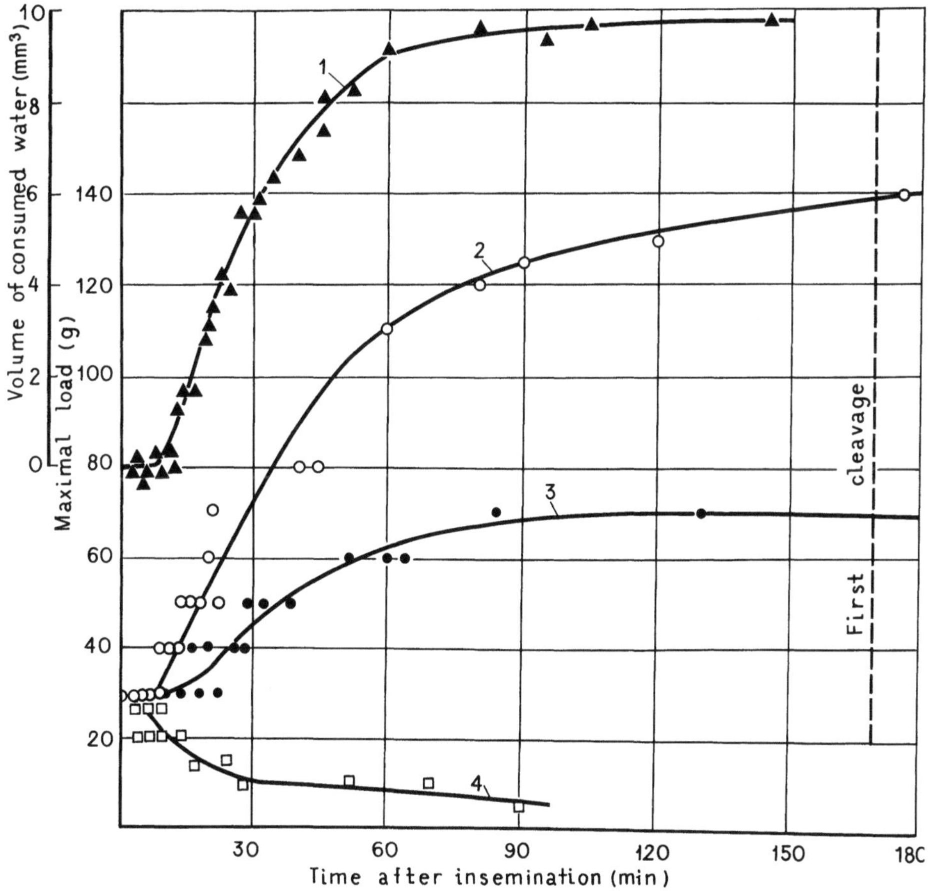

Fig. 30. Water consumption by the fertilized egg (*1*) and hardening of the membrane in the fertilized (*2, 3*) and unfertilized (*4*) eggs of *A. gueldenstaedti* (Zotin 1953c) *1, 2, 4* The eggs in all membranes were studied; *3* the eggs in zona radiata interna alone were studied

a microscope only within 15–20 min, while during the next 20–25 min its width rapidly increases. During this period, the water input in the egg, which began a few minutes after fertilization, rapidly increases as well (Fig. 30; Zotin 1953a). The hydrophilic colloid accumulated in the superficial layer of the cytoplasm swells at the expense of consumed water and passes in the perivitelline space. By 45–50 min after egg activation the volume of the perivitelline space achieves the maximum value (Fig. 36C, Plate I, 2 lat), and the hydrophilic colloid secretion from the egg ceases. At this time the endoplasm becomes more homogeneous, lacunae of the karyoplasm are absent in it, and the egg surface assumes smoother contours.

The hydrophilic colloid expelled by the egg in the perivitelline space differs from the material of the cortical granules in its chemical nature. Unlike the cortical granules, it does not contain mucopolysaccharides.

The first and second phases of the perivitelline space formation are causally interrelated, since in the absence of cortical reaction the hydrophilic colloid accumulated in the superficial cytoplasm is not secreted from the egg.

The biphasic pattern of the perivitelline space formation is inherent not only in sturgeon. Similar patterns of the hydrophilic colloid secretion from the animal part of the egg are also observed in the activated eggs of the *Anura*: *Rana temporaria* and *Bufo bufo* (Dettlaff unpubl. data). In the absence of hydrophilic colloid, activation of the *B. bufo* eggs (matured after the GV removal) by needle pricking is accompanied by flattening of the animal area. It was similar to that observed during normal development simultaneously with the hydrophilic colloid secretion in the perivitelline space. Hence, this flattening is due to the active contraction of microfilaments in the animal area of the egg, rather than to the pressure of the perivitelline fluid.

Nuclear Transformations During Fertilization

In Acipenserids, the spermatozoon penetrates the egg cytoplasm as a whole, together with its tail. Its head is then separated and turns through 180° (Fig. 31, 1A–C), as in many other animal species (Wilson 1925). At the same time, the head starts swelling, the chromatin loosens, and the compact nucleus of the spermatozoon head transforms in a light vesicular male pronucleus with fine-granular chromatin (Fig. 31, 2, 3). A spermaster forms around the sperm centrosome near the pronucleus (Fig. 31, 4). Thereafter, the male pronucleus gradually enlarges and, with the spermaster coming ahead of it, moves towards the center of the animal part of the egg filled with fine-granular yolk, where its encounter with the female pronucleus should take place (Fig. 33, 4, 6).

Detailed cytological studies of fertilization in *H. huso* (Ginsburg 1959) have shown that within the first 20 min after insemination the size of the sperm nucleus remains practically unchanged. Then it starts increasing (Fig. 34), the centrosome becomes surrounded by an aster, spermaster (Fig. 33, 3), and the swelling male pronucleus starts migrating in the depth of the cytoplasm (Fig. 35). The spermatozoon centrosome then divides and an achromatin spindle forms (Fig. 33, 5). At this time the rate of movement of the male pronucleus is maximal, 9 μm/s.

In parallel with these changes of the male nucleus, the II meiotic division is completed, the II polar body extruded and a group of chromosomes remaining in the cytoplasm (a haploid set) transforms into a female pronucleus, as in *A. gueldenstaedti* (Fig. 32). The female pronucleus is a light vesicular nucleus with fine-granular chromatin which is smaller than the male one at all stages (Fig. 34). The female pronucleus moves inward (Fig. 33, 7) and, in the center of the animal cytoplasm, encounters the male pronucleus (Fig. 33, 8).

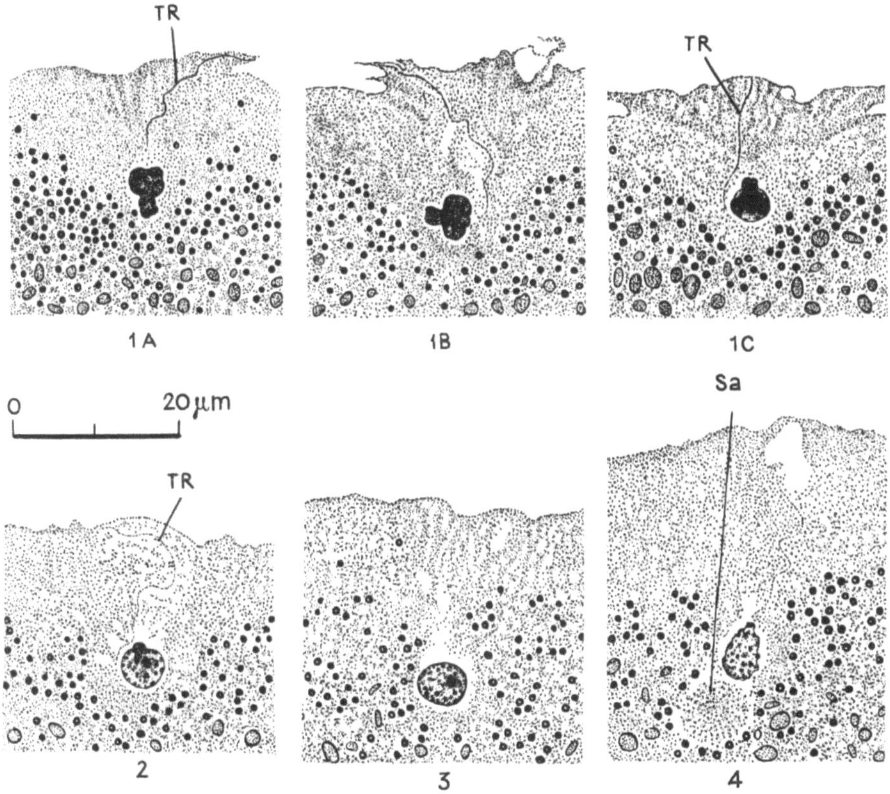

Fig. 31, 1 A–C Formation of the male pronucleus in *A. gueldenstaedti* at 22.4 °C (Ginsburg 1959). **1 A–C** 5 min; **2** 7 min; **3** 10 min, and **4** 15 min after insemination. *Sa* Spermaster; *TR* tail region of the spermatozoon. Sanfelice, iron hematoxylin

For a long time (20–40 min), both pronuclei remain in close contact without their nuclear envelope disintegrating. The spindle rays embrace both pronuclei (Fig. 33, 8). The pronuclei then lose their regular shape and the chromosomes form a distinct group in each of them. Finally, the 1st cleavage spindle forms and paternal and maternal chromosomes become arranged on its equator. At this stage, the normal for the species diploid chromosome number is restored and the process of fertilization completed.

The chronology of nuclear transformations during fertilization in *H. huso* is presented in Table 4.

Turning of the Egg Inside the Membranes, Changes in the Pigment Pattern of the Animal Region, and Establishment of Bilateral Symmetry

As a result of the cortical reaction, a thin (only a few microns thick) layer of the extruded material of the cortical granules appears between the surface of the

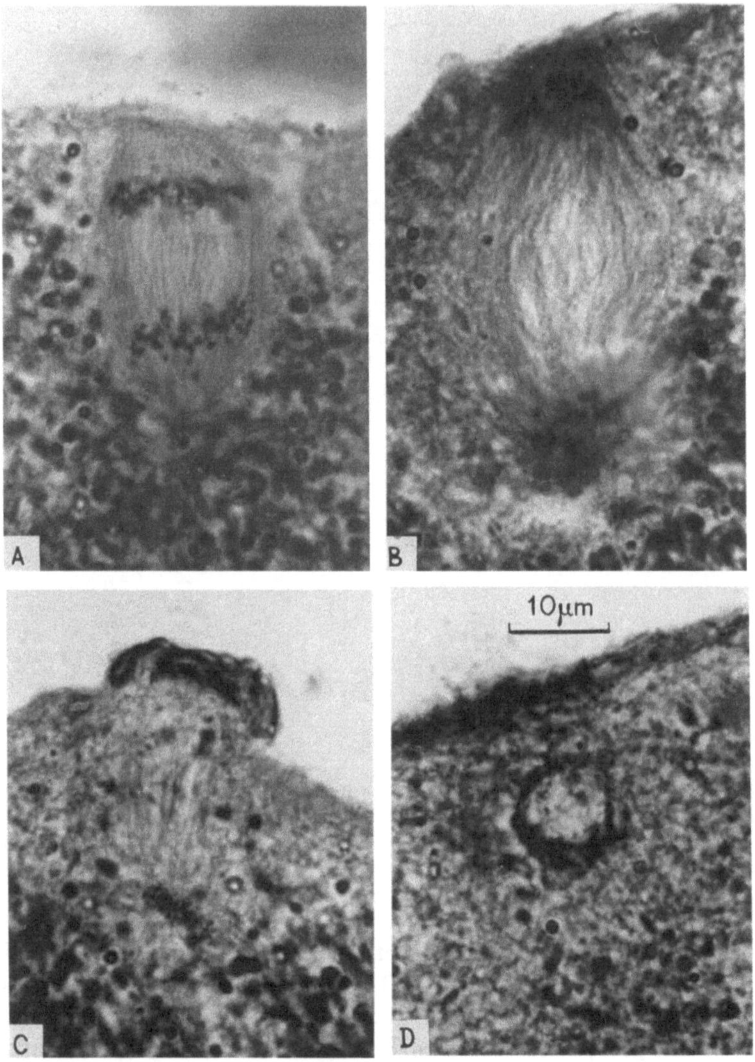

Fig. 32A–D. Completion of the II meiotic division and formation of the female pronucleus in the egg of *A. gueldenstaedti*. **A** Anaphase II; **B** telophase II; **C** extrusion of the II polar body; **D** female pronucleus near the egg surface

Fig. 33. Nuclear transformations during fertilization in *H. huso* (Ginsburg 1959), **1** Spermatozoon penetrates in the cytoplasm (its head is in the fertilization cone); **2** compact male pronucleus; **3** male pronucleus starts swelling and a spermaster forms; **4** male pronucleus increased in size and penetrated deeper into the cytoplasm; **5** spermaster near the pronucleus divided into two asters; **6** and **7** swollen male (**6**) and female (**7**) pronuclei; **8** pronuclei in close contact. *Sa* Spermaster; *ST* sperm trail, trace left by the spermaster and male pronucleus penetrating deep into the cytoplasm

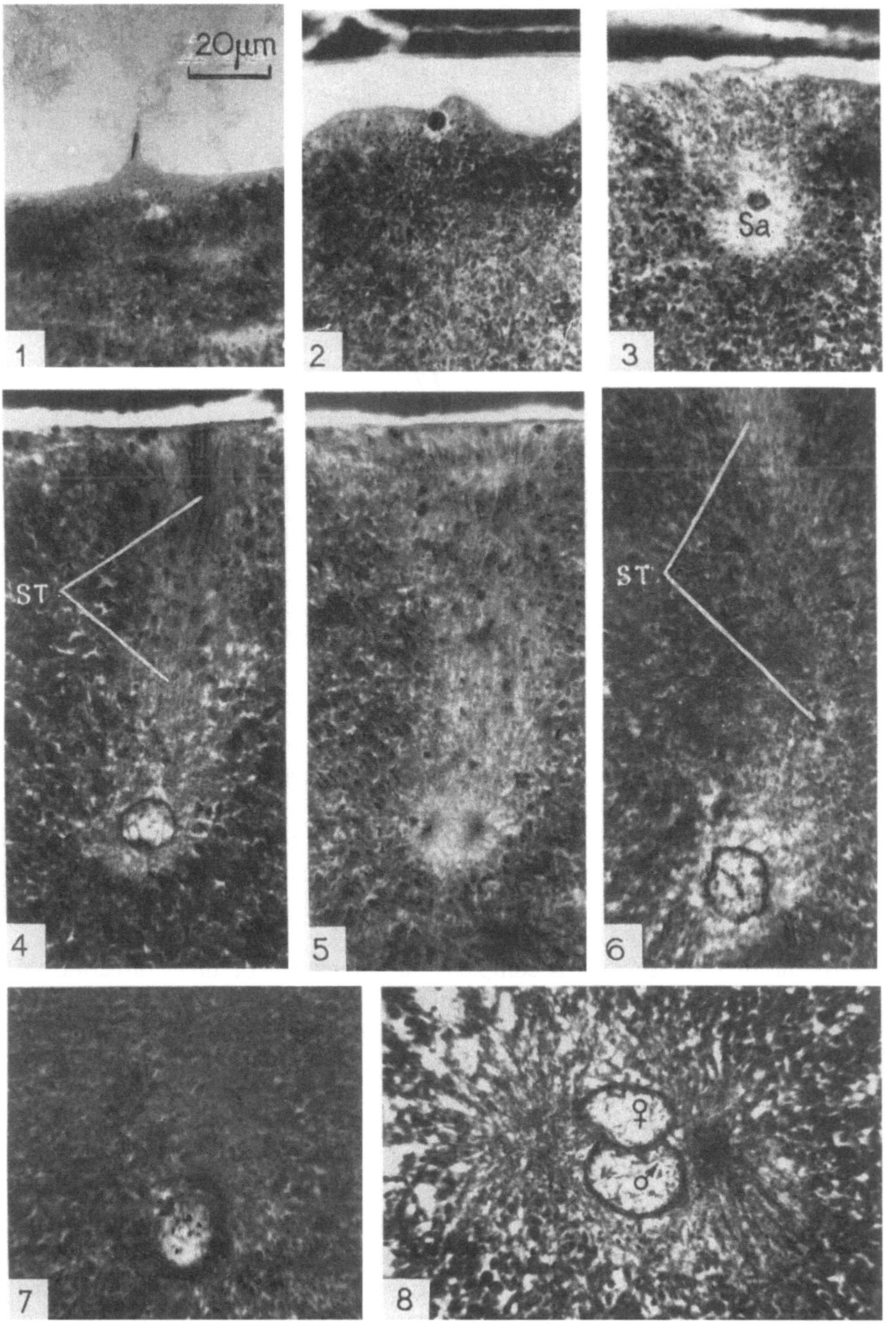
20μm
Sa
ST
ST
♀
♂
1
2
3
4
5
6
7
8

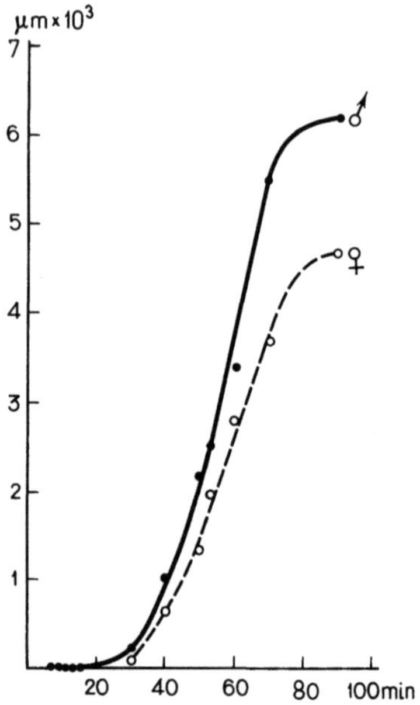

Fig. 34. Changes in the volume of the pronuclei in *H. huso* at 18.4 °C (Ginsburg 1959). *Ordinate*, volume of pronucleus; *abscissa*, time after insemination

cytoplasm and the membranes (Fig. 16C). The egg becomes free inside the membranes and, after attachment to the substrate (see later), begins to turn according to the position of its center of gravity, with the yolk-rich vegetal hemisphere downwards and the animal area upwards (Fig. 36A, B). The turning of the egg begins several minutes after insemination, occurs slowly and gradually and usually ends after 15–20 min.

By the time the turning is completed, a narrow slit becomes discernible (if viewed from the side) in the upper part of the egg, between the cytoplasm and the membranes. The slit then widens due to the hydrophilic colloid secretion from the egg (see earlier), a perivitelline space of considerable size forms, and the animal area of the egg flattens (Fig. 36C; Plate I, 2 lat).

The unique pigment pattern of the animal region, a light polar spot in the center and dark pigmented rings around it, changes after fertilization. The pigment is gradually drawn together, accumulating into the center of the animal area (Plate I, 2an); the light polar spot disappears or, if it was very large (which is often observed in *H. huso*), markedly decreases in size.

After a certain time (1 h after insemination at 20 °C in *A. stellatus* and 3.5 h at 11 °C in *A. gueldenstaedti*), a very light, sometimes absolutely white strip appears at the margin of the animal area, which is almost half the length of the circumference and gradually widens assuming the form of a light crescent. Sometimes the crescent is bright, with distinct boundaries (Plate I, 3 lat and 3an);

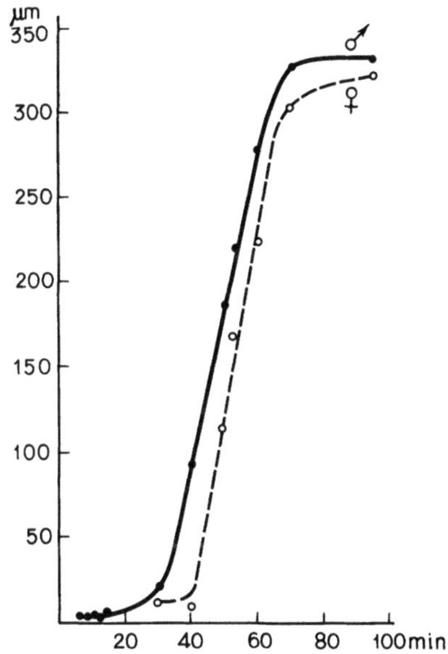

Fig. 35. Migration of the pronuclei in the ooplasm in *H. huso* at 18.4 °C (Ginsburg 1959). *Ordinate*, distance between the pronucleus and the egg surface; *abscissa*, time after insemination

in other cases, the boundaries are less defined, and in some batches it cannot be discerned at all.

Observations have shown that the light crescent indicates the future position of the dorsal blastopore lip, its central part corresponding to the middle of light crescent. These observations were confirmed by experiments with constriction of the eggs of *A. gueldenstaedti* and *A. stellatus* (Fig. 37). The eggs with distinct light crescent at the 1st cleavage division stage were constricted with a fine silk thread or hair along the 1st cleavage furrow. The dorsal lip of blastopore developed in them in strict correspondence with the position of light crescent. In those cases when the 1st furrow and, correspondingly, the ligature superimposed over it, separated the light crescent material, the dorsal lip of blastopore was also divided in two parts, so that when a larger part of the crescent was in one of the halves, a larger part of the dorsal lip developed in this half. In those cases when, after constriction, the whole material of light crescent was in one of the halves, the dorsal lip of blastopore developed only in this half. At the later stages, the middorsal plane of the developing embryo always coincided with the plane crossing the middle of light crescent. Hence, the embryo acquires its bilateral symmetry in the beginning of development, even before the egg starts cleavage.

Formation of the light crescent has much in common with the early changes in amphibian eggs, with which the sturgeon eggs are very similar in structure and in many developmental features. It has been shown for anuran amphibians

Table 4. Chronology of nuclear processes during fertilization in the *H. huso* egg (at 18.4 °C). (After Ginsburg 1959)

Time after insemination		Stage	
min	τ_0	Male	Female
3	0.06	Spermatozoon left the micropyle terminal tubule and lies in the fertilization cone	Chromosomes at the equatorial plate of the II meiotic spindle separate from one another
7	0.14	Spherical pronucleus is formed	
9–15	0.18–0.30		Anaphase II
20	0.4	Pronucleus begins to increase in size and sink deeper in the cytoplasm	Telophase II
30	0.6	Spermaster distinctly seen	Extrusion of the II polar body and formation of pronucleus
40	0.8	Maximum velocity of pronucleus movement, spermaster has divided	Pronucleus is formed and starts to sink deeper in the cytoplasm
53	1.06	Encounter of pronuclei in some eggs	
70	1.4	Pronuclear walls are in close contact in 80% of cases	
80	1.6	Close contact between pronuclei in all cases, prophase chromosomes are distinctly seen	
100	2.0	Breakdown of the pronuclear envelopes	
110	2.2	Metaphase of the 1st cleavage division	
120	2.4	Anaphase of the 1st cleavage division	
130	2.6	Telophase of the 1st cleavage division, chromosome vesicles are formed	
150	3.0	Prophase of the 2nd cleavage division, the 1st cleavage furrow appears on the egg surface	

(Ancel and Vintemberger 1948; Vincent et al. 1986) that before the onset of cleavage the cortical cytoplasm shifts by about 30° with reference to the inner cytoplasm. In species with heavily pigmented eggs, this displacement is accompanied by appearance of a lighter-colored crescent-shaped region below the equator, a so-called gray crescent. The plane of displacement running through the middle of the gray crescent becomes the plane of bilateral symmetry.

Unlike the gray crescent in amphibians, the light crescent in sturgeon is located above the equator. This difference appears to be due to relatively greater yolk reserves in the sturgeon egg: in both cases, the lower boundary of the crescent coincides with the boundary of the vegetal area rich in yolk and distinctive in pigmentation.

In amphibians, the direction of rotation of the cortical cytoplasm is, under natural conditions, determined by the site of the sperm entry but it can be affected by the egg turn under the influence of gravity force (see Gilbert, 1988). In sturgeon, the spermatozoon penetration can play no role whatsoever in the egg

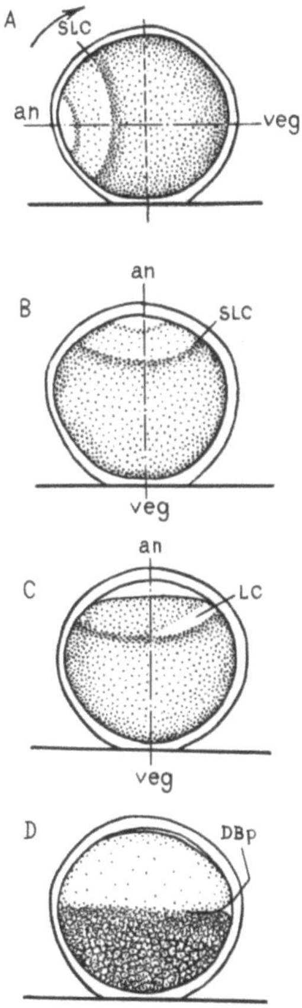

Fig. 36A–D. Position of the egg attached to the slide during: **A** the first few minutes after fertilization; **B** after the animal area turned upwards; **C** at the light crescent stage; **D** at the early gastrula stage (Dettlaff and Ginsburg 1954). *an-veg* Animal-vegetal axis of the egg; *DBp* dorsal lip of blastopore; *LC* light crescent; *SLC* site of prospective light crescent

symmetrization, since in them, unlike amphibians, the spermatozoon penetrates always only near the animal pole. In sturgeon, the direction of displacement of the cortical cytoplasm leading to bilateral symmetry is prepared by the egg turn occuring soon after fertilization (Ginsburg 1953a; Dettlaff and Ginsburg 1954). This was demonstrated by experiments in which eggs turned in a given direction and then the correlation between the direction of the turn and the plane of symmetry was established. One such experiment is depicted in Fig. 38. Unfertilized eggs of *A. stellatus* were placed in a row on a slide submerged in water in such a way that their animal-vegetal axis was oriented perpendicularly to the longer slide side and the animal pole faced from the observed (Fig. 38A). Sperm was then added, the fertilized eggs attached to the slide and turned. The turn direction was determined by the position of the eggs: it took place in the vertical

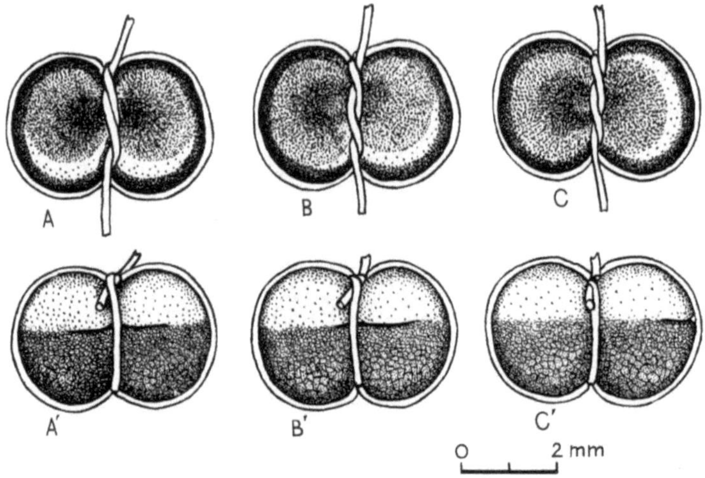

Fig. 37A–C. Constriction of *A. gueldenstaedti* eggs with distinct light crescent along the 1st cleavage furrow (Dettlaff and Ginsburg 1954). **A** 1st cleavage furrow divides the light crescent in approximately equal parts; **B** in unequal parts; **C** the light crescent is entirely in one of the two first blastomeres (upper view). **A'**, **B'**, and **C'** The same embryos at the early gastrula stage (lateral view)

plane running through the animal-vegetal axis of the egg, towards the observer. As a result, the light crescent (Fig. 38B) and later, correspondingly, the dorsal lip of blastopore (Fig. 38C) regularly appeared on the egg side which faced the observed and, thus, became the dorsal side of the embryo. The bilateral symmetry plane running through the middle of light crescent and dorsal lip of blastopore and, later, along the middorsal line of the embryo corresponded to the turn plane. It coincided precisely with the plane of turning or was oriented at a small angle with reference to it: at a directed turning by 90° this angle was less than 45° in 78% of cases, at a turning by 240° — in 92% of cases (Ginsburg 1953a).

Changes in the cytoplasm which determine the site of the light crescent formation are reversible: if the position of the already turned egg is changed within the next 20–30 min, so that it is forced to orient once more with its animal pole upwards, the symmetry plane will be established in accordance with the last turn.

Changes in the Egg Membranes

Following fertilization, the egg membranes become sticky, swell, and harden.

In mature unfertilized eggs of *H. huso*, *A. gueldenstaedti*, and *A. stellatus*, the membranes are cloudy, with individual layers not discernible. If they are placed

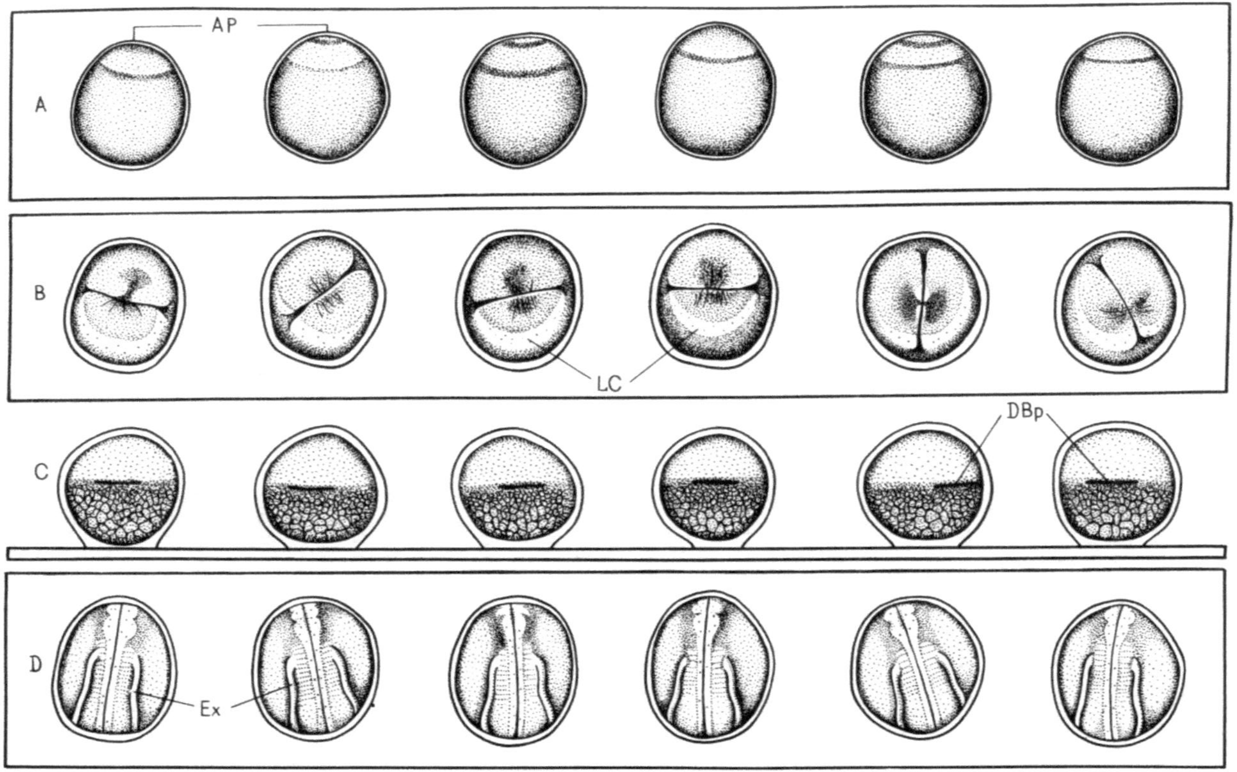

Fig. 38A–D. Development of bilateral symmetry in eggs which have turned in a given direction. The fertilized eggs of *A. stellatus* were placed on a slide, so that the animal-vegetal axis of the eggs was perpendicular to the longer side of the slide (Ginsburg 1953a). **A** Eggs during the first minutes after insemination; **B** after the eggs turned with the animal pole upwards at the stage of the 1st cleavage division; **C** early gastrula; **D** closed neural tube. **A**, **B**, and **D** upper view; **C** lateral view, from the side of the observer. *AP* Animal pole; *DBp* dorsal lip of blastopore; *Ex* rudiment of the excretory system; *LC* light crescent in a Petri dish, they lie there freely and move from their place readily when lightly touched.

Several minutes after insemination (2 to 13 min depending on the species, specific features of a given batch, and temperature), the jelly coat of the egg becomes sticky, and the eggs begin to stick to the bottom of the dish and to one another. Initially, the stickiness of the membranes is slight. Gradually, the membranes become increasingly sticky, and the eggs are firmly attached to the dish bottom. The eggs are usually attached so fast that a considerable effort is needed to detach them from the substrate.

In some batches, the eggs of *A. stellatus* and *A. gueldenstaedti* are weakly attached; these batches which occur both in stimulated females and mature females taken from the spawning ground are an exception.

The stickiness of jelly coat of the fertilized eggs is usually much more pronounced and develops much more rapidly than in unfertilized eggs from the same batch. This difference testifies against the development of adhesiveness solely at the expense of a fine layer of material on the surface of the jelly coat which hydrates when the unfertilized eggs are placed in water, as suggested by Cherr and Clark (1982, 1984a, 1985b). The great stickiness of the jelly coat in the fertilized eggs appears to be related to changes in its properties during fertilization.

After the eggs are attached, the stickiness of the membranes gradually disappears, remaining until hatching only at the site of their attachment to the substrate. Apparently, a substance responsible for the membrane stickiness is gradually washed off with water or changes its properties.

With the appearance of stickiness, the membranes begin to swell. This continues approximately until the 1st cleavage division, when the membranes become three times thicker than before fertilization. In *A. stellatus*, the membranes become 180–200 μm thick and in *A. gueldenstaedti colchicus* — 360 to 400 μm. With swelling, the cloudy membranes clear and become transparent. Now, even with a low magnification their layered structure becomes visible: we see (Plate I, 2 lat) the thin inner membrane zona radiata interna, which is closely apposed to the egg, zona radiata externa which envelops it, and a thicker superficial jelly coat.

These changes in the membranes are due to the effect of some substances secreted by the egg during its activation: if an unactivated egg adheres to the fertilized one, the active factor penetrates the membranes of the former and induces in them, in the area of contact, similar changes.

The jelly coat swells to a much greater extent than zona radiata interna and zona radiata externa. It acquires the consistency of a dense gel while two other envelopes have a leathery consistency and are ruptured upon stretching. By this time, zona radiata interna separates from zona radiata externa which remains closely connected with the jelly coat. Now two outer membranes can be easily removed together.

In parallel with the appearance and strengthening of stickiness and water consumption by the membranes, their mechanical properties undergo changes. Measurements taken with a special device have shown that after fertilization the membranes rapidly harden (Zotin 1961). For example, an unfertilized egg of *A. stellatus* withstands a load of no more than 15–20 g (it is crushed under a heavier load). Fifteen minutes after fertilization, it can already withstand double the load and an hour later — three to four times the load. By the moment of the 1st cleavage division, the strength of the membranes achieves the level which is maximal for the entire embryogenesis. At this time, the egg of *A. stellatus* withstands a load of 60–80 g, the egg of *A. gueldenstaedti* 80–140 g, and the egg of *H. huso* 120–160 g. Hence, the membranes of larger eggs are stronger.

The rapid hardening of the membranes, following the appearance of stickiness and the beginning of swelling, is very important during natural spawning. The eggs spawned by a female are carried away and dispersed by the current, and individual eggs, being slightly heavier than water, slowly settle to the bottom. At this time, their membranes are still soft and delicate but the eggs are not damaged, since they float, suspended in water. With the appearance of stickiness, the eggs are attached to stones or pebble, and here, if the eggs do not fall into a crack, they suffer impacts of sand particles, small stones, and everything that the river current carries with it. Only because of the increased strength of the membranes can the eggs withstand the impacts and blows.

This property of the membranes enables the eggs to endure small collisions with one another and with the walls of a fish-breeding apparatus, which are unavoidable during incubation of the demucilaged (made non-adhesive) eggs.

As far as the other property, the stickiness of the membranes, is concerned, it plays an important role in nature: because of it the embryos remain at the spawning grounds until the hatching of the prelarvae. Here, because of the strength of the current and the specific features of the ground, they are protected against silting and provided with favorable conditions for development. When sturgeon are bred at the hatcheries, the stickiness of the membranes loses the importance it has in natural reproduction. Consequently, in sturgeon culture the eggs are incubated in a free state, having been previously demucilaged.

Whether the incubation of sturgeon eggs deprived of their stickiness can give good results was subject of doubts by the end of the last century (see Dettlaff and Ginsburg 1954). This question was raised again by a group of researchers in the 1950s. These researchers believed that the attachment of the eggs is a prerequisite of the normal embryonic development and must be maintained during the incubation of the eggs for hatchery purposes (Matveev 1951; Sadov 1951; Emel'yanov 1953, 1961; Sadov and Kokhanskaya 1961). Presently, this question, which evoked a lively discussion in its time, has been closed. It was already obvious more than 20 years ago that normal and viable larvae develop from good-quality demucilaged eggs (see Ginsburg and Dettlaff 1969). Particularly demonstrative is the following example. Sexually mature *H. huso* spawners that now migrate in the rivers developed from the fry that had been released from the sturgeon hatcheries after the regulation of the river flow and the cessation of the natural spawning of this species. Hence, these spawners developed from the demucilaged eggs. The biological soundness of such eggs is, thus, unquestionable.

Elimination of Adhesiveness of Egg Membranes

The important element in the biotechnique is the preparation of inseminated eggs for incubation by demucilaging them. The manual method of demucilaging the eggs, which was used at sturgeon hatcheries about 20 years ago, is completely obsolete.

At the present time, demucilaging is done by mechanized means. The apparatus devised by R. K. Latypov–Yu. I. Lytkin and E. V. Orlov are used for this purpose. The first of these is a drum, in which the eggs are mixed during demucilaging with four-bladed rubber mixers. The Orlov apparatus, which has become most widely used, is a metallic container with a double bottom, usually conical in shape, which is connected at the base with an air-feed system receiving air from a compressor (see Mil'shtein 1972). The apparatus is filled with a specially prepared suspension of fine river silt or infusorial flour. The inseminated eggs are placed in it and mixed by passing compressed-air bubbles through the suspension for 40–60 min. The demucilaged eggs are washed with clean water and placed in incubation apparatus. There are a few modifications of the Orlov apparatus (see, for example, Meshcheryakov and Bikbaev 1973).

The use of a chalk suspension for demucilaging the sturgeon eggs has also been proposed (Engel'gardt et al. 1957). Talc and milk are successfully used to demucilage *Cyprinus carpio* eggs. Good results have been obtained when using talc for demucilaging the sturgeon eggs (Soin 1977). These methods are not in use in sturgeon culture, but could be helpful for research. Specifically, *Saprolegnia* develops to a lesser degree on the eggs that have been demucilaged with chalk and talc and the membranes of *C. carpio* eggs remain transparent when demucilaged with a 10-fold diluted cow's milk.

Behavior of Unfertilized Eggs

All of these described changes occur as a result of fertilization of the egg. But, as a rule, some eggs remain unfertilized in all batches. In contrast to the fertilized eggs, the unfertilized ones remain lying with their animal-vegetal axis oriented at random and their shape remains unchanged: the animal region retains convexity and is closely apposed to the membrane. The membranes either do not acquire stickiness at all or become slightly sticky. In the last case the stickiness increases at a much slower rate than in the fertilized eggs. The unfertilized eggs stick to each other weakly and can be easily separated. Their membranes swell slowly, mainly in the vegetal region, assuming wave-shaped contours. The membranes remain cloudy and the layers in them are indistinguishable.

However, together with such eggs, we see among the non-inseminated eggs others that behave like fertilized ones. Their membranes acquire stickiness, swell, and the layers within them become discernible. The eggs turn with their animal pole upwards, and a perivitelline space forms. The pigment pattern also undergoes changes, though less regular than in the fertilized eggs. These eggs are activated, i.e., they began to develop, although they were not fertilized. Here, we deal with so-called virgin or parthenogenetic development. Initially, it has much in common with the normal development. Activated eggs can begin cleavage but it proceeds abnormally and soon ceases, and the eggs gradually die off (see Sect. 2.4).

After insemination, all batches contain, together with the fertilized eggs, a greater or lesser amount of activated and non-activated unfertilized eggs. When the eggs have been inseminated with good-quality sperm, the reasons must be sought in the eggs themselves: at the time the eggs come in contact with the spermatozoa, they either have not acquired fertilizability (they have not yet matured), or for some reason lost that ability (this occurs much more frequently).

Breeding Quality of the Eggs

The eggs vary in quality. This is reflected not only in their fertilizability, but also in subsequent development of the fertilized eggs. When good-quality eggs are inseminated, more than 80–90% of them are fertilized. If these eggs are incubated in favorable conditions, there is almost no loss and the prelarvae have normal structure.

However, it is not always possible to obtain such eggs. Among poor-quality eggs, the percentage of fertilization is lower and among the eggs, that have been fertilized, some develop abnormally and die during incubation or produce inviable prelarvae. Sometimes, serious developmental defects are observed in batches of eggs which still have a high fertilization rate.

Poor-quality eggs are a mixed group, which includes eggs that have been damaged for various reasons. In some cases, these eggs differ distinctly from good-quality eggs in their appearance and properties. These are primarily the batches with many damaged (broken) eggs: the eggs are crumpled, some of the eggs are ruptured and their milky-white contents are leaking out. Such eggs are not taken for insemination. In other cases, the eggs soften and become flaccid. They are easily crushed between the fingers, but do not break on their own. These are "soft" eggs. They are sometimes inseminated and incubated, but one cannot expect good results from the incubation of such eggs. Soft eggs are easily damaged during demucilaging and produce many abnormal embryos, and often great losses are observed, even before hatching. In addition, eggs are encountered which, one can say on the basis of their external appearance, primarily pigment pattern, are undergoing degeneration. Such eggs usually develop abnormally or are entirely incapable of development.

But it also happens that the eggs do not differ from normal eggs in external appearance; however, after insemination only part of them are fertilized. Among the eggs that have not been fertilized, we find both activated and non-activated eggs.

Activation of the eggs, as we have seen, can occur in the body of the female, especially when the water temperature in the holding tank with spawners suddenly changes by the end of their maturation. With a sudden drop in temperature, ovulation is inhibited and eggs in the follicles that have attained maturity become activated. A brief exposure to a high temperature can also activate the matured eggs (see Ginsburg 1968, 1972). Furthermore, the eggs can be activated as a result of injuries inflicted when the eggs are removed for insemination or dried, if they are exposed to the air without being covered by coelomic fluid. They can also be activated when the mature eggs are placed in water before insemination. In this case, activation takes place within a shorter period of time and at a higher rate in batches of more mature eggs than in those of less mature ones. Therefore, washing off the coelomic fluid before the sperm is added (when the wet method of insemination is used) can lead to a considerable drop in the percentage of fertilization. In all of these cases, activation of the eggs before insemination is the reason why they are not fertilizable.

The second group of unfertilized eggs consists of non-activated eggs which do not differ from good-quality mature eggs in external appearance. Usually, retention after ovulation in the body cavity of the female is the reason for their loss of fertilizability. With longer retention (or higher water temperature), the eggs are damaged. In rare cases, the unfertilizability of the eggs of this group may be due to their insufficient maturity (as a rule, immature eggs do not ovulate).

In order to avoid loading the incubation apparatus with poor-quality eggs, attempts have been repeatedly made (see Dettlaff and Ginsburg 1954) to develop methods of assessing the quality of the eggs even before insemination (for batches in which the eggs do not display degenerative changes in the pigment pattern or obvious decrease in turgor). Until now, reliable criteria for such an assessment have not been developed.

However, it is not so much the early rejection of poor-quality batches as the improvement of the hatchery quality of the obtained eggs that is very important to fish culture. Definite progress has been recently made in this direction: the optimum temperatures at which to keep spawners of *A. gueldenstaedti*, *A. stellatus*. *A. ruthenus*, and *H. huso* and the optimum times for obtaining the eggs at the various temperatures have been determined, thus allowing us to considerably reduce the loss of eggs owing to the non-fertilized eggs.

2.4 Cleavage and Blastulation

During cleavage, the sturgeon egg divides into ever smaller cells as a result of successive mitotic divisions. The nuclei are dividing at the maximal rate; the cell cycle is reduced: G_1 and G_2 phases are absent. A cavity arises in the embryo as a result of cleavage and after that it is called blastula.

The period of cleavage is followed by the period of blastulation which is characterized by the decrease in the rate of nuclear divisions, lengthening of the mitotic cycle, and realization of the morphogenetic nuclear function. This period begins at the late blastula stage.

2.4.1 Morphology of Cleavage

Unequal distribution of yolk inclusions in the sturgeon egg provides for unequal cleavage: in the animal part of the egg, which contains small deutoplasmic inclusions at a relatively low concentration, small blastomeres form, while in the vegetal part filled with many large yolk granules and lipid inclusions, the blastomeres are much larger. In accordance with the abundance of yolk inclusions, division of the vegetal blastomeres is completed much later than that of the animal ones, but it is completed in this region as well by the end of the cleavage period. Hence, the cleavage in sturgeon can be regarded as complete unequal cleavage.

The furrow of the 1st cleavage division is laid down at the anaphase of this division: a narrow band of loose cytoplasm oriented perpendicularly to the spindle axis becomes visible in the central part of the spindle (the same is seen during the subsequent divisions, see Fig. 41). The loose band spreads in all directions from the spindle. When it first reaches the cortical layer of the egg, a short light strip appears in the center of the animal area which resembles a crack: this strip signifies appearance of the 1st cleavage furrow on the egg surface. In Acipenserid fish, it appears $3\tau_0$ after insemination (see Table 4); then it deepens and an adjacent pigmented superficial layer of cytoplasm submerges into it. At the same time, the furrow spreads over the egg surface, relatively rapidly in the animal area and ever slower in the vegetal direction.

The 1st cleavage furrow is closed in the vegetal pole region when the 2nd cleavage furrows approach the equator and the 3rd cleavage furrows are laid down in the animal area (Fig. 39). The furrow submerges in the cytoplasm very

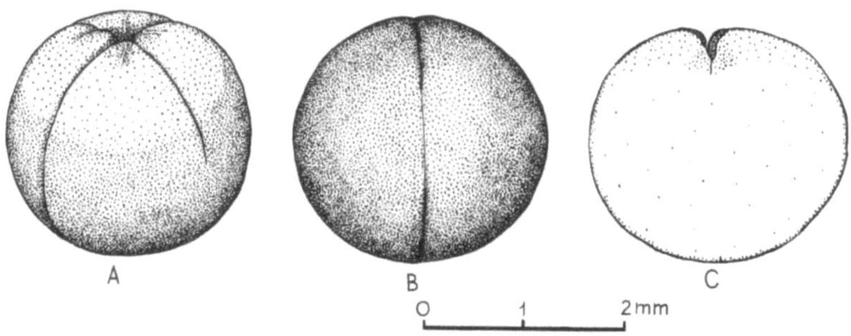

Fig. 39A–C. The egg of *A. stellatus* immediately after the 1st cleavage furrow closure in the vegetal pole region (Dettlaff and Ginsburg 1954). **A** lateral view; **B** view from the vegetal pole; **C** view on section running perpendicularly to the 1st cleavage furrow

slowly. The direction of the 1st cleavage furrow is not strictly oriented with reference to the light crescent and, hence, to the bilateral symmetry plane (Ginsburg 1953a; Fig. 38).

The 2nd cleavage furrows are laid down in the center of the animal area at a right angle to the 1st cleavage furrow and divide the animal area into four more or less equal parts (Plate II, 5an; see p. 240).

The position of cleavage furrows in sturgeon eggs strictly follows the Hertwig rule, according to which the spindle axis is oriented in the direction of the maximum length of the cytoplasm free of inclusions. Hence, the 3rd cleavage furrows are laid down meridianally (Fig. 40A–C), rather than equatorially, as is the case in animals with less yolky eggs, such as amphibians. The cleavage pattern depends also on the shape of the eggs: in spherical eggs the 3rd cleavage furrows are oriented radially (Fig. 40C). Such eggs are often encountered in *A. gueldenstaedti*, whereas in *H. huso* and *A. stellatus* the eggs are usually egg-shaped (Fig. 14). In the egg elongated along the animal-vegetal axis, the animal area assumes, after its turn, an oval shape, and, correspondingly, the 3rd cleavage furrows are laid down almost parallel to the 1st furrow, forming, together with the cleavage furrows laid down earlier, a characteristic pattern (Plate II, 6an; Fig. 40A, B).

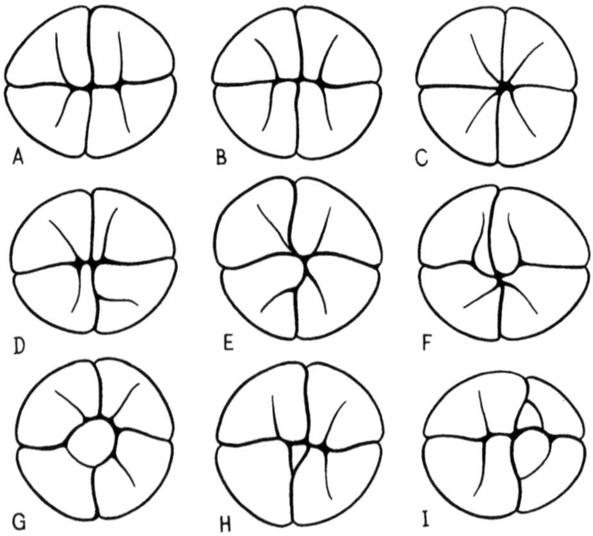

Fig. 40A–I. Pattern of furrows at the stage of 3rd cleavage division in *A. gueldenstaedti* (Ginsburg and Dettlaff 1955). **A, B** Eggs of oval shape, the 3rd cleavage furrows almost parallel to the 1st furrow; **C** globular egg, the pattern is radial; **D–F** significant differences in the shape and sizes of individual blastomeres; **G, H** a small blastomere separated in the center of the animal area; **I** two small blastomeres formed

The pattern formed by the cleavage furrows is usually geometrically not quite regular, and the blastomeres have different shapes and sizes (Fig. 40D–F). Small elongated or spherical blastomeres are sometimes produced already by the 3rd division (Fig. 40G–I).

An even greater diversity in cleavage pattern is observed during the 4th division, when the furrows in the narrower blastomeres are oriented latitudinally, isolating relatively small cells in the center of the animal region, and the furrows in the wider blastomeres are oriented obliquely or radially (Plate II, 7an).

The furrows of each of these divisions appear practically simultaneously in the animal parts of the different blastomeres due to the synchronous division of their nuclei.

A cavity arises in the embryo during cleavage. Already by the stages of 16–32 blastomeres (Plate II, 7 and 8), a fluid begins to accumulate between the blastomeres. This fluid moves the blastomeres apart and small cavities of irregular shapes are formed. Later, a single cavity forms as a result of the merger of these cavities, and the embryo acquires the structure characteristic of the blastula stage (for sections of the embryos at these stages see Dettlaff and Ginsburg 1954). The cavity of a blastula (blastocoel or cleavage cavity) is located closer to the animal pole, at the boundary with the blastomeres which contain a large amount of coarse-granular yolk (see Fig. 45). The size of this cavity gradually increases, while its roof becomes thinner.

At the early blastula stage, the cells of the animal region are still quite large and can be distinguished with a low magnification (Plate II, 11 lat and 11an). Somewhat later, at the late blastula stage (Plate III, 12 lat and 12an; see p. 242), they are already so small that they can be seen only under a microscope. The blastomeres in the vegetal region also become much smaller, but they are still significantly larger than the animal blastomeres. A zone of intermediate size blastomeres called the marginal zone lies between the small animal blastomeres and relatively large vegetal ones (Plate III, 12 lat). It is here, in that part of the marginal zone which corresponds to the light crescent region of the egg before cleavage, that later, during the period of blastulation, changes leading to a new period in development, gastrulation, begin.

2.4.2 Mitotic Cycle During the Period of Synchronous Cleavage Divisions and Its Dependence on Temperature

During the period of synchronous cleavage divisions, the nuclei in various blastomeres are, at any given moment, at the same mitotic phase. Successive phases of the mitotic cycle are depicted in Fig. 41.

The total duration of the cycle during this period was estimated according to the time interval between the onset of metaphases of the 1st and 2nd cleavage divisions. It was shown that the duration of mitotic cycle in Acipenserids equals the interval between the appearance on the egg surface of the 1st and 2nd or 2nd

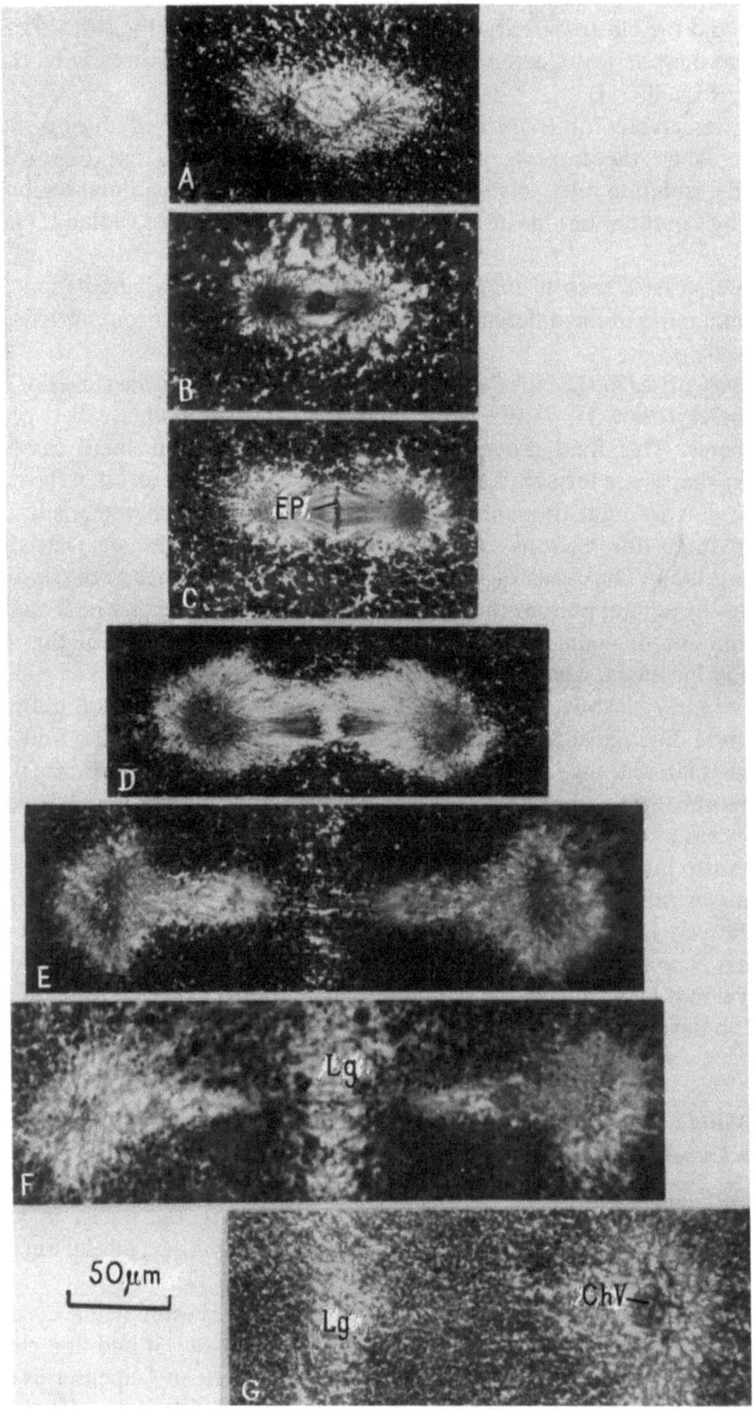
A
B
EP
C
D
E
Lg
F
50µm
Lg
ChV
G

and 3rd cleavage furrows (Ginsburg 1959). The same was established for amphibians, such as *Xenopus laevis* (Rudneva 1972) and *Ambystoma mexicanum* (Rott 1973). This gave us a convenient method of estimating the duration of the mitotic cycle during the period of synchronous cleavage divisions designated as τ_0. The value of τ_0 was estimated in the embryos of *A. gueldenstaedti, A. stellatus, H. huso*, and *A. ruthenus* at different temperatures. The data obtained are given in Fig. 42.

The value of τ_0 was used as a unit for measuring the relative duration of various developmental periods (see Sect. 2.8.2). In particular, this unit was used to estimate the relative duration of individual mitotic phases during this period.

The following criteria for determination of the ranges of subsequent mitotic phases were used.

Interphase and prophase (taken together): from the stage of nucleus consisting of closed chromosomal vesicles adjoined by two centers and up to the moment of the nuclear membrane disintegration and the transition of chromosomes into the cytoplasm. During this period the cleavage furrow of the preceding mitotic cycle appears on the egg surface for the first time.

Prometaphase: from the disintegration of the nuclear membrane up to the moment when kinetochores of the chromosomes arrange strictly in the equatorial plane of the spindle.

Metaphase: from the stage of equatorial plate up to the beginning of rapid movement of the chromosomes to the spindle poles.

Anaphase: from the beginning of rapid poleward movement of the chromosomes (and simultaneous elongation of the spindle) to their transformation into chromosomal vesicles. The central part of the spindle begins to disintegrate during anaphase.

Telophase: from the formation of chromosomal vesicles up to the termination of their movement to the centers.

In *A. gueldenstaedti*, the durations of these phases amounts to 0, 5 for interphase + prophase, about 0.1 for prometaphase, 0.2 for metaphase, 0.14 for anaphase, and 0.06 τ_0 for the telophase (Dettlaff 1963). It was also established that the time interval from insemination to the appearance of the 1st cleavage furrow on the egg surface in sturgeon is 3 τ_0.

Later it was shown (see Dettlaff et al. 1987) that phases of the mitotic cycle in all studied species of poikilothermic animals have practically the same relative (in τ_0) duration: in a sea urchin, four species of teleosts, and two species of amphibians. The interval from insemination up to the appearance of the 1st cleavage furrow in amphibians is, just as in sturgeon, 3 τ_0, while in teleosts it amounts to 2 τ_0.

Fig. 41A–G. Mitotic phases of the 2nd cleavage division in *A. gueldenstaedti* (Dettlaff 1963). **A** Prophase; **B** prometaphase; **C** metaphase; **D** anaphase; **E** early telophase; **F, G** late telophase. *ChV* Chromosomal vesicles; *EP* equatorial plate; *Lg* loosening of the cytoplasm in the region of future furrow

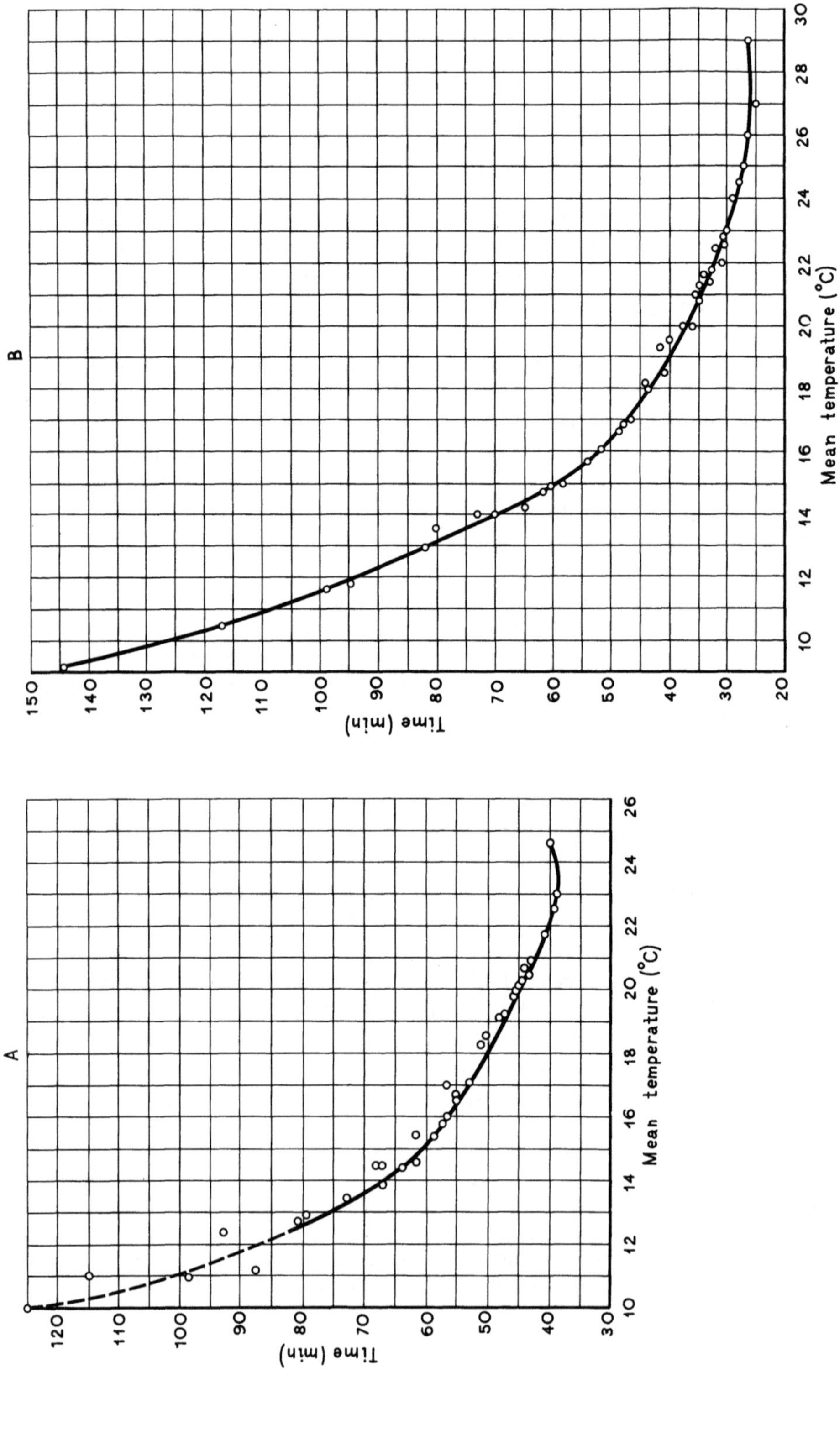
A
Time (min)
Mean temperature (°C)
B
Time (min)
Mean temperature (°C)

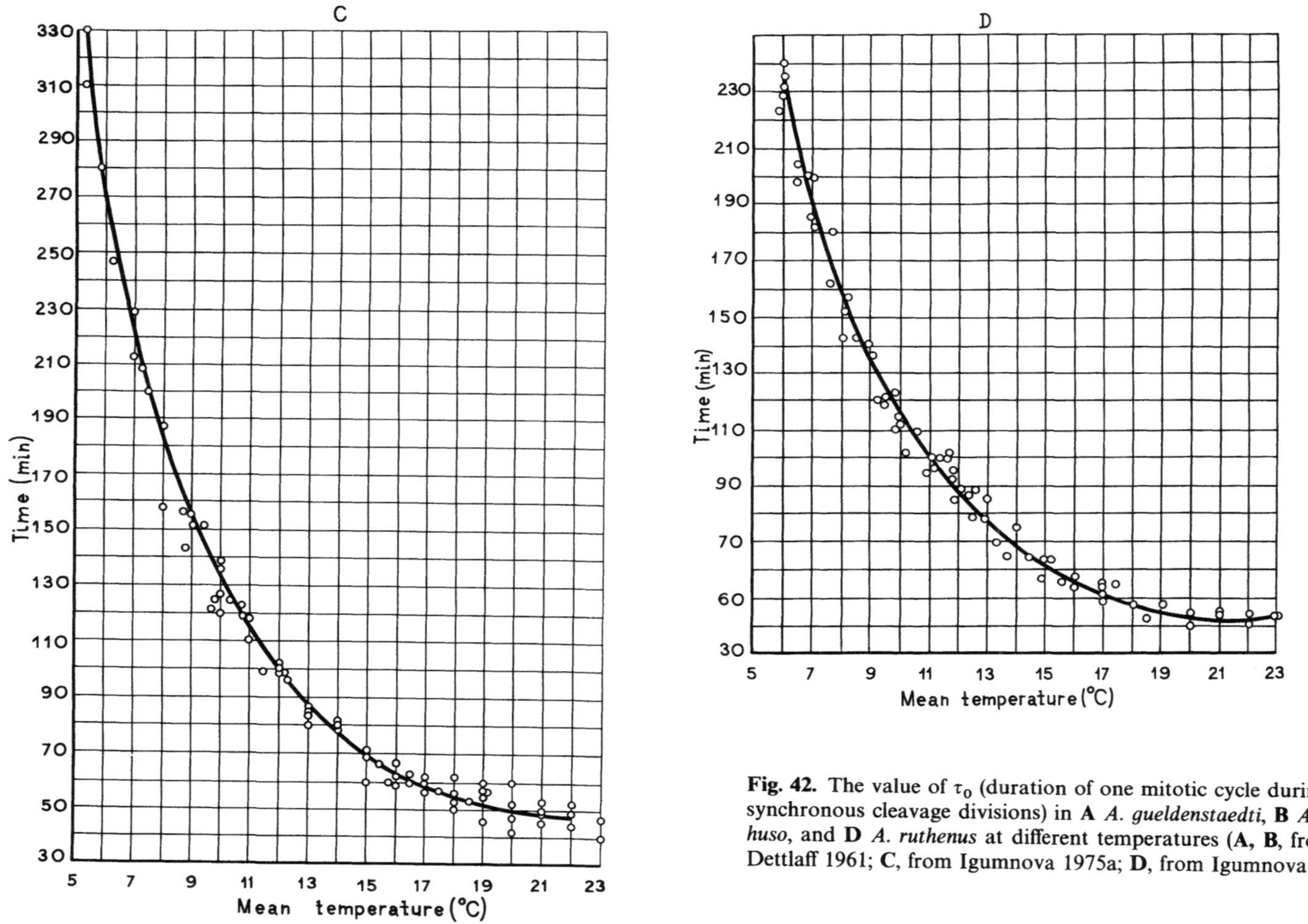

Fig. 42. The value of τ_0 (duration of one mitotic cycle during the period of synchronous cleavage divisions) in **A** *A. gueldenstaedti*, **B** *A. stellatus*, **C** *H. huso*, and **D** *A. ruthenus* at different temperatures (**A, B**, from Dettlaff and Dettlaff 1961; **C**, from Igumnova 1975a; **D**, from Igumnova 1985a)

Synchronous divisions of the nuclei in the animal area of *A. gueldenstaedti* embryos continue for 8 to 10 mitotic cycles (depending on temperature), and, thereafter, desynchronization of their division begins (Chulitskaya 1970).

2.4.3 Desynchronization of the Division of Nuclei During Cleavage and Rearrangement of the Cell Cycle During Transition to Blastulation

The time of desynchronization of the mitosis in the various blastomeres depends on temperature (Chulitskaya 1970). For example, at 23 °C desynchronization in *A. gueldenstaedti* embryos begins during the 8th cleavage division and at 17 °C during the 10th cleavage division. At first, desynchronization involves a small number of cells. The complete desynchronization of nuclear divisions occurs in the period between 14 to 16 τ_0 after insemination. After desynchronization is set in, nuclei at all five mitotic phases can be observed in the embryo at the same time and the percentage of nuclei at different phases of the cycle hardly differs from the relative duration of these phases during the period of synchronous cleavage divisions. This means that during the period of cleavage desynchronization the mitotic cycle preserves the structure it had during the period of synchronous divisions.

During the interval between 17 and 18 τ_0 after insemination, at the 15th cleavage division (stage 12 +), in the animal blastomeres of *A. gueldenstaedti* embryo rearrangement of the cell cycle occurs sharply. From this moment the embryo enters the period of blastulation (called, by Chulitskaya, the "period of asynchronous divisions"). Blastulation is characterized by the decrease in the mitotic index, lengthening of the mitotic cycle, loss of the karyomere structure by the nuclei, increase in the fraction of interphase nuclei, and appearance in them of pyroninophilic nucleoli. At this time, the morphogenetic nuclear function is realized which provides for the transition of the embryo to gastrulation. In the vegetal area, the cell cycle rearrangement occurs earlier than in the animal one but it was not studied in detail.

A similar sharp rearrangement of the cell cycle during transition from cleavage to blastulation was also described in embryos of *Rana temporaria* (Chulitskaya 1970) and *Xenopus laevis* (Newport and Kirshner 1982a, b). The stage of sharp rearrangement of the cell cycle which is accompanied by the onset of the processes of transcription and changes in some cell properties, was called midblastula transition (MBT) by Newport and Kirshner. Since this term is now widely accepted, we are using it for *A. gueldenstaedti*, although in this species the cell cycle rearrangement takes place at the late blastula stage rather than at the midblastula. The period from the MBT to the onset of gastrulation in *A. gueldenstaedti* lasts 2–3 τ_0.

In order to account for causes of the mitotic index fall at the end of the period of cleavage, a hypothesis was advanced suggesting that this is due to exhaustion in the cytoplasm of some substances essential for karyokinesis (Dettlaff 1964). For this hypothesis to be tested, the cytoplasm of mature unfertilized *A.*

gueldenstaedti eggs was injected into the eggs at the stage of 4th cleavage division (Chulitskaya 1970). The introduction of the cytoplasm in a volume of about 5% of the total egg volume led to the delay of the onset of desynchronization of the nuclear division in the animal blastomeres by one mitotic cycle and introduction of the same volume of supernatant of the homogenate of 250–300 eggs, by two cycles. However, these injections exerted no effect on the stage of cell cycle rearrangement (MBT).

The data obtained on sturgeon suggest that factors of desynchronization of nuclear divisions and cell cycle rearrangement are different (for review, see Rott 1980, 1987).

When studying changes in the properties of the cytoplasm in sturgeon embryos during the periods of cleavage and blastulation, the activity of maturation-promoting factor (MPF) was determined in the cytoplasm (Feulgenhauer 1977–see Dettlaff 1988; Ryabova and Dettlaff 1981; Dettlaff and Ryabova 1986). In experiments on *A. stellatus* it was shown that the cytoplasm taken from the egg before the onset of cleavage and the cytoplasm from the blastomeres at different mitotic phases during the first synchronous cleavage divisions, when injected in the fully grown oocytes, induces their maturation (GV breakdown) at a similar rate. At the stages of desynchronization, the MPF-activity of the cytoplasm decreased almost twofold and injections of the cytoplasm taken from the embryos during the period of blastulation induced GV breakdown in single oocytes only. The data obtained suggest that there is a correlation between the mitotic activity of the cells and the MPF content of their cytoplasm.

As to the MPF-activity of the cytoplasm at different mitotic phases during the first synchronous cleavage divisions, no differences were found both in *A. stellatus* and in *Rana temporaria* (Dettlaff and Ryabova 1986). This result does not agree with the data obtained for *Rana pipiens* (Wasserman and Smith 1978) and *Xenopus laevis* (Gerhart et al. 1984), according to which the MPF-activity of the cytoplasm from the blastomeres at different mitotic phases of the first cleavage divisions is different.

2.4.4 Cleavage Defects

The Pattern of Cleavage as a Criterion of Typical Embryonic Development

Some researchers have tried to use variations in the cleavage pattern to develop criteria for estimating the egg quality (Sadov and Kokhanskaya 1953, 1961). They recommended rejecting the eggs which, at the stage of eight blastomeres, were considered to be of poor quality by these criteria.

Sadov and Kokhanskaya believed that the biologically sound embryos have always a geometrically regular cleavage pattern either with furrows of the 3rd cleavage almost parallel to the 1st furrow, or radial, as well as of a mixed type combining these two, while the embryos having eight blasomeres of irregular shapes were considered as abnormal and a source of different deviations and mortality at all developmental stages.

Direct experiments involving individual incubation of *A. gueldenstaedti* and *A. stellatus* embryos with different cleavage patterns (Ginsburg 1954; Dettlaff and Ginsburg 1954) have proved that such a view is unfounded. At the stage of eight blastomeres, we selected embryos with a pattern of furrows far from regular, with blastomeres located in the center of the animal region (Fig. 40G–I), as well as embryos with the regular pattern of furrows (Fig. 40A, B). These embryos were incubated, and the hatched prelarvae were reared up to the beginning of active feeding. We found no regular differences between these two groups in morphology, rate of growth, or viability.

Asymmetry of the cleavage pattern at the stage of eight blastomeres and differences in the shapes and sizes of individual blastomeres occur in all batches of eggs. They express the natural variability of the cleavage pattern, and are by no means indicative of the poor quality of the eggs. This variability is based on the differences in the shape of the eggs: e.g., it has been shown that a slight deformation of the eggs prior to cleavage by pressing them between two slides leads to a change in the pattern of furrowing but does not cause developmental defects (Dettlaff and Ginsburg 1954).

Disturbance of Cleavage in the Fertilized Eggs

True cleavage defects, which, as a rule, include defects caused by polyspermic fertilization (occurring to some extent in almost all batches of eggs), should be distinguished from variations in the form and distribution of furrows in a cleaving egg. As we have already mentioned, normally only one spermatozoon enters the sturgeon egg. However, several spermatozoa can sometimes enter the egg through different micropyles. During polyspermic fertilization, all of the spermatozoa which entered the cytoplasm become involved in development, and three, four, and even more blastomeres appear at once during the 1st cleavage division (Fig. 43C–E). During the 2nd cleavage division, new furrows in the polyspermic eggs are laid down not regularly: most of the blastomeres divide into two parts but some of them divide into three parts (see Fig. 43C′), while the others do not divide at all (see Fig. 43D′, E′). During the 1st division, the dispermic eggs do not differ from the normal monospermic ones: one furrow is laid down, but during the 2nd division in each of the first two blastomeres two furrows appear and the animal area divides in six blastomeres (Fig. 43B and B′). Later, the polyspermic eggs divide simultaneously with the normal ones from which all of them, without exception, differ starting from the stage of the 2nd cleavage division in the presence of an excessive number of blastomeres (cf. Fig. 43A′ and B′–E′). The polyspermic eggs develop abnormally; the embryos developed from them die mostly even before incubation is completed and only few of them develop into inviable larvae of abnormal structure: shortened, bent,

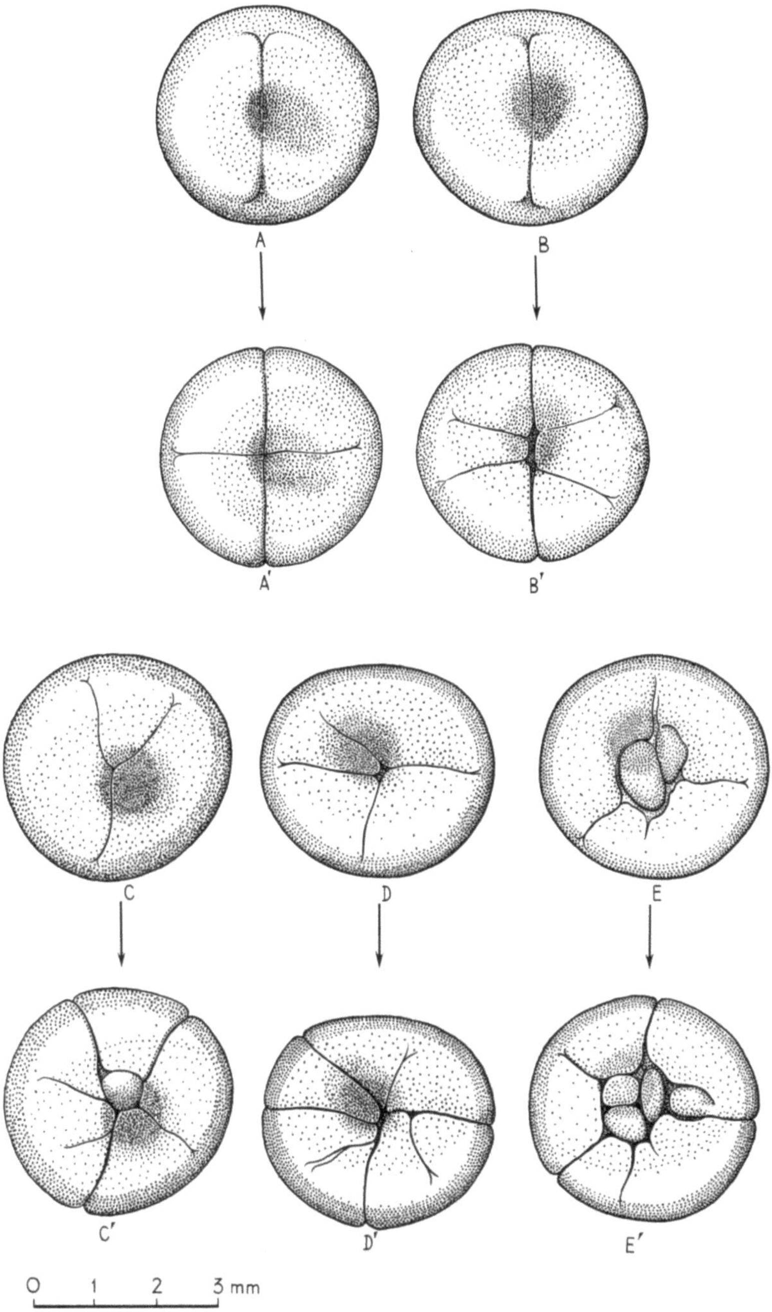

Fig. 43A–E. Cleavage pattern in polyspermic eggs of *A. gueldenstaedti* (Ginsburg 1968). Eggs at stages of the 1st (**A–E**) and 2nd (**A′–E′**) cleavage divisions. **A, A′** Normal monospermic egg; the rest are polyspermic with underdeveloped heads, dropsy. Single larvae, probably dispermic, do not have sharp structural defects at the stage of hatching but even they display later various defects and all of them are delayed in growth and development (Ginsburg 1953b, 1963, 1968; Ginsburg et al. 1963).

Developmental defects of the polyspermic eggs in sturgeon are very similar to those observed under experimental polyspermy in the frog (Brachet 1910, Herlant 1911). A cytological study has shown that these defects are due to the involvement of additional sperm (haploid) nuclei in development.

When inseminating sturgeon eggs of good quality with the semi-dry method and at the correct dosage of sperm, polyspermic eggs amount to no more than 4–6% as a rule. Given sperm overdosage, especially in cases of low-quality eggs, the fraction of these eggs can be much greater (up to 20%).

Profound cleavage abnormalities occur when the ooplasm is damaged by temperature (substantial disparities in time of furrowing, distortions of cleavage pattern). Sometimes, a greater or smaller part of the egg remains undivided (mosaic cleavage). This is due to the fact that exocytosis of the cortical granules was not realized in some regions of the egg surface and the furrows cannot run over these regions. When the eggs with mosaic pattern of cleavage are incubated, we get a high percentage of markedly deformed embryos. The undivided part of the egg not only does not participate in embryonic development itself, but also acts as a mechanical barrier against the normal movement of cell sheets during gastrulation. Such developmental abnormalities are especially common in *H. huso* when the female matures at water temperature close to the upper limit of the optimal zone and ovulated eggs are detained in the body cavity, having been obtained too late.

Parthenogenetic Cleavage

Eggs activated without fertilization usually divide. Cleavage in such eggs begins much later than in the fertilized eggs and continues to lag considerably. However, if activation has already taken place in the body of the female, the cleavage of the activated eggs can begin sooner than in the fertilized eggs. In some cases, it starts even before the eggs are collected, i.e., in the ovary or body cavity of the female (Dettlaff and Ginsburg 1950).

Parthenogenetic cleavage proceeds irregularly (Fig. 44). The furrows are laid at different times and, frequently, a large region of the egg does not divide at all. In some cases, cleavage ceases early, after the appearance of only a few furrows; in other cases, it continues up to the stage of several dozen of blastomeres (Fig. 44C). This is the limiting stage, beyond which parthenogenetic development never proceeds; the embryos never start gastrulating and slowly die off. The cell boundaries gradually disappear and the embryos acquire a whitish color with free patterns (Fig. 44D). The dying off of unfertilized eggs (activated and dividing ones, as well as non-activated ones) takes place at the time when the fertilized eggs of the same batch are gastrulating.

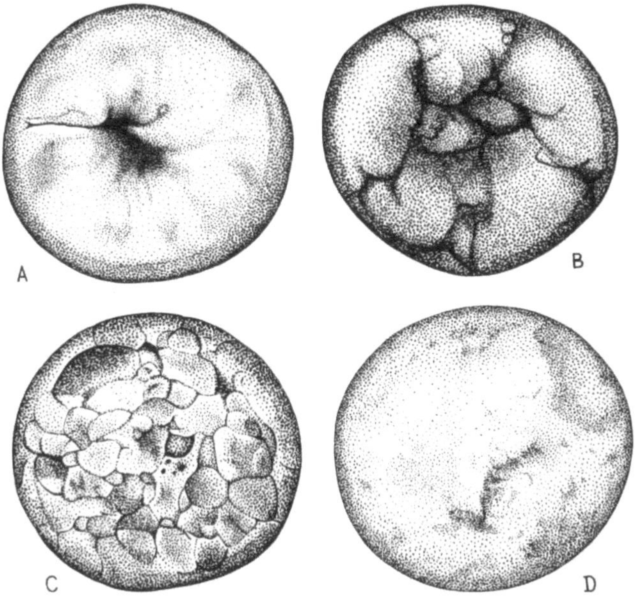

Fig. 44A–D. Parthenogenetic cleavage: **A** beginning and **B** somewhat more advanced; **C** limiting stage of cleavage of unfertilized eggs; and **D** a dying unfertilized egg (Ginsburg and Dettlaff 1969)

2.4.5 Percentage of Fertilization

Determination of the percentage of fertilization is of great importance to fish culture and is carried out for all incubated batches of eggs.

Having determined the percentage of fertilization, a fish breeder decides whether it is worthwhile to continue incubating the given batch or whether it would be better to discard it if the percentage of fertilization is very low. Furthermore, the percentage of fertilization makes it possible to carry out preliminary calculations of the forthcoming losses and abundance of prelarvae in the given batch, since unfertilized egg constitute the main source of the losses in batches of good-quality eggs under normal fertilization and incubation conditions.

Beginning with the 1st cleavage division stage, we can already clearly distinguish between the fertilized and unfertilized eggs, but it is better to determine the percentage of fertilization at the 2nd division stage. First of all, we can safely detect all the fertilized eggs at this stage, whereas eggs dividing with some delay can remain undetected when the sample is fixed at the 1st division stage. At the 2nd division stage, it is also possible to determine the number of inviable polyspermic embryos which cannot be fully detected before this stage. It is undesirable to determine the percentage of fertilization at later stages of cleavage, since at this time activated unfertilized eggs usually start dividing and it becomes more difficult to distinguish between them and the fertilized eggs. A certain skill is required for this (parthenogenetically developing eggs are characterized by atypical cleavage patterns and by a significant lag in development, as indicated earlier). If, for some reason, it was not possible to determine the percentage of fertilization at the 2nd division stage, it is best to postpone doing so until much later, i.e., to the gastrula stage, when all the unfertilized eggs (both activated and non-activated) degenerate and differ distinctly from the gastrulating embryos.

When examining a sample fixed at the 2nd division stage, it is important to pay attention to the embryos with more than four blastomeres. These embryos, as we have aleady said, are inviable polyspermic embryos. With good-quality eggs and the proper biotechnique of artificial insemination (see Appendix F), the number of these embryos should not exceed 4–6%. If there are considerably more of them, fish breeders should turn their attention to the conditions under which the spawners are kept during maturation, since the cortical reaction in poor-quality eggs spreads more slowly than in the normal eggs, and, as a result, they are less protected against penetration by supernumerary spermatozoa.

Appendix G provides information as to how the time best suited for taking a sample to determine the percentage of fertilization is established and how the sample is examined.

2.5 Gastrulation

Blastulation is followed by the period of gastrulation. It is characterized by complex cell translocations referred to as morphogenetic movements. As a result three germ layers arise: outer, ectoderm; inner, endoderm, and lying between them middle layer, mesoderm. During gastrulation, embryonic material acquires localization corresponding to the adult body plan. Interactions between newly positioned parts of the embryo determine the direction of their subsequent development.

2.5.1 Reorganization of the Embryo During Gastrulation

Gastrulation begins with the formation of a narrow heavily pigmented strip on the prospective dorsal side of the embryo, where, before the onset of cleavage, the light crescent was located. This strip forms at the boundary between the marginal zone cells and large vegetal blastomeres, roughly at the equator level (stage 13, Plate III). Then within the dark strip a groove is formed and a sheet of marginal zone cells adjacent to it invaginates. It moves towards the equator and, having reached the site of invagination, rolls over the edge and disappears inside. Thereupon new portions of cell materials approach the site of invagination and, once inside the embryo, move towards the animal pole, underlying the outer cell sheet (Fig. 45A, B). At the same time, the groove spreads

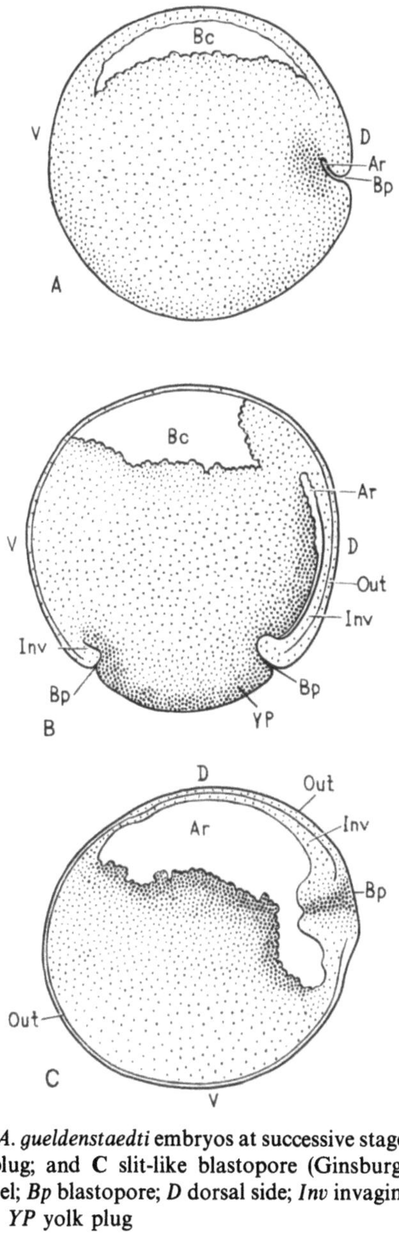

Fig. 45A–C. Sections of *A. gueldenstaedti* embryos at successive stages of gastrulation: **A** early gastrula; **B** large yolk plug; and **C** slit-like blastopore (Ginsburg and Dettlaff 1955). *Ar* Archenteron; *Bc* blastocoel; *Bp* blastopore; *D* dorsal side; *Inv* invaginated cell sheet; *Out* outer cell sheet; *V* ventral side; *YP* yolk plug

transversely and its lateral ends are prolonged all along the boundary between the marginal zone and the vegetal region (stages 14 and 15, Plate III). The edge of invagination shifts over the surface of the embryo in the direction of the vegetal pole, the superficial cell sheet overgrowing the dark vegetal material (epiboly). From the beginning of invagination, a narrow slit-like cavity forms between the invaginating cell sheet and yolk-rich vegetal blastomeres. This is the cavity of primitive gut, or archenteron. It widens markedly only at the end of gastrulation (Fig. 45A–C). The archenteron cavity communicates with the outside through an opening, primitive mouth or blastopore. The edge of invagination which arises first on the dorsal side of the embryo is called the dorsal lip of blastopore. Lateral lips form later and lastly the ventral lip of blastopore.

The stage of onset of invagination when the blastopore looks like a short slit is called early gastrula (stage 14, Plate III). As invagination spreads transversely, the slit elongates and becomes crescent-shaped, in accordance with the spherical shape of the embryo. At the midgastrula stage (stage 15, Plate III), invagination spreads already to half of the circumference, the lateral lips of the blastopore are formed, and the light animal material covers much more than half of the egg surface. When invagination spreads over the whole circumference of the embryo, the blastopore is closed into a ring (stage 16, Plate III). This ring surrounds the region of large-cellular heavily pigmented material of the egg's vegetal hemisphere, which is not yet covered by the outer cell sheet, called yolk plug. Correspondingly, stage 16 is called the stage of the large yolk plug.

As epiboly proceeds, the light animal material moves upon the yolk plug. Epiboly is accompanied by a marked stretching of the outer cell sheet.

At the end of gastrulation, the blastopore edges are drawn together still closer (see the stage of small yolk plug or stage 17, Plate III). The lateral lips of blastopore approach each other, the yolk plug is drawn inside, and the blastopore edges are closed. However, communication of the archenteron with the outside through the narrow slit between them persists for a long time. This stage (stage 18, Plate III) may be called the stage of slit-like blastopore. It represents a transition to the next developmental period. By this time, a shallow groove running along the midline from the upper part of the slit-like blastopore forwards, to the head end, becomes discernible. It forms at the site where ectoderm grows together with the notochord material. At this time, the inner layer of ectoderm thickens on both sides of the groove: this is the initial stage of the neural plate formation, similar to that described in the *Anura* (Dettlaff 1983). It is noteworthy that in normally developing embryos the neural plate forms and becomes visible from outside only after the blastopore has been closed, during neurulation (see Sect. 2.6).

At the beginning of gastrulation, the blastocoel continues to increase in size, while the epithelial layer covering it from outside becomes ever thinner (Fig. 45A, B). At the stages of midgastrula and large yolk plug, the blastocoel looks like a dark spot in the animal area distinctly seen through the thin roof (Plate III, 16an). While following development of the embryo at the subsequent stages, one can detect a gradual decrease in size of the blastocoel occurring as the invagination and epiboly proceed. By the moment of the blastopore closure, the blastocoel disappears (Fig. 45C). Its traces in the form of a small dark patch can be seen even at the neurula stage on the ventral side of the embryo.

The gastrula wall above the archenteron is formed by two cell sheets, superficial and invaginated ones. It is much thicker than the blastocoel roof which consists of the covering sheet alone (Fig. 45B). However, the dark pigment on the surface of the vegetal material forming the archenteron floor can still be seen through this wall as a dark projection on the dorsal side of the embryo, whose outer boundary outlines the archenteron cavity (Plate III, 15d, 16d). In the lateral parts of the embryo, the dark projection turns into a fairly narrow dark band which shows the archenteron boundary in this area.

It can be seen on sagittal sections of the embryo that, at the stage of large yolk plug, the invaginating cell sheet moved far forwords in the dorsal region but the archenteron cavity is still relatively small. At the end of gastrulation it markedly increases in size (Fig. 45C), thus displacing the center of gravity of the embryo. The massive, yolk-rich ventral wall of the archenteron, which previously faced to the side, now sinks down and its relatively thin roof moves upwards. Hence, at the end of gastrulation the embryo turns its dorsal side upwards inside the membranes.

2.5.2 The Fate Map by the Beginning of Gastrulation and the Dynamics of Morphogenetic Movements

Translocations of individual cell regions, which are located on the surface of the embryo by the beginning of gastrulation, and their fate can be established, using the method of vital spot staining technique developed by Vogt (1925). These studies were made on the embryos of four species of Acipenserid fish (Dettlaff and Ginsburg 1954; Ballard and Ginsburg 1980).

The jelly coat with the zona radiata externa were removed, leaving the zona radiata interna in place. Small bits of agar or squares of cellophane impregnated with Nile blue sulfate or neutral red dye were pressed upon the zona radiata and held with a strip of tinfoil or a glass bridge while the dye passed into the nearest cells. The dyed cells remained recognizably marked during two to four days. By that time, most of the organs have become visibly differentiated (Fig. 46). The final location of cells was charted according to the position and shapes of particular spots.

As a result, the fate map of the various rudiments by the beginning of gastrulation was charted for *A. stellatus* and *A. gueldenstaedti* (Fig. 47). The results obtained for the other Acipenserid species fully conform with this fate map.

It can be seen that presumptive endoderm, which later lines the anteriormost part of the archenteron, is located directly above the equator and to 20° latitude. The band of this material embraces 360° of the embryo's surface. Above this band there is a band of marginal cells from 20 to about 30° latitude, which produce notochord and somite mesoderm. In subsequent development, they together with the marginal cells of the embryo's ventral half, form skeleton,

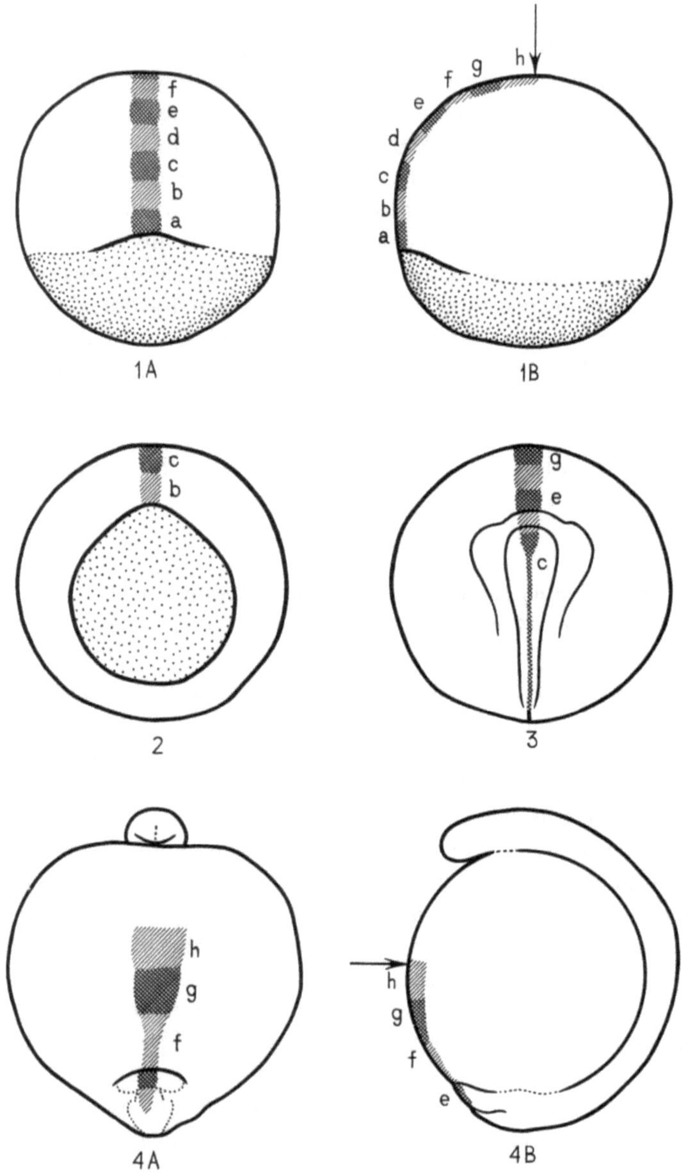

Fig. 46. Changes in the position of colored spots printed on the surface of *A. stellatus* embryo at the beginning of gastrulation. The alternating spots were printed with small bits of agar impregnated with Nile blue sulfate or neutral red (Dettlaff and Ginsburg 1954). **1** Stage 14 (**A**, dorsal view; **B**, lateral view); **2** stage 16, view from the vegetal pole; **3** stage 21, dorsal view; **4** stage 26 (**A**, ventral view; **B**, lateral view). The *arrow* indicates the position of material from the animal pole region

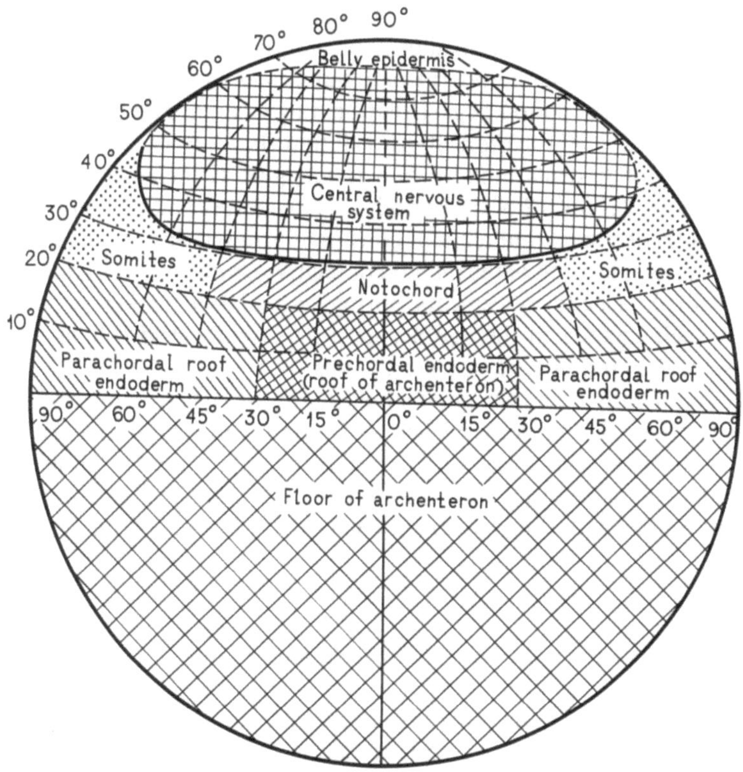

Fig. 47. Fate map of *A. stellatus* and *A. gueldenstaedti* colchicus, dorsal view, stage 12 (Ballard and Ginsburg 1980). The dorsal lip of blastopore will appear later along the 0° parallel, from 15° left to 15° right of the zero meridian

muscles, circulatory, excretory, and genital systems. The territory of prospective ectoderm is located above the marginal zone band. It includes the material of neural plate (forming organs of the central nervous system) and the material of covering epithelium (whose major part is located on the ventral side of the embryo). The cells of the vegetal hemisphere, lying below the equator, are the material of endoderm; they form the digestive system and its derivatives.

In the region of the dorsal lip of blastopore, invagination begins earlier and proceeds at a higher rate than in the other regions of blastopore. Invagination of the prechordal plate material (to 20° latitude) is completed soon after the onset of stage 14. The material from 20 to 30° latitude moves inside the embryo before stage 16 begins. Invagination of the material over the whole blastopore is completed before the blastopore is closed.

Interesting results have been obtained in studies of the morphogenetic movements of the axial organ rudiments with another method (Ignatieva 1979; Dettlaff et al. 1987). The size of these rudiments in *A. stellatus* embryos during gastrulation were measured in a relative unit, angle degrees. It was shown that movement of the anterior edge of invaginating chordamesoderm material towards the animal pole begins after about a quarter of the period of gastrulation. The entire chordamesoderm material comes inside the embryo for the second quarter of the period of gastrulation, the maximal rate of this movement being 42° for 0.2 of the period of gastrulation. Epiboly continues during the whole period of gastrulation. Its rate is unequal: it increases, attaining the maximal value, 26° for 0.2 of the period of gastrulation, and then gradually, but insignificantly decreases and remains constant during the last 0.4 of the period of gastrulation. Application of the method of relative characteristic of duration and rate of morphogenetic movements during gastrulation makes it possible to compare their kinetics in the various species of Acipenserid fish, as well as in Acipenserid fish and other lower vertebrates.

We have already noted the great similarity in the egg structure and cleavage pattern between the Acipenserid fish and amphibians. This similarity is also expressed in the process of gastrulation, the changes in the morphology of the embryo, morphogenetic movements, and the fate map in Acipenserids being especially similar with those in *Anura* (Ballard 1981; Dettlaff 1982).

2.5.3 Inductive Interactions During Gastrulation

In Acipenserid fish, as in the other vertebrates, the chordamesoderm induces the neural plate formation in the gastrula ectoderm. In order to study this phenomenon, the first experiments (Ginsburg and Dettlaff 1944; Dettlaff 1946) were followed by a large series of experiments involving the removal and transplantation of the material of the dorsal lip of blastopore, which later forms the archenteron roof, in *A. stellatus*, *A. gueldenstaedti*, and *H. huso* (Dettlaff and Ginsburg 1954).

Operations were performed at stages 13 and 14. When the central part of the dorsal lip of blastopore was removed, no neural plate formed and the axial organs were absent at the later stages. When the removed material of the dorsal lip of blastopore was grafted onto the ventral side of another embryo at the same stage, the transplant invaginated and lined the ventral ectoderm from inside. The latter developed, at the place of contact with the graft, into the neural plate and, later, into a more or less complete nervous system and its derivatives. Starting from the stage of crescent-like blastopore (stage 14), the ability of ventral ectoderm to form the neural plate (when in contact with the material of the archenteron roof) was rapidly decreased and disappeared altogether by the midgastrula stage (stage 15).

In sturgeon, as in amphibians, there are regional differences in the inductive action of chordamesoderm (Ignatieva 1979; Fig. 48). In order to reveal these differences, a central region of the dorsal lip of the blastopore apposed to the edge of invagination was removed in *A. gueldenstaedti* embryos at successive stages of gastrulation and transplanted on the ventral side of a recipient embryo at the early gastrula stage (stage 14).

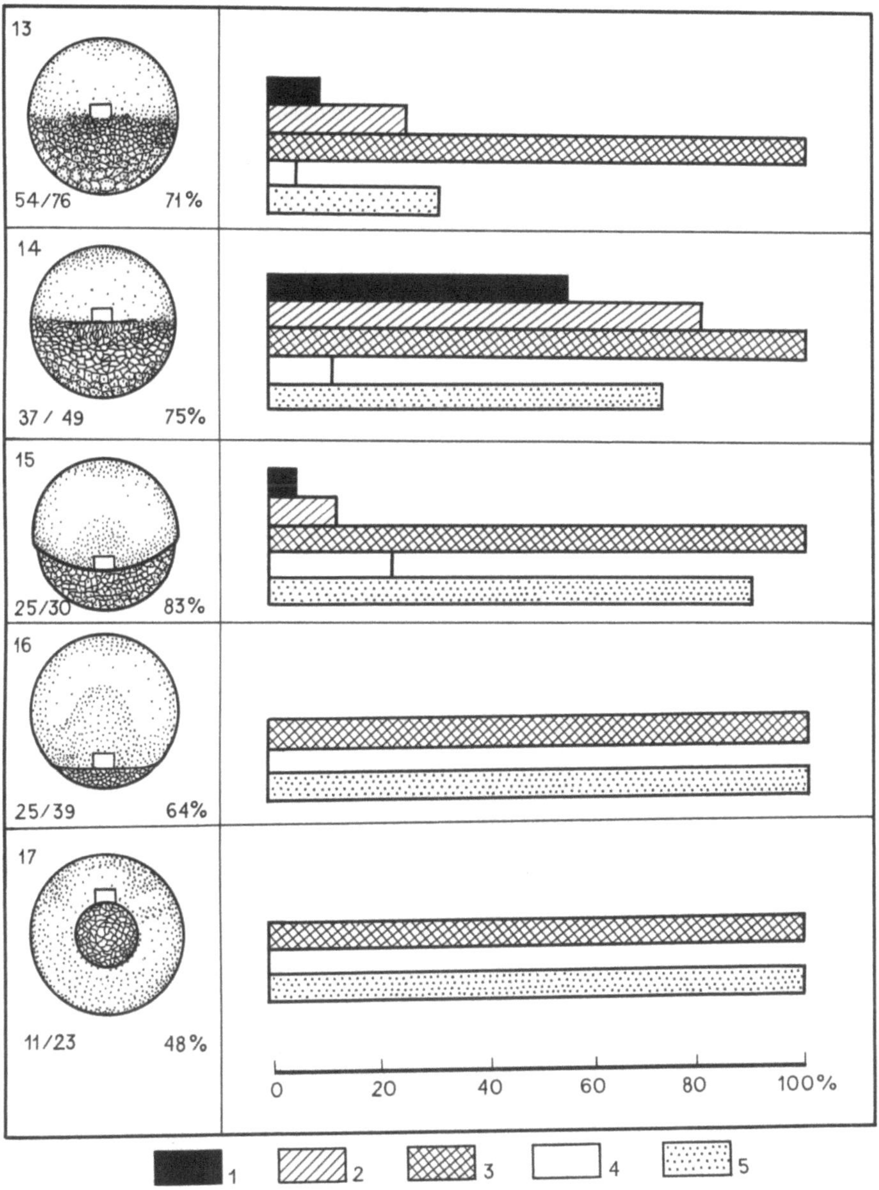

Fig. 48. Regional differences in the inductive properties of the material of chordamesoderm in embryos of *A. gueldenstaedti* (Ignatieva 1979). In the *left part* a schematic representation of donor embryo is given and a region of the dorsal lip of blastopore is *outlined*, which was grafted onto the abdominal wall of recipient embryo at the early gastrula stage. *Numerals* below the drawings designate: *left*, ratio of the number of recipient embryos with inductions to the total number of studied embryos; *right*, percentage of inductions. The *right part* of the diagram represents frequency of inductions of the various parts of the central nervous system and cases of notochord differentiation (in percent of the total number of embryos with inductions). *1* Brain regions in front of the level of auditory vesicles; *2* myelencephalon with auditory vesicles; *3* spinal cord; *4* unidentified neural structures; *5* notochord

The material of prechordal plate together with the notochord material taken before the beginning of invagination (stage 13) induced predominantly the spinal cord, while brain regions rarely developed. The grafts taken at stage 14 induced the brain structures at a high rate. When the material of the dorsal lip of the blastopore of the midgastrula (stage 15) was transplanted, the number of archencephalic and deuterencephalic inductions was markedly reduced. The transplants taken at stages 16 and 17 induced the spinal cord only. Hence, in sturgeon embryos regional differences in the inductive properties of different regions of the chordamesoderm material appear after the beginning of gastrulation. They are established, apparently, by the midgastrula stage. At the same time, this material acquires the ability to differentiate according to its fate to the full extent.

2.5.4 Gastrulation Defects

Most of the defects observed in sturgeon embryos arise as a result of disturbance of morphogenetic movements during gastrulation. Such defects are observed in the embryos with abnormal cleavage (usually, due to polyspermic fertilization), but they can also arise in the normally dividing embryos under unfavorable conditions of development.

If epiboly of the vegetal hemisphere of the blastula is inhibited at the early stages of gastrulation, the embryo, as a rule, soon dies. When such a disturbance is not very serious and a more or less large yolk plug remains naked by the moment of the neural plate formation, the embryos can live much longer and reach the stage of hatching. Many of these embryos can liberate themselves from the membranes but they usually have some structural defects. However, if only a small plug remains by the beginning of neurulation, it is either drawn inside or rejected, and at later developmental stages the embryos may have no marked defects.

There are gastrulation disturbances of another type which affect not epiboly of the vegetal blastomeres, but invagination. Pronounced defects of this type arise in the embryos with incomplete cleavage (due to unfavourable conditions of maturation). Various deformed embryos develop depending on how large the undivided part of the egg is and in what part of the embryo it comes to lie: with different degree of under-development of the anterior body regions or bifurcation of the axial organs (see Sects. 2.6.3 and 2.7.2).

Defects of gastrulation are also observed under unfavorable conditions of incubation: too high or too low temperature, oxygen deficiency (overloading of the incubation apparatus, insufficient water current, silting, or, more rarely, low oxygen content of water). Similar defects can be induced by increased salinity, but this happens very rarely in sturgeon culture. If, because of insufficient demucilaging, the eggs are agglutinated in clusters, the embryos in the center of such a cluster have usually yolk plugs of much larger size than in those on the cluster surface. The same is true of the embryos covered with silt or sand.

In good-quality eggs, gastrulation proceeds simultaneously and at its end only single embryos with large yolk plugs of irregular shape can be seen (such embryos usually develop from polyspermic eggs). If good-quality eggs develop in embryos with yolk plugs of varying sizes, this indicates insufficiently favorable conditions of incubation, and attention must be paid to this.

2.6 Development of the Embryo from the End of Gastrulation to the Onset of Heart Beating

This period is characterized by intensive processes of organogenesis. Also, the rudiments of the main systems of organs (nervous, muscle, circulatory, excretory, digestive) do not perform definitive functions. As earlier, the embryo is immobile, feeds on yolk, and breathes with the entire body surface.

2.6.1 Neurulation

At the beginning of this period, the material of the germ layers divides into rudiments of individual systems of organs. The most characteristic changes in the external appearance of the embryo at these stages consist of the formation of a neural plate on the dorsal side and its closure in a neural tube, a rudiment of the central nervous system (neurulation).

We have already mentioned that at the end of gastrulation the inner layer of ectoderm begins to thicken on the dorsal side of the embryo, in the region of the archenteron roof. After the blastopore is closed, this process continues, the dorsal side of the embryo slightly flattens, and the neural plate outlines become more and more distinct. Its anterior, widest region is the rudiment of two primary brain vesicles, the prosencephalon and mesencephalon, while the narrower middle region is the rudiment of the third brain vesicle, the rhombencephalon. This part of the neural plate is bordered by horseshoe-shaped neural folds. In the posterior, still narrower part of the neural plate, which corresponds to the trunk region of spinal cord, the neural folds form somewhat later than in its anterior part.

At the end of gastrulation, simultaneously with the thickening of the inner layer of ectoderm, its outer layer intergrows with the notochord rudiment, which underlies it along the middorsal line. A shallow groove appears in this place. This groove, called the neural middorsal groove, runs from the slit-like blastopore forwards, to the level of presumptive mesencephalon, i.e., to the anterior end of the notochord rudiment. In the region of the mid-dorsal groove, the neural plate remains thin (later it forms the floor of the neural tube). The inner layer of ectoderm markedly thickens on both sides of the groove; this region of ectoderm later forms the lateral walls of the neutral tube.

During neurulation, the neural folds appear all along the neural plate. They rise a little and become more distinct (Plate IV, 20, 21; see p. 244, 245). At the same time, the folds markedly thicken and are subdivided into the inner and outer parts in the anterior region, at the level where the neural plate has its maximum width. Simultaneously with the rising, thickening, and backward spreading of the neural folds, they are also drawn together. The central part of the neural plate somewhat deepens, its edges rise a little, bend, and gradually approach each other. Finally, the edges are closed and the neural plate is converted into a tube. The neural plate closes earliest in its anteriormost part, in the region of prospective prosencephalon. Then it closes in the posterior part from where the closure gradually spreads forwards. And, finally, the neural plate closes in its widest part, in the region of prospective mesencephalon.

We distinguish five stages of neurulation: an early neurula stage when the contours of the head part of the neural plate and neural fold around it are first outlined (Plate IV, 19); a stage with wide neural plate when its all contours and the neural folds in the head region are distinct (Plate IV, 20); a stage when the neural folds begin to draw together and the excretory system rudiment first appear (Plate IV, 21); a late neurula stage when the neural folds in the trunk region of the neural plate are drawn together (Plate IV, 22), and finally a stage of the closed neural tube (Plate IV, 23).

Application of the method of vital spot staining (Dettlaff and Ginsburg 1954; Ballard and Ginsburg 1980) has shown that in the anteriormost part of the neural plate the closure proceeds along the boundary of the neural plate and the entire material of the neural fold remains outside. In the greater part of the head region of the neural plate, the boundary of closure runs roughly in the middle of the folds, at the level of a groove separating the outer, thickened and an inner, narrower parts of the folds. The outer part of the folds later forms the visceral skeleton: mandibular, hyoid, and branchial arches. If it is removed at the midneurula stage, the mandibular and branchial arches do not develop, or develop but are reduced in size. The inner part of the folds forms neural ganglia.

During neurulation, the neural plate elongates. The slit-like blastopore, which was opened outside at the posterior end of the neural plate, loses (after the neural plate closes in a tube) its communication with the outside and opens in the neural tube cavity. Hence, the blastopore changes into a neurenteric canal connecting the gut cavity with the lumen of the neural tube.

During neurulation, the growth of endoderm towards the dorsal side of the embryo, which had already started at the end of gastrulation, is completed and the archenteron transforms into a gut cavity. A large amount of yolk in sturgeon eggs provides for a great size of intestinal tube. The latter has a very thin roof, lateral walls thickening downwards, and a massive floor (Fig. 49). The intestinal cavity is small and located in the upper part of the intestinal tube. The yolk-rich endodermal cells of the intestinal tube floor are bulky and, as a result, the endoderm in sturgeon embryos resembles the yolk sac in teleosts. However, in the latter, the yolk is not involved in cleavage and fills the gut cavity. By the end

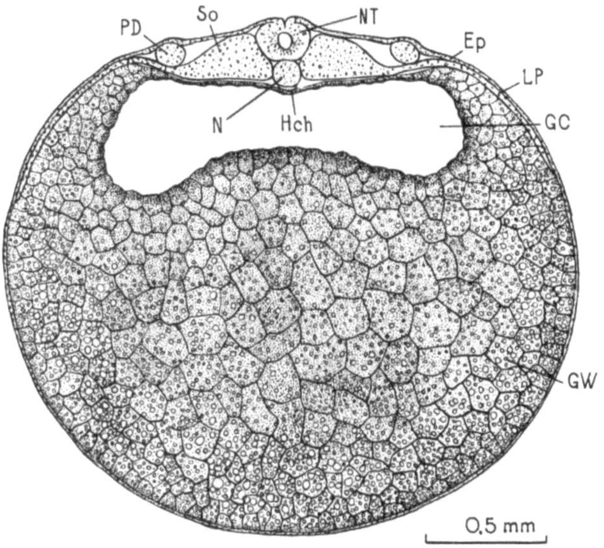

Fig. 49. Cross-section of an *A. stellatus* embryo (in the trunk region) at stage 23 (Dettlaff and Ginsburg 1954). *Ep* Covering epithelium; *GC* gut cavity; *GW* gut wall; *Hch* hypochord; *LP* lateral plate; *N* notochord; *NT* neural tube; *PD* rudiment of pronephric duct; *So* somite

of neurulation, the gut rudiment is a closed structure which communicates with the neural tube cavity alone.

During neurulation, the middle germ layer, mesoderm, becomes subdivided. A long dense cellular strand is separated on the dorsal side of the embryo along the midline in mesoderm under the neural plate. This strand develops in a notochord (Figs. 49, 50). The notochord acquires a great compactness and elasticity and is the main skeletal element of the embryo. The notochord is underlain by the hypochordal plate formed by the outer layer of a bilayered chordamesoderm rudiment.

The dorsal part of mesoderm lying on both sides of the notochord begins to be segmented, i.e., it becomes subdivided into a series of primary segments, or somites. Segmentation begins in the anterior part of the trunk region and gradually progresses backwards. In the end of neurulation, eight or nine pairs of somites can be dimly seen through the epidermis on both sides of the closing neural tube.

Parts of the mesoderm located laterally to the material of somites remain unsegmented. These are lateral plates consisting of two thin, closely adhering layers: outer, parietal, and inner, visceral. A cavity appears later between them, the secondary body cavity or coelom. A part of the mesoderm material between the rows of somites and the lateral plates forms the excretory system of the embryo. Its rudiments have already been developed by the middle of neurulation as two short strands which are seen through the epidermis on the dorsal side as light strips (Plate IV, 21). Gradually, the rudiments of the excretory

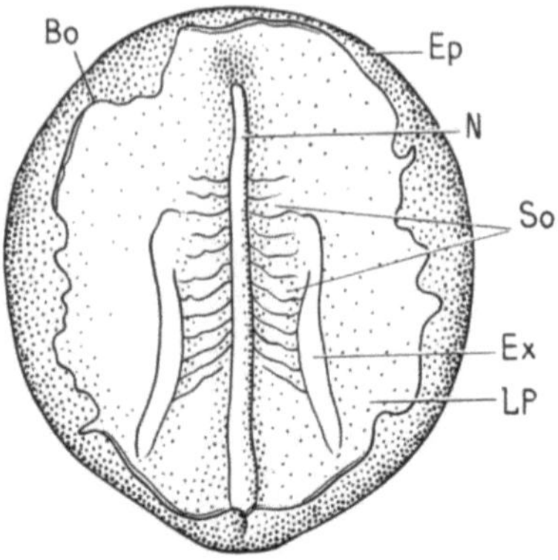

Fig. 50. Embryo of *A. gueldenstaedti* at the stage of closed neural tube (stage 23). Dorsal view after removal of the covering epithelium and the neural tube (Ginsburg and Dettlaff 1955). *Bo* Boundary of the exposed area (after removal of epithelium and neural tube); *Ep* covering epithelium; *Ex* rudiment of excretory system; *LP* lateral plate; *N* notochord; *So* somites

system elongate, spread into more posterior body regions, and bend somewhat (Plate IV, 22, 23d; Fig. 50). At these stages, they are compact mesodermal strands without any signs of differentiation.

The processes of neural plate and neural tube formation were studied in *A. ruthenus* by Salensky (1878). He was wrong in assuming that the neural plate is already formed before the end of gastrulation, at the stage of large yolk plug, and the neural folds appear first in the posterior part of the embryo, while the head part of the neural plate forms later. Indeed, the reverse is observed: the anterior part of the neural plate and bordering anterior parts of the neural folds form first and, only afterwards, do these structures gradually arise in the posterior part of the embryo. As to the relationship between gastrulation and neurulation, normally, the neural plate becomes distinct only after epiboly has been completed and the yolk plug has disappeared (see Sect. 2.6.3).

Inductive Interactions During Neurulation [Determination of the Internal Ear (Labyrinth) Taken as an Example]

In sturgeon, as in amphibians, the neural tube material exerts an inductive effect on the covering epithelium, when in contact with the latter.

The right half of presumptive rhombencephalon was removed in *A. stellatus* at different stages of neurulation (Ginsburg and Dettlaff 1944). In some cases, a partial regulation of the brain was observed and a small auditory vesicle (rudiment of the membranous labyrinth) developed on the operated side. The weaker the regulation, the smaller the vesicle, and when there was no regulation, the vesicle was absent. These first experiments on a small number of embryos suggested that the labyrinth determination is stimulated by the material of the rhombencephalon rudiment.

Later, the labyrinth determination was studied in detail in embryos of *A. gueldenstaedti*, *A. stellatus*, and *H. huso* (Chulitskaya 1961a,b, 1962). The covering epithelium from the region containing the auditory vesicle material was explanted at successive stages of neurulation in a modified Holtfreter solution. Auditory vesicles developed only in those explants which were taken from the embryos after they achieved the stage of approachment of the neural folds, i.e., after the contact was established between the covering epithelium of the auditory region and the presumptive rhombencephalon. In the explants taken at some subsequent stages, the frequency of auditory vesicle formation, their size, and completeness of differentiation (frequency of sensory areas formation) increased. These results suggest that in sturgeon, as in amphibians (Yntema 1955; Jacobson 1966; Ginsburg 1989), the presumptive rhombencephalon is the inductor of the labyrinth.

In amphibians, the action of this inductor is preceded by the first phase of the labyrinth determination when the inductive effect is exerted by the mesoderm underlying the ear ectoderm. As a result of its inductive influence, the ectoderm transplanted into a foreign environment (on the ventral side of a host embryo) forms in the *Urodela* small hardly differentiated auditory vesicles at a low rate and in the *Anura* larger vesicles, many of which have sensory areas, at a much higher rate (Ginsburg 1950, 1989). In sturgeon, no signs of determination in the labyrinth material have been found at the stages preceding the inductive action of the presumptive rhombencephalon (Chulitskaya 1962).

These data suggest that labyrinth determination in the *Acipenseridae* and *Urodela* is similar according to an important feature: the covering epithelium of the ear region acquires stable organ-specific properties and, correspondingly, the ability to form auditory vesicles with sensory areas in a foreign environment or upon explantation only after the contact with the presumptive rhombencephalon has been established, while in *Anura* this occurs already by the beginning of neurulation.

Differences in the timing of stable determination of the labyrinth in the *Acipenseridae* and *Urodela*, on the one hand, and the studied species of the *Anura*, on the other, correlate with those in relative duration of the periods from insemination to the beginning and end of gastrulation (Table 5). As a result, the embryos of *Acipenseridae* and *Urodela* at the morphologically similar stages of neurulation are formed by cells of younger cell generations than those of the *Anura* and have a lesser total duration of interkinetic cell states (Dettlaff 1964; Dettlaff et al. 1987). Such a correlation is found in amphibians not only with respect to the labyrinth, but also to the other derivatives of ectoderm (Ginsburg 1989). Unfortunately, for sturgeon we do not have data about determination of any derivatives of ectoderm, except the labyrinth.

Table 5. Duration (in τ_0) of the period from fertilization to the begining and end of gastrulation in Acipenserids and amphibians

Order	Species	Duration to the Beginning of gastrulation	Duration to the End of gastrulation	Reference
Acipenseriformes	*Acipenser gueldenstaedti*	19.4–20.2		Krenig (Chulitskaya) (1960)
		19.5	38.0	Ginsburg and Dettlaff (1991)
	A. stellatus	19.6–20.3		Krenig (Chulitskaya) (1960)
		20.0–20.8	37.9–39.1	Dettlaff and Dettlaff (1961)
	Huso huso	20	38	Igumnova (1975a)
Urodela	*Triturus vulgaris*	16	< 36	Berman et al. (1987)
	Pleurodeles waltl	18.7	34.7	Gallien and Durocher (1957; see Dettlaff and Rudneva 1973)
	Ambystoma mexicanum	18–19	33.3–35	Skoblina (1963)
		17	33–35	Bordzilovskaya et al. (1989)
Anura	*Rana temporaria*	24.9–25.4	51.1–52.2	Chulitskaya (1965)
		24.6–25.9	48.6–52.2	Ryabova et al. (1984)
	R. terrestris	23–24	50–51	Skoblina (cited in Dettlaff 1964)

2.6.2 Development of the Embryo from the Closure of the Neural Folds to the Onset of Heart Beating

After neurulation is over, the sturgeon embryo has, basically, its adult body plan (Fig. 49). The outer epithelial layer represents the covering epithelium, the neural tube is located on the dorsal side under the epithelium, with a notochord and hypochord beneath, and the intestine closely adheres to them downwards. On both sides of the neural tube, there are somites, rudiments of the excretory system and lower, between the epithelium and the intestinal wall, lateral plates. The anterior part of the neural tube is widened, this is the brain rudiment. The rudiment of the posterior body regions (common rudiment of the posterior trunk and tail regions) has not yet formed.

Processes of morphogenesis continue after the neural tube has been formed: the body shape of the embryo changes and the rudiment of the posterior body regions is separated, parts of the brain form, rudiments of the sense organs (olfactory sacs, eyes, and membranous labyrinths) appear; rudiments of pituitary, epiphysis, and hatching gland also appear; visceral arches develop; pronephros appears and differentiates; heart and blood vessels form; branchial pouches and liver diverticulum are laid down. Tissue differentiation begins: muscle fibers form, notochord cells are vacuolized, blood cells appear.

At these stages, the embryo is immobile and does not change its position inside the membranes.

Changes in the body shape of the embryo are characterized by a number of specific features related to a large amount of yolk in the cells of the gut wall and, correspondingly, the relatively large size of the ventral body region, which brings the embryos of the *Acipenseridae* close, in their external appearance, to those of the *Holostei* and *Teleostei*. The head region, as in the embryos of these groups, remains for a long time spread on the surface of the body and only gradually separates from it by means of the head body fold. In sturgeon, this fold appears at the end of the period under description (Plate V, 27, 28 lat; see p. 247). The rudiment of the posterior body regions, like the head, separates by means of body fold formation, but somewhat earlier (Plate IV, 25 lat; Plate V, 26 lat).

Changes in the body shape of *A. stellatus*, *A. gueldenstaedti*, and *H. huso* embryos are very similar. Once the neurulation is over, they start from a gradual tightening of the axial mesoderm towards the middorsal plane. As a result, the somites, initially low and wide, become higher and, together with the neural tube, rise above the rest of the body surface. This can easily be seen if one compares their width and height at consecutive stages (Plate IV, 25t; Plate V, 26–28t).

At the same time, the axial organs elongate, especially in the head region. Undivided axial mesoderm, apposed to the posterior end of the neural tube, is drawn together towards the middorsal line, forms there a thickening and rises a little over the embryo's surface, forming, together with the end of the neural tube, the first rudiment of the posterior body regions. This rudiment separates from the ventral region by means of the caudal body fold. If the membranes are removed from the embryos at these stages, we can see (Plate V 26–28 lat) that the rudiment of the posterior body regions separates from the ventral region wall. At first, this rudiment looks like a wide and flattened lobe (Plate IV, 25t) and, then, becomes narrower and higher (Plate V, 27t) and rod-shaped (Plate V, 28t). A dark spot is discernible at its end which is an extension of the cavity of the tail gut posterior end at the site of its connection with the neural tube cavity.

It is noteworthy that at the same developmental stages the rudiment of the posterior body regions in *A. stellatus* is somewhat shorter than in *A. gueldenstaedti* and *H. huso*, and in the latter slightly shorter than in *A. ruthenus*. There are differences also in differentiation of this rudiment. By the end of this period (stage 28), it represents a rod-shaped, spherical in cross section, structure in *A. stellatus*. In *A. gueldenstaedti* it is not yet flattened but a thickening of the epithelium (rudiment of fin fold) is outlined on its dorsal side, while in *H. huso* its flattening began and a low fin fold is present.

Unlike the rudiment of the posterior body regions, the head region, as already mentioned, remains spread on the surface of the embryo for a long time. It rises up markedly only after the heart rudiment formation (Plate V, 28 lat).

It is noteworthy that the spherical body shape of the embryo at these stages is due to the fact that it is surrounded by strong membranes, that do not yield to stretching. After they are removed, the body elongates, the head rises up still more and the rudiment of the posterior body regions separates from the ventral region (Plate IV, 25 lat; Plate V, 26–28 lat).

The integuments of the embryos of *A. stellatus* and *H. huso* at these stages are weakly pigmented and fairly transparent. In *A. gueldenstaedti colchicus* they are pigmented more intensely, the pigment pattern being unequal in different parts of the body. After the neural tube closes, large pigment spots of varying sizes and in varying numbers appear on the ventral side of the embryo. The brightness of these spots varies both in different batches of eggs and in the embryos from the same batch. Despite more intensive pigmentation, the integuments of *A. gueldenstaedti* embryos are also transparent enough. Because of this, many of the later described changes in the internal structure of the embryos of these species can be followed by simple observations of intact embryos.

The widened head end of the neural tube begins to elongate by the end of neurulation. Simultaneously with elongation, it divides first into two regions and, then into three primary brain vesicles due to unequal growth of its walls: forebrain, or prosencephalon, midbrain, or mesencephalon, and hindbrain, or rhombencephalon. The prosencephalon gives rise later to telencephalon with olfactory lobes and diencephalon with optic processes, epiphysis and hypophysial infundibulum. Mesencephalon does not become subdivided, while rhombencephalon gives rise to myelencephalon, or medulla oblongata and rudimentary metencephalon, or cerebellum. Relatively small longitudinal cell strands (neural crest) develop from the material of neural folds where they close over the neural tube. Later, they give rise to ganglia and ectomesenchyme forming the visceral skeleton. Initially, the brain vesicles differ slightly from each other: they have similar wall thickness and cavity size, but their differentiation soon begins.

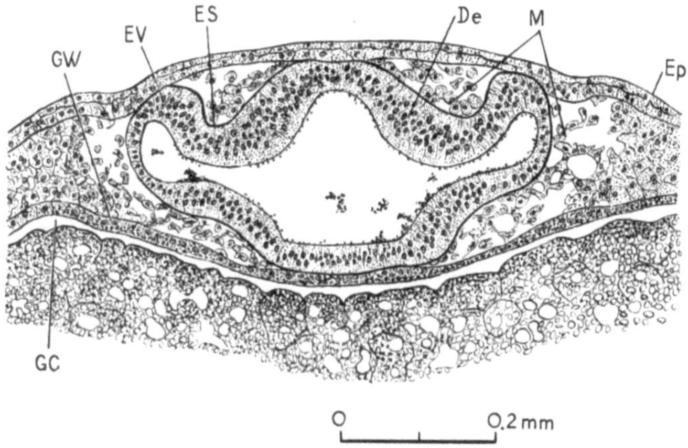

Fig. 51. Cross section of an *A. stellatus* embryo at the level of eye vesicles, stage 26 (Dettlaff and Ginsburg 1954). *De* Diencephalon; *Ep* bilayered covering epithelium; *ES* eye stalk; *EV* eye vesicle; *GC* gut cavity; *GW* gut wall; *M* mesenchyme

In the posterior part of prosencephalon, the cavity widens and becomes egg-shaped; its lateral walls then form two hollow processes, eye rudiments. Optic processes gradually grow laterally and upwards, approaching the epithelium covering the dorsal side of the body. At first, the brain wall directly passes into the wall of broad optic processes. But, then, their walls are gradually drawn together, elongate and become thinner at the site of connection of the optic processes with the brain. In such a way the optic processes transform into optic vesicles (Plate V, 26h), which are connected with the brain by an initially massive (Fig. 51) and then thin optic stalk (into which optic nerves grow at the later stages).

After the optic vesicles have grown up to the covering epithelium, their contact wall with it markedly thickens (rudiment of retina). By the end of this period (stage 28), the inner layer of the epithelium draws together and thickens in the region of contact with the eye rudiment (the onset of lens rudiment formation).

Rudiments of olfactory sacs appear anterolaterally to the anterior end of prosencephalon at a stage when the optic processes have already formed (between stages 24 and 25). They look at first like thicknenings of the epithelium inner layer (olfactory placodes). Later, these thicknenings, together with the covering them outer epithelium layer, deepen, and olfactory pits arise, clearly visible from outside.

By the end of this period, an evagination forms in the posterior part of the thin floor of the prosencephalon, the first rudiment of infundibulum. A thickening of the inner layer of epithelium forms near the hatching gland rudiment (Fig. 52) whose cells later move to the forebrain base and form the epithelial part of hypophysis. Finally, a small spherical evagination forms (stage 28) in the forebrain roof at its boundary with the midbrain—a rudiment of epiphysis.

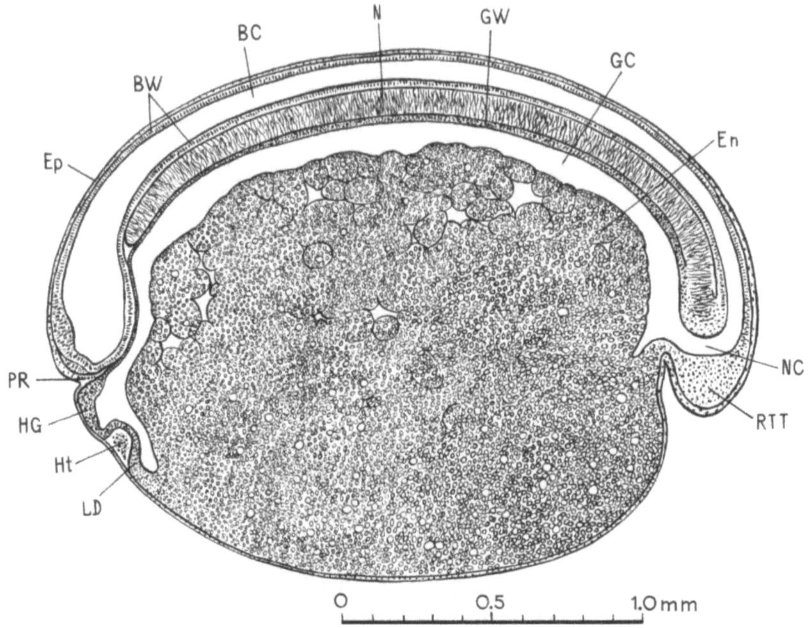

Fig. 52. Section of an *A. stellatus* embryo in the sagittal plane, stage 26 (Dettlaff and Ginsburg 1954).*BC* Brain cavity; *BW* brain wall; *En* yolk-rich endodermal cells; *Ep* covering epithelium; *GC* gut cavity; *GW* gut wall; *HG* hatching gland rudiment; *Ht* site of heart rudiment formation; *LD* liver diverticulum; *N* notochord; *NC* neurenteric canal; *PR* ectodermal part of the pituitary rudiment; *RTT* common rudiment of posterior trunk and tail regions

In the mesencephalon, the walls thicken, while the cavity narrows down.

In the rhombencephalon, which gradually passes into the spinal cord, the cavity, on the contrary, markedly increases. At the same time, its roof becomes much thinner. The increase of the hindbrain cavity can be easily followed by simple observation, since this cavity is distinctly seen from outside (Plate IV, 24h; Plate V, 26 and 27h). At the beginning of this period, auditory placodes (rudiments of membranous labyrinths) appear on both sides of the hindbrain. At first, they are large thickenings of the inner layer of the covering epithelium. Gradually, the middle part of these thickenings deepens, while the edges are drawn together. As a result, two vesicles are formed (auditory vesicles, Fig. 53) which have characteristic elongated pear-like shape. With their narrow end they face the epithelium, with whose inner layer they remain connected until the end of the period under description. The walls of the auditory vesicles are fairly thick, but their cavity is still small. It increases only at the end of this period (stage 28). At this time, the auditory vesicles can hardly be seen through the covering epithelium. They can be seen much better at the later stages, when their cavity markedly increases and the walls become thinner.

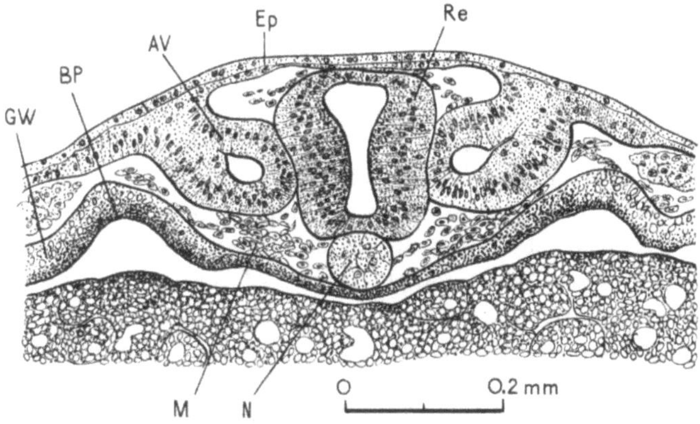

Fig. 53. Cross-section of an *A. stellatus* embryo at the level of auditory vesicles, stage 26 (Dettlaff and Ginsburg 1954). *AV* Auditory vesicle; *BP* endodermal branchial pouch; *Ep* covering epithelium; *GW* gut wall; *M* mesenchyme; *N* notochord; *Re* rhombencephalon

Soon after three primary brain vesicles have formed, a light plate of semilunar shape becomes visible in front of the forebrain (Plate V, 26h). Its curvature repeats that of the forebrain wall to which it adheres. This plate is the rudiment of hatching gland. It is formed by a region of the foregut wall which closely adjoins the epithelium (see Fig. 52) and intergrows with the latter to such an extent that they cannot be separated in live embryos. The hatching gland rudiment borders frontally with the head fold and posteriorly with the deepening of the epithelium at the place where the ectodermal part of the hypophysis rudiment forms (see Plate V, 26h). During this period, the hatching gland rudiment lies in front of the forebrain and only at the end, as a result of undercutting the head by the body fold, does it begin to displace downwards.

Simultaneously with the neural tube closure, visceral arches begin to develop. Thickenings of the outer parts of the neural folds at the level of midbrain, which have already formed at the midneurula stage, increase and embrace the forebrain laterally, like two wings. As the head region elongates and brain vesicles separate, the rudiments of visceral arches grow forwards and, bypassing the optic processes, approach the edges of the hatching gland rudiment from the left and right (Plate V, 26h). Behind the 1st pair of arches, which is the rudiment of the mandibular arch, the 2nd pair of arches soon appears (between stages 24 and 25). It extends from the hindbrain, embraces the midbrain and a part of the forebrain and, at the level of the optic processes, adjoints from the outside the anterior pair of arches. This is the rudiment of hyoid, or sublingual arch. Behind the 2nd pair of arches, shorter branchial arches, directed forward and laterally, develop at the level of hindbrain. When viewing the embryo from the dorsal side, the visceral arches look like light dense structures. At the stages where the head is not yet raised up, the visceral arches lie in the same plane and are seen throughout, but their boundaries are hardly visible (Plate IV, 24, 25h; Plate V, 26, 27h).

It can be seen in sections that the visceral arches are dense accumulations of mesenchyme forming directed flows of cells. The first two pairs of arches are directed forward, to the brain, and the rest are located between endodermal branchial pouches. We have already noted that in those cases, when certain parts of the neural folds were removed at the midneurula stage, some of these arches were absent or sharply reduced in size at the later stages. Hence, the visceral arches are formed in sturgeon, as in amphibians (Hörstadius 1950), by the ectomesenchyme cells arising from certain part of the neural folds.

Notochord is the main skeletal element of the embryo, which becomes increasingly strong. Its cells elongate and are vacuolized.

At the beginning of the period under description, the number of somites markedly increases and then constantly rises due to the rapid spreading of segmentation of the mesoderm from the head towards the tail. Only the mesoderm of the end of the rudiment of posterior body regions remains undivided and here the notochord is not separated from it. Further increase in the number of segments proceeds at the expense of this material. In sections, the somites are compact triangular structures (see Fig. 49).

At the same time as the increase in the number of somites, their differentiation proceeds: individual mesenchyme cells emigrate from them, which surround the notochord and neural tube. This is skeletogenic mesenchyme. These cells emigrate from that part of the somites which faces the notochord. A greater part of the somite is a myotome in which differentiation of the muscle fibers soon (stages 26–27) begins. Differentiation of the myotomes, as well as segmentation of undivided axial mesoderm, progresses backwards.

Rudiments of the excretory system also differentiate at these stages. This process was studied in detail in *A. ruthenus* by Pavlenko (1916) and Sawadski (1926b). In the Acipenserid species studied by us, the excretory system develops in a similar way.

All the changes next described are clearly seen when viewing the embryos from the dorsal side (Plate IV, 21–25d; Plate V, 26–28d), since the excretory systems rudiments are clearly seen from outside.

The rudiments of the excretory system can first be seen in sturgeon embryos in the second half of the period of neurulation (stages 21–22) on both sides of the neural plate. These are narrow dense strands of mesodermal cells located at the external boundary of the somites. These strands are positioned at a certain angle to the neural plate and diverge backwards. Later, the rudiments elongate and, at the same time, differentiate. The anterior part thickens and becomes subdivided (stage 24 and subsequent stages) in the pronephros rudiment, consisting of six pronephric tubules and the collecting part of the pronephric duct. The rest of the rudiment forms a greater part of the pronephric duct. At first, the rudiments of the excretory system are organized like dense cellular cords, but at stage 25, a cavity appears in them, almost simultaneously all over their length. The first renal tubule is located at the level of the 3rd pair of somites.

The distal ends of the pronephric tubules unite to form the collecting part of the pronephric duct. At the site of the transition of this part of the pronephric

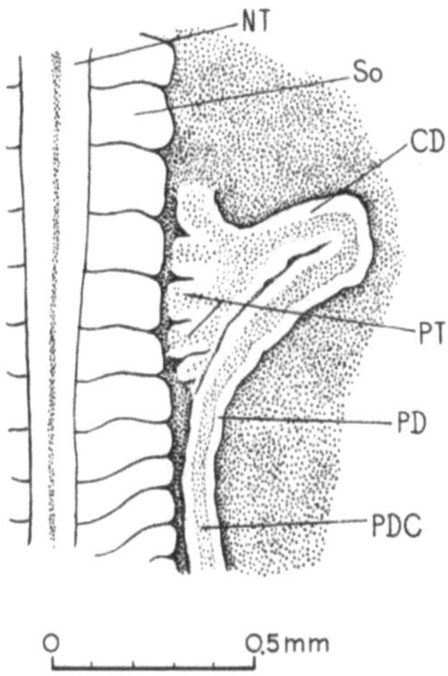

Fig. 54. Structure of pronephros in an *A. stellatus* embryo, stage 26. Covering epithelium removed (Detlaff and Ginsburg 1954). *CD* Collecting part of the pronephric duct; *NT* neural tube; *PD* pronephric duct; *PDC* cavity of the pronephric duct; *PT* pronephric tubules; *So* somites

duct into its main part, a characteristic, ever-increasing and distinctly visible (from outside) loop forms which is directed forwards and laterally (Fig. 54). The other end of the renal tubules, at first connected with the somites, opens: in the 1st tubule — into the widened region of the coelomic cavity, and in all others — into relatively small chambers (pronephridial chambers). Two posterior tubules and adjoining chambers are greatly reduced.

Simultaneously with the differentiation of the anterior region, the pronephric ducts grow backwards. At the stages of heart formation, before the onset of beating, these ducts reach the ectodermal cloacal invagination and open into it. Hence, the pronephric ducts communicate with the environment even before the onset of heart beating, when the gut does not yet have such communication.

As the shape of the somites changes (decrease in width and increase in height), the pronephric ducts gradually displace towards the middorsal plane.

The changes in the region of lateral plates occur as follows. After the rudiment of the excretory system has separated, the dorsal ends of the lateral plates remain for a while connected with the somites (Fig. 49) by means of mesenchyme strands described in *A. ruthenus* by Salensky (1878). This connection is then lost and the separated ends of the lateral plates grow under the myotomes. At the same time, the lateral plates spread very slowly forwards and downwards, onto the ventral side of the embryo. The lateral plates split into two thin layers of mesenchyme cells closely adjoining to each other. During the period under description, a small cavity between them (onset of the formation of secondary body cavity, or coelom) appears only in the head end and in the region of pronephros where it receives the first pair of renal tubules forming funnels on their ends.

While moving forwards, the lateral plates embrace the brain (with adjoining visceral arches) as two wing-like processes tapering towards their anterior ends. These processes gradually elongate, until their ends meet in front of the head, in front of the hatching gland (Plate IV, 25h). Somewhat later, the cavity between the two layers of the lateral plates spreads from the posterior part of the head forwards (it can be seen as a dark band in the head region of the embryo if viewed from above); finally, it reaches the anterior end of the lateral plates. Now, the right and left cavities are separated only by two adjoining walls of the lateral plates (Plate V, 26h). The heart rudiment forms in this place.

Individual mesenchyme cells emigrate from the inner (visceral) layer of the lateral plates at the place of their junction, while the rest of these plates thickens and caves (Fig. 55). The edges of this hollow are then drawn together and, by closing, form a short massive tube. This tube embraces another, at first bifurcated, thin-walled tube formed by the earlier emigrated mesenchyme cells. The thin-walled tube forms the endothelial lining of the heart cavity, endocardium, while the thick-walled one forms the muscle layer of the heart, myocardium.

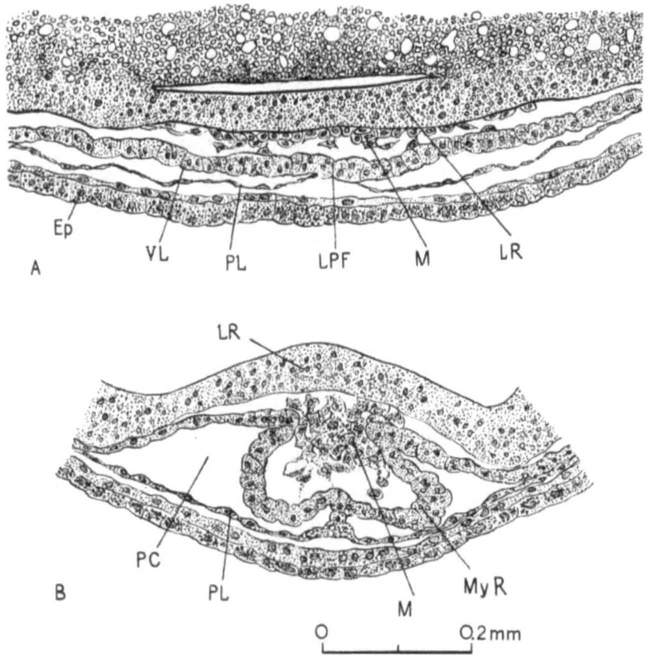

Fig. 55A, B. Sections of *A. gueldenstaedti* embryos in the region of the heart rudiment formation **A** at stage 26 and **B** at stage 26/27 (Dettlaff and Ginsburg 1954). *Ep* Bilayered covering epithelium; *LPF* region of fusion of the lateral plates; *LR* liver rudiment; *M* mesenchyme forming the endocardium rudiment; *MyR* myocardium rudiment; *PC* pericardial cavity; *PL* parietal layer of the lateral plate; *VL* visceral layer of the lateral plate which forms the myocardium rudiment

The coelomic cavity surrounding the cardiac tube forms the pericardial cavity.

The cardiac tube elongates relatively rapidly and, in the beginning of the next period, begins to bend in an S-shape. At the same time, first weak, hardly noticeable and then ever more distinct contractions of the heart muscle appear which signify the beginning of the next period.

From the beginning, the cardiac tube is slightly bifurcated at its anterior end. Later these bifurcated ends form the ducts of Cuvier which appear to be laid down simultaneously with the heart.

Blood vessels also develop simultaneously with the heart. For example, the blood vessels of the head can be seen on both sides of the brain, a wide aorta forms under the notochord, a fairly dense network of blood vessels develops in the body wall, so-called yolk sac. We describe the position of the main blood vessels in the next section when dealing with stages at which blood circulation takes place.

Large blood vessels and the heart are laid down almost empty of blood: only single blood cells occur in the heart.

At the stages when the heart rudiment forms, blood cells develop from the mesenchyme cells of the lateral plates on the ventral side of the embryo. They accumulate in great numbers in foci of their formation (blood islands). In addition, single blood cells also occur. Endothelial walls form around the aggregates of blood cells and, later, individual islands fuse into a network of the yolk sac vessels.

Formation of the blood cells on the ventral side of the embryo can sometimes be seen from outside in defective embryos in which the heart and blood circulation are absent. By the stage of hatching, these embryos have on the ventral side a large red field, an accumulation of red blood cells.

We consider next the differentiation of the digestive system. During the period under description and the next one, the intestinal tube does not communicate with the outside by either end (Fig. 52). Near the base of the rudiment of posterior body regions, an ectodermal deepening forms on the ventral side by the end of this period (stage 28). It becomes closely apposed to the intestine process but their common wall is yet not perforated. The ectodermal deepening, which is the rudiment of the most posterior part of the intestine (cloaca), receives pronephric ducts.

A tail gut appears and gradually elongates along with the formation and growth of the rudiment of posterior body regions. During the whole period, it retains its communication with the neural tube cavity (neurenteric canal).

The intestinal cavity widens relatively slightly; it is somewhat wider in its pharyngeal region. Four pairs of pouch-like deepenings develop in the upper and lateral walls of the pharyngeal region. These are endodermal branchial pouches directed towards the epithelium (see Fig. 53); two anterior pouches come into contact with the nascent thickenings of the inner layer of epithelium relatively early (stage 26). At these stages, the outer layer of epithelium does not yet deepen. This occurs during the next developmental period, while perforation of the walls separating the cavities of endodermal branchial pouches and ectodermal branchial grooves as well as the formation of gill clefts take place only after the embryos are liberated from the membranes (Salensky 1880; Schmalhausen 1951). Finally, a process forms in the anterior part of the intestine, near the heart rudiment. This is the liver diverticulum (see Fig. 52).

Five stages can be distinguished in this period on the basis of structure of the rudiments of brain, visceral arches, heart, pronephros and the degree of separation of the common rudiment of the posterior trunk region and tail. These stages are as follows: a stage when optic processes appear and the pronephros rudiment forms (Plate IV, 24); a stage of approachment of the lateral plates and formation of a thickening in the region of the rudiment of posterior body regions (Plate IV, 25); the stage of fusion of the lateral plates and formation of the heart rudiment there, as well as the onset of separation of the rudiment of posterior body regions (Plate V, 26); the stage of the short cardiac tube (Plate V, 27); and the stage of the elongated straight cardiac tube (Plate V, 28).

2.6.3 Developmental Defects

Developmental defects arising during this period are most frequently due to disturbances of epiboly. As a result, the neural plate forms in the presence of a yolk plug of a greater or lesser size (Fig. 56). If the yolk plug is large (Fig. 56A), a shortened and deformed neural plate forms, the embryo develops abnormally and soon dies. With a smaller yolk plug (Fig. 56B), developmental defects are less significant. Only if the yolk plug is very small, can it be drawn in or rejected and the embryo undergoes normal development.

Developmental defects include underdevelopment of the anterior body regions. A shortened neural plate forms in these embryos and later they are found

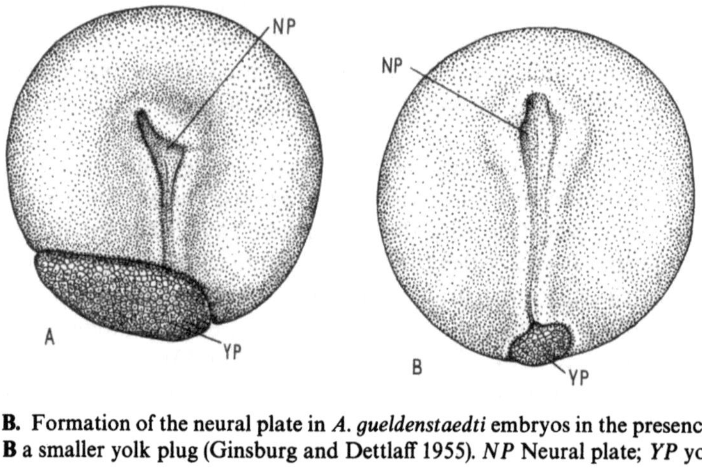

Fig. 56A, B. Formation of the neural plate in *A. gueldenstaedti* embryos in the presence of **A** a large and **B** a smaller yolk plug (Ginsburg and Dettlaff 1955). *NP* Neural plate; *YP* yolk plug

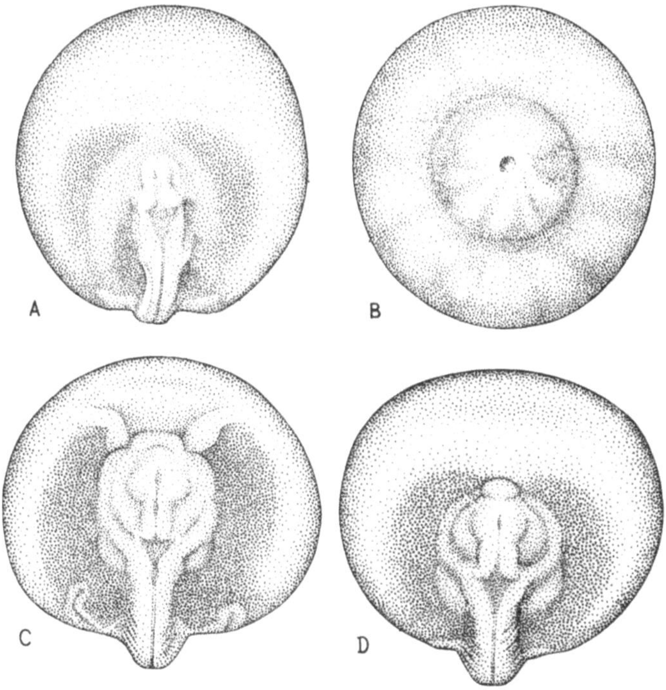

Fig. 57A–D. Abnormal embryos of *A. gueldenstaedti* at stages 26 and 28 (Ginsburg and Dettlaff 1975). **A, B** Deformed embryos at stage 26 with underdeveloped anterior body regions (**A** prosencephalon absent; **B** head and anterior trunk axial organs absent, only a rudiment of the posterior trunk and tail regions developed); **C, D** deformed embryos at stage 28 with defects of the heart rudiment (**C** two heart rudiments developed; **D** heart rudiment absent)

to have no anterior regions of the head (Fig. 57A), no head, or even no head and anterior part of the axial body rudiments (Fig. 57B). Defects of this type will be dealt with in greater detail in the next Section (see Sect. 2.7.2).

Sometimes, embryos occur with bifurcated axial rudiments. Such embryos have an area of naked endoderm on both sides of which the left and right halves of the neural plate form (spina bifida). Thereafter an asymmetrical neural tube, one row of somites, one pronephros, and one pronephric duct develop on each side. The defect may be expressed to a different degree and cover one or another region of the embryo depending on the stage of gastrulation disturbance. Under favorable conditions of incubation, the defective embryo with the endoderm protruding as a cupola continues to develop for some time but, as a rule, dies during the period in question.

Sometimes (very rarely), we observe cases of duplication when the embryo has two heads or a bifurcated tail.

We frequently encounter a unique complex defect where the head, trunk, and heart develop abnormalities at the same time. The head in such embryos does not separate from the yolk sac at all, or does so to a lesser degree than during normal development. The posterior body region is markedly shortened. The heart is split into halves, sometimes two hearts are formed (Fig. 57C). In some cases, there is no heart (Fig. 57D). Defects in the development of the heart are caused by disturbances affecting the fusion of the lateral plates. As already mentioned, the right and left lateral plates are normally drawn together in front of the head and, fusing, form a single rudiment of the cardiac tube. In cases of incomplete fusion, a bipartite heart forms. If no fusion takes place, two independent cardiac tubes develop, or a heart does not form at all.

This complex defect occurs in cases when the embryos are exposed to unfavorable environmental factors at the end of gastrulation and during the neural plate formation, or just before the heart rudiment formation.

2.7 Development of the Embryo from the Onset of Heart Beating to the Time of Hatching

2.7.1 Morphology of the Embryo

This period is characterized by intensive growth of the embryo, continuation of differentiation, and appearance of definitive functions for some systems of organs. The heart starts to function and blood circulation begins. Muscle contractions appear: starting as hardly perceptible jerking, they gradually, by the end of the period, assume the character of swimming movements. The nervous system begins to function: the embryo responds to external stimuli (mechanical, chemical) by muscle contractions and, at the end of the period, swimming movements. The onset of the excretory function of pronephros has as yet not been established; Sawadski (1926b) noted that in *A. ruthenus* the pronephros functions on the day of hatching.

During this period, the gas exchange realized by the entire body surface is complemented by gas exchange mediated by the yolk sac vessels and, later, segmental vessels of the posterior trunk region and tail.

With the beginning of blood circulation and appearance of muscle contractions, the oxygen consumption increases (Korzhuev 1941). Later, when the embryo assumes a free way of life, it also rises markedly. The increase in the level of respiration appears to provide for a more intensive growth of the embryo.

Besides great similarity in the morphology of developing embryos in studied Acipenserid species during transition to this period, there is a distinct difference between them: the heart starts to beat at different stages. The first, barely perceptible contractions of the heart in *A. stellatus* begin at a stage, when it has the shape of an elongated, but still straight tube (Plate V, 28; see p. 246). In *A. gueldenstaedti*, the first contractions begin at the stage of the sigmoid heart, when the posterior body region begins to flatten and lie on its side (Plate VI, 29). In *H. huso* and *A. ruthenus*, the first contractions of the heart are detected even somewhat later when the heart is distinctly S-shaped and the posterior body rudiment is flattened and lies on its side. These differences are separated by several hours and are apparent in embryos developing at different temperatures. With respect to these differences, two moments should be considered. First, the intensity of metabolism may be related to the size of the organism: in the smaller size, it is higher. Second, *A. stellatus* spawns at temperatures higher, on average, than *A. gueldenstaedti*, *A. ruthenus*, and *H. huso*. The smaller size of the *A. stellatus* embryos and the higher temperature at which they develop provide for their relatively higher demand of oxygen. On the other hand, the oxygen content of water decreases at a higher temperature. Hence, this demand can, apparently, be satisfied by an earlier transition to intensive forms of gas exchange. In the embryos of *H. huso* and *A. ruthenus*, which develop at the lowest temperature, the heart starts to beat somewhat later than in *A. gueldenstaedti*. Circulation in the network of yolk sac vessels begins with the onset of heart beating.

During the period from the beginning of blood circulation to hatching, the outlines of the embryo undergo marked changes. Separation of the head, which had hardly begun at the end of the preceding period, progresses rapidly. Successive stages of this process can be clearly seen in the embryos from which the membranes were removed while they were still alive (Plate VI, 29–31 lat; plate VII, 32–33 lat; see pp. 248–251), since inside the membranes the head, like the posterior body rudiment, remains pressed to the ventral region. As a result of progressive deepening of the anterior body fold, the head becomes more and more separated from the ventral region and its lower surface increases. At the same time, the hatching gland, which was previously located in front of the brain, becomes fully displaced onto the lower surface of the head. The flattened head becomes spherical and this leads to relative displacement of the other organs: olfactory pits come to lie on the anterior surface of the head, the eyes now face laterally, rather than upwards, the first visceral arches bend laterally and downwards. Through the covering epithelium, we can see brain regions, olfactory sacs, eyes in which the lens and, just before hatching, a dark pigment spot gradually become discernible; on both sides of the medulla oblongata relatively large thin-walled auditory vesicles can be seen. A mouth recession, often hardly descernible, is outlined at the lower surface of the head, at its base, just before hatching. Two small vertically running epithelial folds are formed by the time of hatching on both sides of the head in the elongated branchial region. These are ectodermal branchial grooves.

The posterior trunk and tail regions undergo marked changes. In the preceding period, the common rudiment of these regions was rod-shaped and strongly bent. During the period in question, these regions considerably elongate, straighten and flatten. As the posterior trunk region elongates, the pronephric duct also becomes elongated. The latter opens into the cloaca, a pouch-like deepening of the covering epithelium (Plate VI, 29 lat). At this site later, after hatching, the anal opening will break through.

The fin fold, a median fold of integument which is the rudiment of all unpaired fins, forms on the dorsal and ventral sides of the posterior trunk and tail regions. Only that part of the fin fold, which is located on the ventral side of the posterior trunk region, in front of the cloaca, is later reduced. During the period in question, the fin fold rapidly widens (Plate VI, 29–31; Plate VII, 32–34; Plate VIII, 35B; see p. 248–252).

As the posterior trunk and tail regions flatten, they lie on their sides and, being pressed by the membranes to the yolk sac, bend onto the ventral side of the embryo. The tail tip gradually reaches the heart (Plate VI, 31 v), head (Plate VII, 32h) and moves onto the head (Plate VII, 33–34 h; Plate VIII, 35B, d).

The position of the tail tip is a convenient criterion for determination of developmental stage during this period. It should be remembered however that there are certain differences between the embryos of various Acipenserid species in terms of this feature: in *A. ruthenus*, the posterior trunk region together with the tail is relatively longer than in *H. huso*, *A. gueldenstaedti*, and, especially, *A. stellatus*. One can judge how close the embryo is to hatching by the position of the tail at the late stages of embryogenesis.

The hatching starts: in *A. stellatus*, when the tail tip reaches the posterior boundary of auditory vesicles or goes beyond it (Plate VIII, 35A, d), in *A. gueldenstaedti* and *H. huso* when it reaches the loop formed by the anterior part of the pronephric duct (Plate VIII, 35B, d), and in *A. ruthenus* when the tail tip goes even farther, (Plate VIII, 35C, d).

In some cases, long before hatching the membranes stretch, more frequently in *A. gueldenstaedti*; thereafter, the position of the tail tip cannot serve as a criterion of developmental stage.

The degree of straightening of the posterior trunk region and tail can serve as the second common criterion for determining the stage of the embryo in this period. Since to use this criterion, we have to take the egg membranes away, this criterion will be used predominantly for research, when it is necessary to determine the stage very precisely. At these stages, the embryo can easily be liberated from the membranes by two forceps with sharpened tips.

The turn of the posterior trunk and tail regions on to their sides is realized by muscle contractions, and the embryo changes, from time to time, the position of these regions, by lifting them to the vertical position or turning them again either to the right or to the left. At the stage when heart begins to assume the sigmoid shape, the embryo first responds to pricking by weak muscle jerks. At this stage, the tail tip begins to lie on its side. At stages, when the tail tip approaches the head, the embryo is already capable of responding to pricking by moving the posterior body regions and head to the left or to the right, it can even now and then turn from one side to another. The movements of the posterior body regions start slightly earlier than those of the head. Inside the membranes, the tail and the major part of the posterior trunk region already lie on their sides and the embryo can change their position. Starting from when the tail approaches the head and moves onto it, the movements of the embryo inside the membranes become more active. If the embryo is released from the membranes at this time, it "swims on the spot", bends the head to the left and right, performs the oscillatory movements of the tail but does not move, since the heavy yolk sac holds it in place. Somewhat later, the embryo liberated from the membranes turns from one side to another but still cannot move. Only just before hatching is the embryo capable for forward movement.

While moving inside the membranes, the embryo mixes the perivitelline fluid, thus improving the conditions of respiration. If the egges are incubated under unfavorable conditions (at a too high or too low temperature or oxygen deficiency), the motility of the embryos markedly reduces.

The ventral region of the embryonic body preserves, during a greater part of this period, a spherical shape, in accordance with the shape of egg membranes. Only just before hatching, due to the weakening of the latter under the influence of the hatching enzyme (see below) and frequently occurring stretching of the membranes, can the shape of yolk sac change: the yolk sac becomes egg-shaped in *A. stellatus* and more elongated in *A. gueldenstaedti* and *H. huso*. Characteristic, especially for *A. gueldenstaedti*, straightening of the lower surface of the abdomen and its anterior wall, which usually takes place during the first hours after hatching, can occur also before hatching.

In accordance with various intensities of egg coloration, the embryos of *A. stellatus* and *H. huso* before hatching are relatively light, while those of *A. gueldenstaedti colchicus* are usually pigmented rather heavily. In the latter species, the pigmentation of the embryos is often unequal. Mosaic pigmentation is frequent: one part of the embryo may be relatively light, while the other, on the contrary, heavily pigmented. The distribution of light and pigmented region varies. Sometimes, one half of the embryo (right or left) is dark, the other is light. There may also be some other combinations: for example, the anterior and posterior parts can be pigmented in different ways. The pigment pattern is preserved at all successive developmental stages and can be used as a natural marker when observing the development. Mosaic pigmentation also occurs in *A. stellatus* and *H. huso* but in these species it is relatively weaker and occurs less frequently. It is noteworthy that the mosaic pattern of embryonic pigment does not affect the pigmentation of prelarvae determined by the development of melanophores.

Let us consider changes in the internal structure of the embryos during this period.

The nervous system of the embryos continues to differentiate and its changes at these stages do not differ from those in *A. ruthenus* described by Salensky (1878). A fold appears in the brain roof which separates the hindbrain and is the rudiment of cerebellum. The hollow process in the diencephalon floor, which forms the infundibulum, increases in size. Neuromeres can be seen in the wall of rhombencephalon. The number of cells in the brain walls increases, the nerve cells begin to differentiate: hardly discernible at first but large enough by the moment of hatching a zone of white matter (marginal veil) appears. A chain of spinal ganglia forms along the spinal cord, nerve roots differentiate.

The sense organs undergo relatively slight changes. Olfactory pits deepen slightly more. Two large olfactory sacs form which open outside by round apertures. In *A. stellatus*, by hatching stage the sensory epithelium of the olfactory sacs is connected with the brain by the nerves (Dragomirov 1953b). In the brain regions adjacent to the sensory epithelium of the olfactory sacs, nerve cells begin to differentiate earlier than in the neighboring regions, and a considerable zone of white matter forms by the moment of hatching.

The anterior wall of the optic vesicles gradually deepens, and they transform into optic cups; the optic stalk becomes thinner. The lens rudiment increases in size and gradually separates from the covering epithelium. It separates completely from the epithelium in *A. stellatus* when the tail tip approaches the head and in *A. gueldenstaedti* and *H. huso* when the tail moves beyond the head. Shortly before hatching, a small cavity appears in the center of the lens. Definitive pigment starts to accumulate in the thin outer layer of the optic cap in a small group of the cells facing the iris. Its presence was described by Salensky (1878, 1880) in the just-hatched larvae of *A. ruthenus*. The amount of pigment rapidly increases. A small more or less dark spot can first be seen slightly above the center of the eye. Usually, it becomes distinct by the beginning of the hatching period. However, when the eggs of *A. persicus* were incubated in Sadov's tray in a bright dispersed light (at mean temperatures of 21.4–22.1 °C), the pigment spot in the eye could be seen already at stage 34, 10 to 14 h before hatching.

Marked changes are observed in the circulatory system. The first visceral vessels to form are two mandibular arches of the aorta (see Schmalhausen 1952) which, at the anterior notochord end, connect with the dorsal aorta roots and form sinus cephalicus. Paired cerebral vessels run from it forwards and ventrally. A pair of hyoid arches and the first pair of branchial arches of the aorta are also partially laid down before hatching. The dorsal aorta carrying blood from the head to the tail runs under the notochord, reaches its end, bypasses it and passes into the caudal vein. Segmental vessels form in septs, between the myomeres, by which blood from the aorta passes to the posterior cardinal veins. The latter are being formed and consist of separate parts; they are located parallel to the dorsal aorta at the base of mytomes. At the hind gut end, the tail vein passes into the subintestinal vein and carries blood along the lower wall of the intestine. At the tail base it goes onto the yolk sac wall and connects with a dense network of yolk sac vessels. Vessels from the back (from segmental vessels and parts of cardinal veins) also lead onto the yolk sac wall. At first the yolk sac vessels are clearly seen only on its dorsal side. Then their number rapidly increases and they become seen also on the ventral side. By this time almost the entire surface of the posterior half of the yolk sac represents a continuous network of vessels, or sinuses, deepened in a mass of yolk-rich cells. At the beginning of this period, the number of blood islands rapidly increases on the ventral and lateral walls and hemopoiesis is intense. The blood cells have a spherical shape and contain nuclei and yolk granules.

The vascular network of the ventral region wall and segmental vessels developing somewhat later fullfill the function of gas exchange not only in the embryos but also in prelarvae during the first period of free life, until their transition to gas exchange through the gills. The importance of the vascular network in the yolk sac wall in Acipenserids for gas exchange was first stressed by Kryzhanovskii (1934, 1949). Blood is collected from the network of the yolk sac vessels in two yolk veins which, in their lower parts, near the heart, enter the ducts of Cuvier. The latter are two large vessels running from the back (from the area of pronephros) to the heart, along the yolk sac wall, parallel to the pericardial cavity boundary. The ducts of Cuvier collect blood from the venous system of the head and anterior trunk regions.

The lateral plates continue to grow in the ventral direction and soon close up.

No marked changes occur in the excretory system.

Significant changes are observed in the anterior and posterior regions of the digestive system. The head body fold, which appeared at the end of the preceding period, continues to deepen and under-cuts the head, so that it projects over the surface of the yolk sac. The mouth invagination is laid down much later, before hatching, as a small triangular depression of the covering epithelium. Perforation of the mouth opening in this area takes place after hatching.

In the pharyngeal region, the two first pairs of branchial grooves are outlined by the hatching stage. Perforation of the gill clefts takes place after hatching.

In the posterior part of the digestive system, which elongates as the tail grows, there are two parts differing in their development. These are the hindgut, which terminates in the area of the cloaca, and its continuation, the tail (postanal) gut, which reaches the tail tip at the hatching stage. The tailgut still communicates with the cavity of the spinal cord (neurentheric canal). It is a transient structure and is gradually reduced after hatching. The hindgut cavity tapers backwards and finally disappears, so that the posteriormost region of the gut adjoining the cavity of the uniting pronephric ducts has no lumen. The hindgut wall rapidly differentiates, separately from the yolky endodermal cells which are not involved in its formation. There is a small number of endodermal cells freely lying in the hindgut cavity.

In the middle part of the intestine, in the yolk sac area, the intestinal epithelium by the hatching stage has been formed in the dorsal part of the intestine and in its anterior part (in front of the liver diverticulum). The lower wall of the intestine has not yet separated from the adjacent yolk-rich endodermal cells. The cavity in the middle part of the intestine is almost filled with endodermal cells whose boundaries gradually disappear. These cells are then used for feeding of prelarvae. Hence, despite the fact that Acipenserids are characterized by complete cleavage of the egg, part of the cellular material does not participate in the formation of embryonic tissues. From this viewpoint, it is justified to some extent to call the ventral region of the Acipenserid embryos the yolk sac.

Seven stages are identified in this period. Various criteria have been used for their brief characterization.

It should be borne in mind that the characteristics of developmental stages of the membrane-enclosed embryo according to the position of the tail tip is given for the example of *A. gueldenstaedti*. This characteristics is true also for *H. huso*, but in *A. stellatus* the relative length of the posterior trunk and tail regions is somewhat less and in *A. ruthenus* is greater than in *A. gueldenstaedti*. Besides, given the convenience of this criterion, one should not forget that in some batches of eggs the membrane stretches in the second half of the described period. Thereafter the position of the tail tip cannot be used as a criterion of the developmental stage.

Specific features of the structure of the embryo are safe criteria which can easily be identified after liberation of the embryo from the membranes: firstly, degree of straightening of the rudiment of the posterior trunk and tail regions, and the width of the fin fold.

The following stages have been identified: stage 29, when the cardiac tube has an S-shaped flexure and has started to beat (Plate VI, 29); stage 30; when the rudiment of the posterior trunk and tail regions begins to straighten and the tail tip approaches the heart in the embryo inside the membranes (Plate VI, 30); stage 31, when straightening of the posterior trunk and tail regions is markedly advanced and the tail tip reaches the heart in the embryo inside the membranes (Plate VI, 31); stage 32, when the posterior trunk region is straightened, the tail is slightly bent, and in the embryo inside the membranes the tail tip moves slightly on the head (Plate VII, 32): stage 33, when the tail is almost straight and in the embryo inside the membranes the tail tip moves slightly onto the head (Plate VII, 33); stage 34, when the fin fold becomes markely wider and the embryo, liberated from the membranes, is capable of slow forward motion (Plate VII, 34); stage 35, the stage of hatching of single prelarvae, when the embryo, liberated from the membranes, is capable of rapid forward motion; if the membranes are not yet stretched, in the embryos the tail tip reaches the posterior boundary of the auditory vesicles in *A. stellatus* and the loop formed by the anterior part of the pronephric duct in *A. gueldenstaedti* and *H. huso*; in *A. ruthenus*, the tail tip goes still farther (Plate VIII, 35).

2.7.2 Structure of Atypical Embryos at the End of Incubation

Abnormal embryos often cannot free themselves of their membranes; the less abnormal ones hatch out, but usually with some delay compared with the normal embryos. This is why abnormal embryos are still found in their membranes on the bottom of the incubator together with degenerating unfertilized eggs by the end of the hatching period.

If the incubation conditions are favorable, only single deformed embryos occur in batches of good-quality eggs, and up to 25–30% in those of poor-quality eggs. If the eggs are kept in unfavorable conditions, the percentage of deformed embryos increases drastically. For instance, when *A. persicus* eggs were incubated at a high temperature (at times, higher than 26 °C), more than 90% of the embryos had distinct structural defects.

When studying deformed embryos of various structure, we are sometimes amazed that embryos with profound defects do not die by the end of the incubation period. Meanwhile, if we liberate these deformed embryos from their membranes and place them in water, many of them will continue to live for a long time, until there is some yolk left in the intestine. Left in their membranes at the bottom of the incubator together with the. dead eggs, they soon die.

As already mentioned (see Sect. 2.6.3), underdevelopment of the anterior regions of the body is the usual type of developmental defect. One can select a whole series of embryos with all the grades of these defects. In the least abnormal embryos, the anterior regions of the brain are simply reduced. With significant underdevelopment of these regions of the brain, we may find that in place of two olfactory sacs and two eyes, only one of each organs has developed in a central position. Further, we find embryos in which the anterior region of the head is absent altogether (Fig. 58A), i.e. there is no prosencephalon, diencephalon, olfactory sacs, or eyes; the head begins from the mesencephalon or directly from

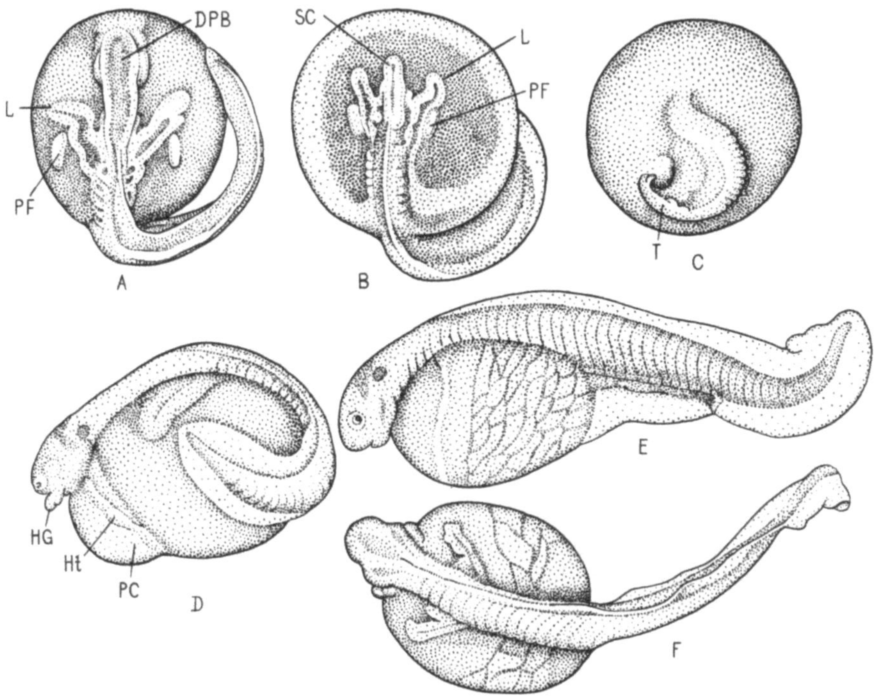

Fig. 58A–F. Abnormal embryos of *A. gueldenstaedti* at stage 35 (Ginsburg and Dettlaff 1955). **A** Prosencephalon and diencephalon absent; **B** headless embryo; **C** only defective posterior trunk and tail regions developed; **D** a deformed embryo with incompletely separated head, bent shortened posterior trunk and tail regions and edema of the pericardial cavity; **E, F** shortened embryos with bent tails. *DPB* Defective posterior regions of the brain; *HG* hatching gland; *Ht* heart; *L* loop formed by the anterior part of the pronephric duct; *PC* pericardial cavity; *PF* pectoral fin rudiment; *SC* spinal cord; *T* tail myelencephalon, on both sides of which we find the ear vesicles. Other embryos do not have the myelencephalon either (Fig. 58B), i.e., the whole head is missing. In this case the loops formed by the pronephric duct on both sides of the spinal cord and rows of muscle segments occupy the anteriormost position. The embryos sometimes have no anterior trunk region either, i.e., only an abnormal posterior trunk region and tail extend from the yolk sac (Fig. 58C). Most of the deformed embryos of this type have shortened and bent tails and many of them have no heart.

Embryos with underdeveloped anterior regions of the body move their tail spasmodically, but usually are unable to swim. Embryos without the anterior region of the head and more deformed ones remain immobile and even do not respond to needle pricking.

Embryos with more or less complete doubling of the axial organs are sometimes encountered; these embryos may have two heads, or two heads and two trunk regions on one yolk sac. Abnormal heart structure, frequently accompanied by shortening of the tail and incomplete elevation of head, is a common type of defect. In some embryos, the heart is shaped like a long, narrow tube, while the pericardial cavity is highly inflated (due to a pathological accumulation of fluid, edema) (Fig. 58D). In others, the heart is bifurcated, or there are two hearts, and in still others, there is no heart at all. Embryos of this type at early stages of development can be seen in Fig. 57C, D.

We also encountered embryos with less significant defects, i.e., they are shortened, bent, with edema of the pericardial cavity expressed to a different extent, as well as embryos with various defects of the fin fold (Fig. 58E, F).

The majority of the deformities are caused by disturbances at the earliest stages of development, i.e., during oocyte maturation and fertilization. In turn, these disturbances determine the abnormal cleavage and gastrulation and the entire subsequent development. The exceptions are various types of distortions which may also appear at the later stages of development, in many cases as a result of muscle paralyses (which occur under unfavorable conditions, primarily when the temperature is too high at the end of the incubation period).

2.7.3 Duration of the Hatching Period and the Means by Which the Embryos are Liberated from Their Membranes

In various Acipenserid species (*H. huso*, *A. gueldenstaedti*, *A. stellatus*, *A. ruthenus*), the embryos are liberated from their membranes at similar stages of development. These stages are identical for the batches of eggs incubated at different temperatures within the range of spawning temperatures (Dettlaff and Ginsburg 1954; Ignatieva 1979). The first embryos to hatch (stage 35, Plate VIII) often do not yet have a pigment spot in the eyes; rudiments of the pectoral fins in these embryos are hardly discernible or totally indistinguishable, and their blood is colorless or yellowish. The embryos that hatch at the stage of mass hatching (stage 36, see Sect. 3.2) always have a pigment spot in the eyes, clearly defined rudiments of the pectoral fins, and pink or even reddish blood. As a result of the movement of the embryos inside the membrane, the shape of the ventral region changes and becomes more elongated.

During the hatching period, the embryos continue to grow inside the membranes. Therefore, a progressive increase in the size of hatching prelarvae is observed.

At first, single prelarvae hatch (hatching of single prelarvae stage), and then the number of prelarvae swimming in the incubator gradually increases. It can be said that the hatching period begins when several dozen prelarvae are seen in the incubator. The number of prelarvae usually increases rapidly after that, and mass hatching of prelarvae sets in. The hatching period ends when only deformed embryos, degenerating unfertilized eggs, and dead embryos remain on the bottom of the incubator.

The duration of the hatching period in different batches of eggs and in the same batch at different levels of incubator loading can vary significantly. For instance, in *H. huso* (Igumnova 1979), the period from the hatching of single prelarvae to the end of hatching constitutes 15–30% of the interval from insemination to the onset of this period. The length of hatching period poses considerable difficulties for fish culture, and hence, it is important to know how the embryos are liberated from their membranes and what conditions favor simultaneous hatching of the prelarvae.

The liberation of the embryos of their membranes is caused by an enzyme secreted from the hatching gland on the lower surface of the head (Zotin 1953b; Ignatieva 1957a). In species whose embryos have especially strong egg membranes (*H. huso, A. gueldenstaedti*), a certain decrease in their strength takes

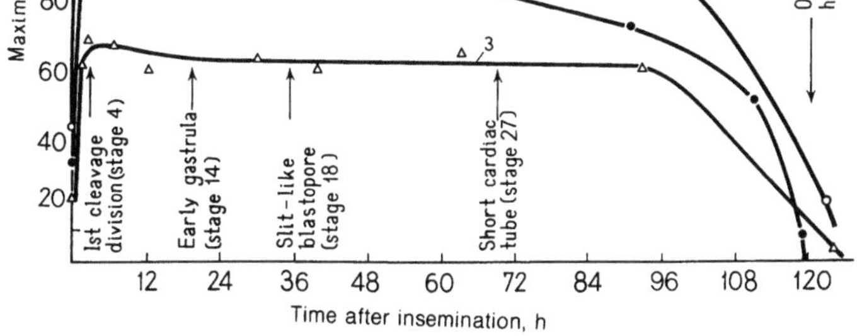

Fig. 59. Hardening of egg membranes in *H. huso* (*1*), *A. gueldenstaedti* (*2*), and *A. stellatus* (*3*) during development (Zotin 1953c). Time plotted on abscissa refers to *H. huso* embryos

place gradually during embryogenesis, starting from the early cleavage when the membranes are the strongest (Fig. 59, 1 and 2). In *A. stellatus*, the strength of the membranes decreases during development insignificantly or not at all (Fig. 59, 3). Shortly before hatching, the strength of the membranes sharply diminishes in all three species.

It is not clear what the mechanism of the decrease in the strength of the egg membranes is in the course of development. Its sharp decrease before hatching is due to the influence of the hatching enzyme. This enzyme is able to dissolve the egg membranes at all stages, starting from the uncleaved egg. It is not species-specific, shows the highest activity at 40–45 °C and is inactivated in the range of 45 to 50 °C.

Testing of the enzymatic activity of the head at successive developmental stages allowed to study changes in the accumulation of the enzyme in the gland and to find out the stage of its secretion into the perivitelline space (Zotin 1953b). As seen in Fig. 60, the enzyme appears in the gland after the onset of heart beating, and then its content increases rapidly up to the stage directly preceding hatching (Fig. 60, 2). At this stage, the gland secretes the enzyme into the perivitelline fluid, the enzyme activity of which increases rapidly (Fig. 60, 3), while the activity of the gland decreases. The strength of the egg membranes quickly diminishes with the appearance of the enzyme in the perivitelline fluid

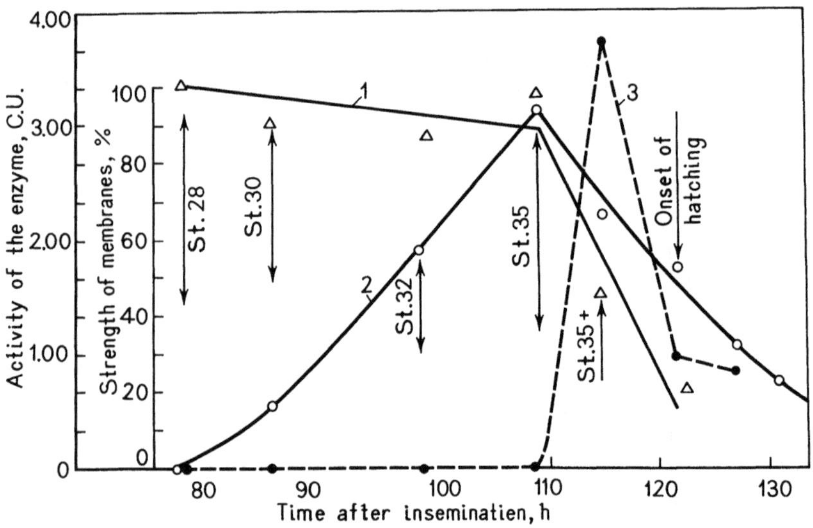

Fig. 60. Relationship between the changes in the strength of egg membranes, accumulation of the enzyme in the hatching gland, and its secretion in the perivitelline fluid in *A. gueldenstaedti* embryos (Zotin 1953b). *1* Strength of the membranes at successive developmental stages expressed in per cent of their strength at stage 28 (studies started at this stage); *2* enzymatic activity of the hatching gland in conditional units; *3* enzymatic activity of the perivitelline fluid in the same units

(Fig. 60, 1). These experiments convincingly demonstrate that the decrease in the strength of the membranes before hatching is due to the influence of the enzyme which accumulates, for a certain period of time, in the gland and, before hatching, is secreted into the perivitelline space.

When studying the sections of the heads of the embryos from the same batch taken a few hours before hatching of the first prelarvae, at the stage of enzyme secretion onset, we could see in the cells of the hatching gland, near the plasma membrane, large clusters of drops of the secretory substance which were intensively stained with acid fuchsin. In some embryos, there were many such drops, while in the others only a few cells contained some drops: almost all material was secreted by the moment of fixation.

Secretory granules and vacuoles in the cells of the hatching gland were earlier described by Sawadski (1911) who studied the development of this gland in detail in *A. ruthenus*. However, he did not see the discharge of the contents of these organelles and, therefore, he believed, like Salensky (1880), that the hatching gland is an organ of attachment, similar to the sucker of the *Holostei*.

Moving around in the weakened membranes, the embryo breaks them. Its head or tail appears first, and then the whole embryo emerges. If the tail is freed first, the embryo moves it actively from side to side, and then gradually emerges from the membranes. Sometimes, it does not manage to do this at once, and then it has to swim around for a while with the membranes covering its head and yolk sac. If it hatches headfirst, the embryo pushes against the membrane with its curved tail and pokes its head through the opening first, and the trunk region next. In cases where the opening in the membrane is small, the yolk sac becomes extremely deformed, i.e., it acquires the shape of a figure eight, and then becomes pear-shaped. However, the yolk sac quickly regains its normal shape once the embryo is outside.

The secretion of the hatching enzyme and muscle activity, which are of primary importance to the liberation of the embryo from the membranes, greatly depend on environmental conditions. They are stimulated by the improvement of aeration, the movement of the water and by the impacts to which the embryos are exposed (Dettlaff and Ginsburg 1954; Ignatieva 1958). It is highly significant that the effect of the stimuli does not result in premature hatching, and the embryos never emerge from the membranes before the stage characteristic of the onset of hatching. Clearly, the hatching gland in the Acipenserid embryos acquires the ability to expel the accumulated enzyme only at a specific stage of development, thus determining the constancy of the hatching stage in these fish.

In the absence of conditions favorable to the discharge of the enzyme, the latter is secreted slowly, the membranes stretch and the embryo straightens out partially, so that the pressure exerted on the membranes by its convoluted body lessens. In such cases, the embryos can remain inside the membranes for many hours. The hatching period is prolonged by other factors, i.e., poor quality of the eggs and unfavorable environmental conditions during the incubation period, which lead to a substantial disparity in the developmental stages.

As for conditions of simultaneous hatching, we should emphasize that the conditions ensuring normal development of eggs during almost the entire incubation period may prove to be inadequate during the hatching period when a rapid water current and vigorous mixing of the eggs in the incubator are necessary.

In particular, the Yushchenko incubators widely used for the incubation of demucilaged sturgeon eggs do not ensure sufficiently simultaneous hatching, which can last for many hours. Some kind of auxiliary devices should obviously be used for more intense mixing of the eggs during the period preceding hatching.

The hatching conditions in the tray incubators of the Sadov and Kokhanskaya system (Sadov and Kokhanskaya 1961; Emel'yanov and Sytina 1967) are even less favorable. In many cases, hatching in them begins with a very substantial delay; in *A. persicus*, the hatching stage at optimal temperatures (17–18 °C) is characterized by pigmented eyes, the presence of nasal bridge and gill filaments on the branchial arches (Sadov and Kokhanskaya 1961), i.e., by features that normally appear no sooner that at stage 39 of prelarval development. When eggs of fairly good quality are incubated in trays, hatching often lasts for a long time, i.e, up to 2–3 days (Emel'yanov and Sytina 1967).

2.8 Duration and Rate of Embryogenesis in Various Acipenserid Species and Their Dependence on Temperature. Chronology of Development

2.8.1 Duration of Embryogenesis as Expressed in Astronomical Time Units

Over many years of study of the development of Acipenserid species, numerous data were obtained about timing of some transient and clearly defined developmental stages in the embryos of *A. gueldenstaedti*, *A. stellatus*, *H. huso*, and *A. ruthenus*, which were incubated at the hatcheries on the Kuban', Don, Volga, and Kura at different temperatures. These data were used to plot graphs of the duration of development in astronomical time units, hours and minutes, as a function of temperature (see Fig. 62, 81). These graphs served as a basis for studying the temporal patterns of development and predicting the timing of various developmental periods at different temperatures. In the studied species, the spawning temperatures coincide only partially. Therefore, the curves of timing of similar developmental stages as a function of temperature are displaced with reference to one another along the temperature scale (Fig. 61). In the zone of spawning temperatures, common for different species, these curves cross each other, i.e., at certain temperatures the embryos of different species develop at the same rate. At a temperature below the crosspoint, species adapted to colder temperatures develop more rapidly, while above the crosspoint those

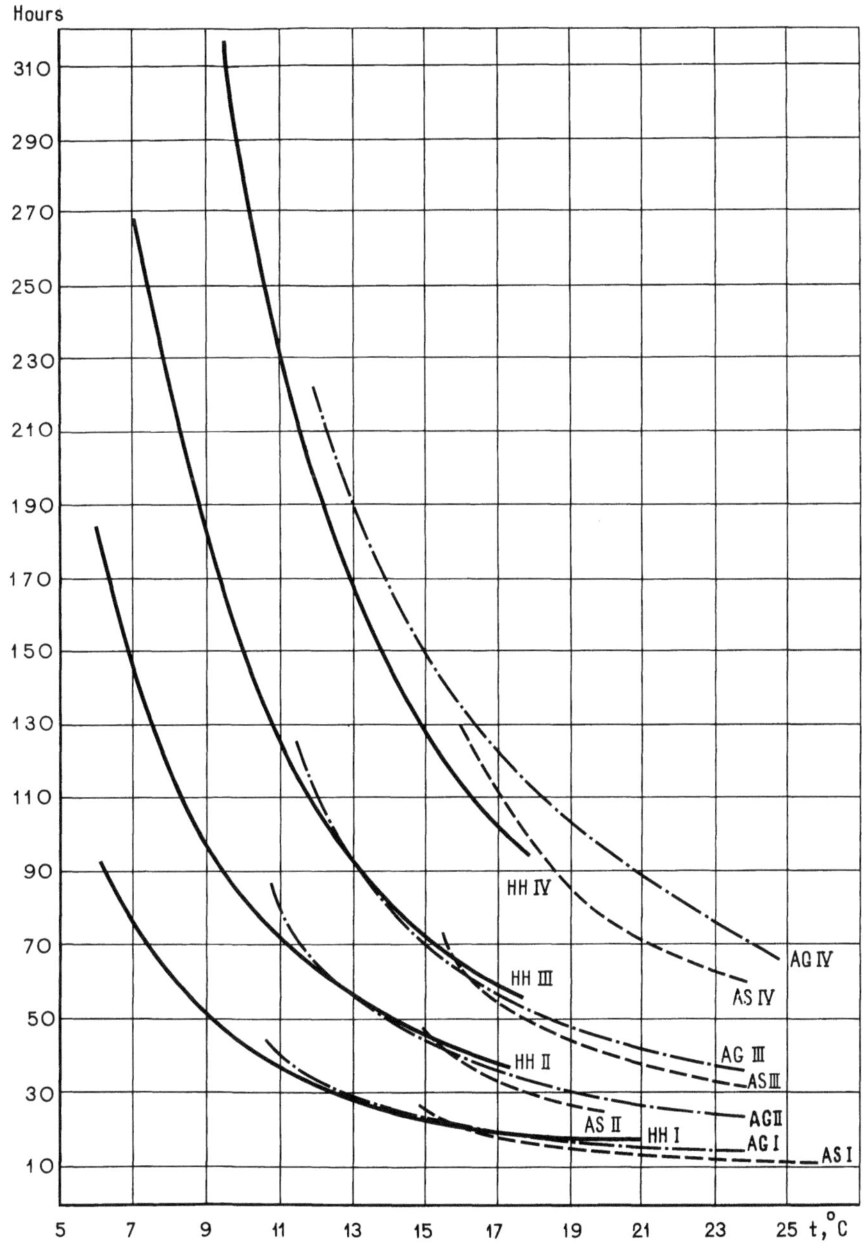

Fig. 61. Duration of embryogenesis in *A. gueldenstaedti* (*AG*), *A. stellatus* (*AS*), and *H. huso* (*HH*) from insemination of eggs to the onset of stages 13 (*I*), 18 (*II*), 26 (*III*), and 35 (*IV*) at different temperatures. The curves for *A. gueldenstaedti* and *A. stellatus* are given from Dettlaff and Dettlaff (1961) and for *H. huso* from Igumnova (1975a)

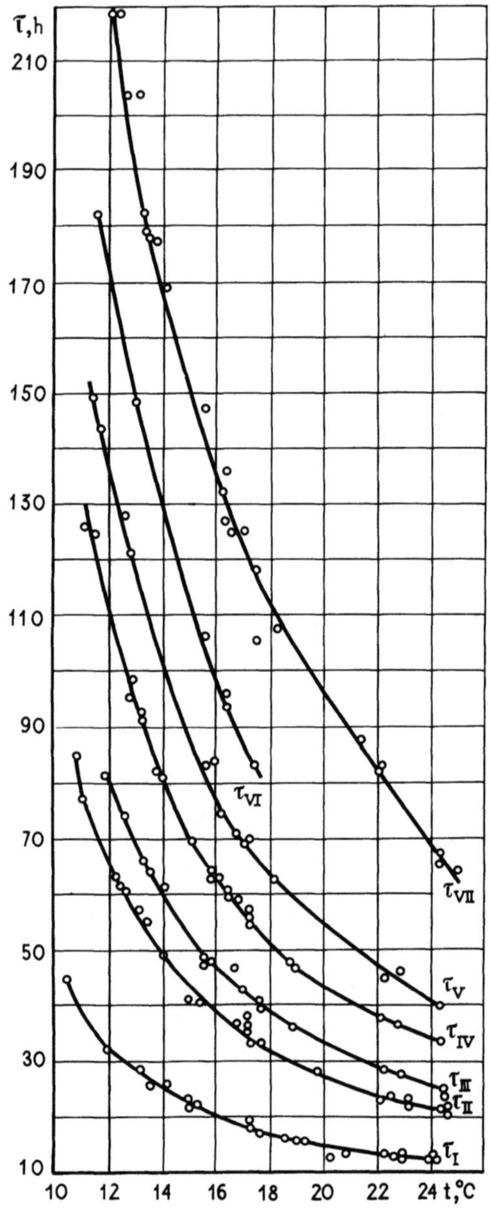

Fig. 62. Duration of the development of *A. gueldenstaedti* embryos from insemination of eggs to the onset of gastrulation (τ_I), end of gastrulation (τ_{II}), closure of neural folds (τ_{III}), fusion of lateral plates in the region of heart rudiment formation (τ_{IV}), onset of heart beating (τ_V), the stage when the tail tip touches the head (τ_{VI}), onset of hatching (τ_{VII}) (Dettlaff and Dettlaff 1961). *Ordinate*, time, h; *abscissa*, temperature, °C

adapted to warmer temperatures develop faster. For example, at a temperature of 15 °C the embryos of *A. stellatus*, *A. gueldenstaedti*, and *H. huso* develop at the same rate. At temperatures below 15 °C the embryos of cold-preferring *A. gueldenstaedti* develop more rapidly, and at 17 °C and more the embryos of warm-preferring *A. stellatus* develop at a higher rate. The embryos of *A. gueldenstaedti* and *H. huso* develop at the same rate at 13–15 °C, at 10–11 °C the embryos of *H. huso* develop more rapidly and at temperatures above 15 °C at a slower rate than those of *A. gueldenstaedti*. The spawning temperatures for *A. ruthenus* and *H. huso* coincide, while the embryos of *A. ruthenus* develop at all temperatures slightly more rapidly than those of *H. huso*. It is evident that the data expressed in astronomical time units characterize the duration of development of a given species only at a given temperature. They do not allow forecasting of the timing and duration of certain developmental periods at the other temperatures in the given species or moreover, their comparison with different animal species. Such prognoses and comparisons become possible if the relative duration of developmental periods is determined in dimensionless units.

2.8.2 Relative Duration of Embryonic Periods

It was shown (Dettlaff and Ginsburg 1954) that the duration of different developmental periods in *A. gueldenstaedti* and *A. stellatus*, when expressed in fractions of the total duration of embryogenesis, rather than in hours, remains constant at different optimal temperatures (Table 6). It was also shown (Dettlaff 1953) that the duration of the period from fertilization to the appearance of the 1st cleavage furrow on the egg surface remains constant, when expressed in the number of intervals between the appearance of the 1st and 2nd cleavage furrows, later designated by symbol τ_0 (Dettlaff and Dettlaff 1961). In *A. gueldenstaedti*, *A. stellatus*, and *H. huso*, the duration of this period equals $3\tau_0$ at different temperatures optimal for each species (Dettlaff 1953).

The observations carried out contradict the widely accepted concept that different developmental periods depend on temperature in different ways (Bělehrádek 1934; Hayes 1949; Svetlov 1978). In order to confirm the data suggesting a similar dependence of various developmental periods on temperature, additional investigations were carried out. The results of some of these investigations are presented in Figs. 62–64.

Figure 62 demonstrates the duration of periods from fertilization to the onset of seven developmental stages in *A. gueldenstaedti colchicus* at different temperatures. If the data presented are used to plot curves in logarithmic coordinates (Fig. 63), these curves are located almost parallel one to another, thus suggesting the proportional changes in the duration of various developmental periods upon changes in temperature. Figure 64 shows the duration of the same developmental periods in different units, number of τ_0, number of periods from fertilization to the onset of gastrulation, and in fractions of the total duration of embryogenesis (from fertilization to the hatching of single prelarvae).

Table 6. Relative duration of certain developmental periods in *A. gueldenstaedti colchicus* and *A. stellatus* in percent of the total duration of the period from fertilization to hatching of single prelarvae. (Dettlaff and Ginsburg 1954)

Developmental periods	*A. gueldenstaedti colchicus*			*A. stellatus*			
	Mean temperature, °C						
	12.1	14.0	16.5	16.0	18.0	20.0	23.0
Fertilization and cleavage (stages 1 to 13)	14.4	15.2	15.5	15.0	15.7	15.6	15.9
Gastrulation (stages 13 to 18)	15.4	14.5	13.6	14.2	14.2	16.0	16.9
Neurulation (stages 18 to 23)	6.7	5.7	6.0	6.3	5.2	5.2	5.2
Period from the end of neurulation to the fusion on lateral plates (stages 23 to 26)	15.3	12.4	12.4	14.2	14.7	13.5	12.8
Total (stages 1 to 26)	51.8	47.8	47.5	49.7	49.8	50.3	50.8
Period from the fusion of lateral plates to hatching of single prelarvae (stages 26 to 35)	48.2	52.2	52.5	50.3	50.2	49.7	49.2

The curves in Fig. 64 are straightened in the central zone of optimal temperatures for all studied developmental periods and run more or less parallel to the abscissa axis. At a suboptimal high temperature (22–24 °C), the proportion of changes in the duration of a more thermosensitive period of early cleavage and less thermosensitive later developomental periods is affected. This occurs because at 22–24 °C the duration of the period of early cleavage begins to decrease more slowly with the increase of temperature than the duration of later periods.

The proportionality of changes in the duration of various developmental periods in the zone of optimal temperatures (at the other favorable conditions) is preserved not only during development of the embryos and prelarvae, but also during oocyte maturation. Hence, the duration of processes with a different nature and realized at different levels of the embryo organization changes proportionally with the change of temperature. It has also been shown that this pattern previously established for the Acipenserid embryos, is true for other poikilothermic animals (see Dettlaff et al. 1987).

The proportionality of changes in the duration of various developmental periods and processes upon changes in temperature is very important for poikilothermic animals: it is an essential condition for preserving the integrity of a developing organism and its adaptation to varying environmental conditions (Dettlaff 1981, 1985). A study of this pattern allowed us to develop the method of

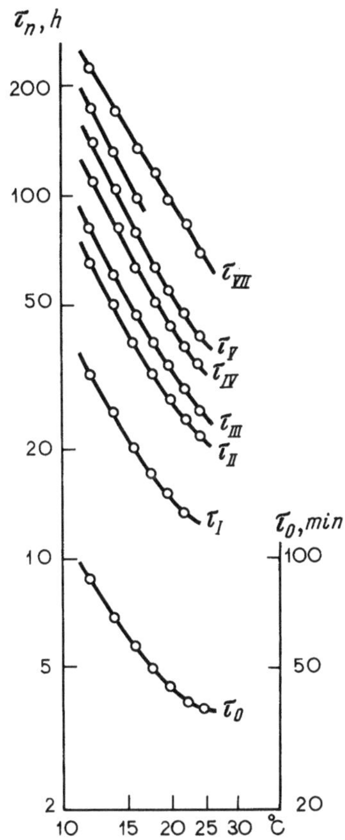

Fig. 63. Logarithm of the duration of developmental periods τ_I–τ_{VII} as a function of the logarithm of temperature (Dettlaff and Dettlaff 1961). *Ordinate*, time, h; *abscissa*, temperature, °C

characterization of the development duration, or biological time of development, in dimensionless values.

2.8.3 Method Characterizing the Duration of Various Developmental Periods in Dimensionless Values

This method consists in measuring the duration of any developmental period designated as τ_n by using the duration of one mitotic cycle during synchronous cleavage divisions, τ_o, at the same temperature as a time unit. The quotient of division of τ_n by τ_o shows by how many times τ_n is greater (or lesser) than τ_o and, therefore, the obtained value is dimensionless. This method of estimating the duration of developmental periods allowed us to "liberate" time from its

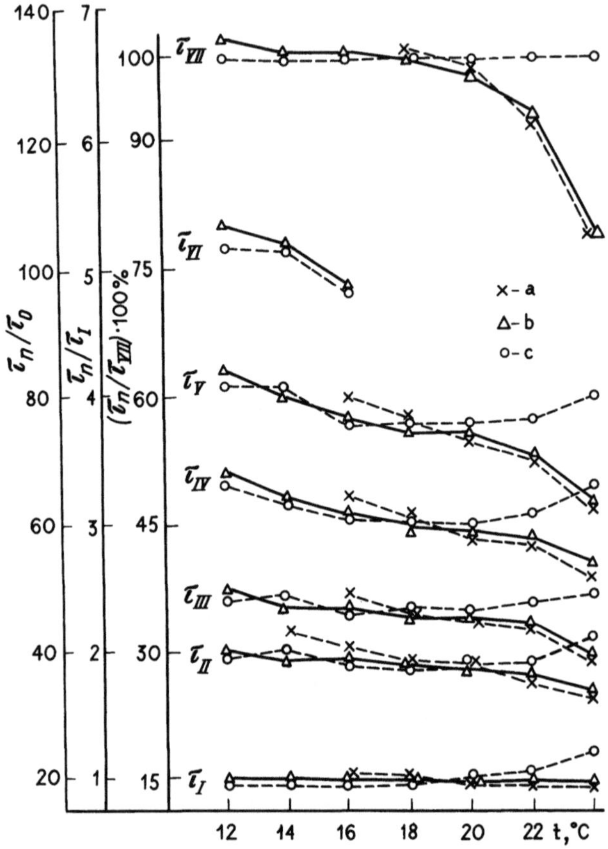

Fig. 64. Duration of the various developmental periods in *Acipenser gueldenstaedti* in dimensionless units: (*a*) in the number of periods of one mitotic cycle during the first cleavage divisions (τ_n/τ_o), (*b*) in the number of cleavage periods (τ_n/τ_I), (*c*) in percent of the total duration of development from insemination to the onset of hatching ($\tau_n/\tau_{VII} \times 100$) (Dettlaff and Dettlaff 1961). *Ordinate*, duration of different developmental periods expressed in different relative units (*a–c*); *abscissa*, temperature, °C. For designations see Fig. 62

dependence on temperature, since the value of τ_n/τ_o is the same in the zone of optimal temperatures at different temperatures. Also, it liberated time from the species-specific dependence of the developmental rate from temperature as, it is the same for τ_n and τ_o. Finally, it liberated us from the astronomical time units, dimension inherent in both numerator and denominator of the fraction τ_n/τ_o, which is reduced.

This method of estimating the relative duration of developmental periods allowed us to overcome the main difficulties which were blocking the path of studying the temporal patterns of development. Since τ_o represents the minimal cell cycle and is characterized by a complex of features, similar in all animals which pass the period of synchronous cleavage divisions in their development, this time unit is comparable in most animals. This allowed us to use the proposed method of characterization of the relative development duration for prognostication of the time of development and the comparison of temporal patterns in representatives of various species, genera, and families. This method cannot be applied for studying the development of mammals and some animal species with the mosaic type of development, in which the development lacks the period of synchronous cleavage divisions.

Graphs of the values of τ_0 as a function of temperature were plotted and the values of τ_n/τ_0 at one of the optimal temperatures for each given species were determined in order to characterize the relative duration of various developmental periods. The results of application of this method for studying the development and predicting the timing of various stages in Acipenserid fish at different temperatures have been widely used in this book. The detailed description of the method and the results of studies of the temporal patterns of development in various poikilothermic animals are given by Dettlaff et al. (1987), Dettlaff (1991), and Dettlaff and Vassetzky (1990, 1991).

2.8.4 The Age of Embryos and Chronology of the Embryogenesis

It is not enough to estimate the age of the embryos of Acipenserid fish, as well as the other poikilothermic animals, using astronomical time units, since at different temperatures the embryos reach the same developmental stage for different periods of time. This estimation only allows to repeat observations at the same temperature. Therefore, until recently, the embryos at the same developmental stages were considered as being of the same age for lack of more adequate age criteria. However, this estimation of age is not correct, since in the embryos of different animals gastrulation can begin at different stages of cleavage (a particular case of heterochrony). Therefore, the embryos at the same developmental stages can consist of cells from different generations (Dettlaff 1956). In order to compare the characteristics of age between different animals, the age should be determined in hours and minutes at one of the optimal temperatures and then recalculated in the number of τ_0 at the same temperature.

Table 7 presents the data about timing of most of the developmental stages in *A. gueldenstaedti colchicus* in hours and minutes at 18 °C and in the number of τ_0. In order to obtain these data, a batch of good-quality eggs with a high percentage of fertilization was incubated at a hatchery, and samples of 100 to 200 embryos were fixed at short intervals. When the samples were taken, the temperature was measured and the mean temperature for a time interval between the stages was calculated. The duration of this interval was then calculated in the number of τ_0 at the same temperature and, using the curve of τ_0 vs. temperature dependence, the absolute timing of this stage at 18 °C was determined.

The data presented in Table 7 about the relative durations of the periods from fertilization to the onset of successive developmental stages allow to

Table 7. Chronology of embryonic development in *A. gueldenstaedti colchicus*

Stage	Time from insemination		Specific features of stage
	h and min at 18 °C	Number of τ_o	
1	0	0	Egg at the moment of fertilization
	~ 0.50	~ 1	Light polar spot disappeared; perivitelline space formed
3	~ 1.40	~ 2	Pigment accumulation in animal region excentric; light crescent formed
4	2.55	3.5	Animal region divided by the 1st cleavage furrow
5	3.45	4.5	Animal region divided into 4 blastomeres
6	4.35	5.5	Animal region divided into 8 blastomeres
7	5.25	6.5	Furrows of 4th cleavage division appeared on the surface of animal blastomeres
8	6.15	7.5	Furrows of 5th cleavage division appeared on the surface of animal blastomeres
9	7.3	9	7th cleavage division; furrows divide vegetal area completely
10	8.2	10	Blastocoel begins to form; nuclear divisions in animal region are still synchronous
11	10	12	Early blastula; blastomeres in animal region clearly discernible under low magnification; desynchronization of nuclear divisions in animal region
12	12.3	15	Late blastula; individual cells in animal region are indiscernible at low magnification
12 +	14.1	17	Rearrangement of cell cycle in blastomeres of animal region; drop in mitotic index
13	16.15	19.5	Beginning of gastrulation; highly pigmented band formed at equator level in region of future dorsal lip of blastopore
14	17.05	20.5	Early gastrula; invagination of material through dorsal lip of blastopore over a small area
15	22.55	27.5	Middle gastrula; animal material covers 2/3 of surface of embryo; blastopore closed in ring
16	25.00	30.0	Yolk plug of considerable size
17	27.50	32.5	Small yolk plug
18	31.40	38.0	Gastrulation completed, blastopore slit-like

19	32.30	39.0	Early neurula; neural folds around head region begin to show
20	33.45	40.5	Wide neural plate; neural folds around head region of neural plate clearly defined
21	34.45	41.7	Neural folds begin to draw together; rudiments of excretory system begin to show for the first time
22	36.4	44	Late neurula; neural folds in trunk region drawn together, rudiments of excretory system longer
23	37.30	45	Neural tube closed; suture along the line of neural folds fusion clearly discernible
24	—	—	Eye processes formed; thickening appears in anterior parts of rudiments of excretory system
25	—	—	Lateral plates reach anterior end of head, their tapering ends come together in front of hatching gland rudiment; common rudiment of posterior trunk and tail regions appears as a thickening at the posterior end of the body
26	50.00	60.0	Lateral plates fuse, and rudiment of heart forms at the site of their fusion; body fold undercuts rudiment of posterior trunk and tail, making it project freely over the yolk sac surface
27	53.20	64.0	Short, tube-like heart rudiment formed
28	57.30	69.0	Heart rudiment as straight elongated tube; trunk muscles do not yet contract in response to stimulation
29	60.00	72.0	Heart tube assumed S-shaped flexure, begins to beat; trunk muscles twitch in response to pricking
30	62.05	74.5	Tip of tail of embryo in membranes approaches heart; posterior trunk region and tail begin to straighten out
31	—	—	Tip of tail of embryo in membranes reaches heart; embryo can move head and tail
32	78.00	93.5	Tip of tail of embryo in membranes touches head
33	—	—	Posterior trunk region and tail straighten out completely after removal of membranes
34	—	—	Embryo removed from membranes can move forward slowly
35	104–106.40	125–128	Hatching of individual prelarvae; embryo can move forward quickly after removal of membranes
36	—	—	Mass hatching

predict their timing at different optimal temperatures. Using this, the number of τ_0 corresponding to the duration of the period in question should be multiplied by the value of τ_0 at the desired temperature. The value of τ_0 should be determined from the curve of its temperature dependence in *A. gueldenstaedti* (Fig. 42A).

It is noteworthy that the relative duration (expressed in the number of τ_0) of the periods from insemination to the onset of successive developmental stages in *A. stellatus* coincides with the corresponding values for *A. gueldenstaedti* over the whole embryogenesis, as well as at later stages, up to stage 40, and with those for *H. huso* and *A. ruthenus* only until the middle of embryogenesis. Therefore, the data about the relative duration of various developmental periods in *A. gueldenstaedti* can be used for predicting of the duration of corresponding periods in embryogenesis of *A. stellatus* and in the first half of embryogenesis in *H. huso* and *A. ruthenus*. Thus, the number of τ_0 characterizing the duration of any developmental period in *A. gueldenstaedti* should be multiplied by the value of τ_0 in *A. stellatus*, *H. huso*, or *A. ruthenus* (Fig. 42B–D) at the desired temperature.

It should be borne in mind that different embryos in the same batch of eggs do not develop quite synchronously: embryos at two successive developmental stages can be encountered in the same sample. The data given in Table 7 were obtained for one or two dozen embryos in the sample, most advanced in their development. The synchrony of development in different batches of eggs varies. In synchronously developing batches of eggs, a considerable proportion of embryos pass simultaneously to the next stage.

Some deviations from the time predicted by the data presented in this chapter can be due to the fact that the rate of development is influenced, besides temperature, by the other conditions of development, such as loading of incubator and flow pattern.

Chapter 3 Development of Prelarvae

3.1 Characteristics of the Prelarval Period of Development

The period from hatching of the embryo to the beginning of active (exogenous) feeding is at least as long as the embryonic development. Characteristic features of this period include the replacement of embryonic adaptations and functions by the definitive ones. This change involves alterations in the relationships of the developing embryo with the environment. This affects its behavior such as response to light, substrate and water current. During this period several traits are formed, which have taxonomic value and characterize adult fish of various species: the subpharyngeal fold is formed (or not formed), differences arise in the length and location of barbels, position of the dorsal fin, structure of mouth, lower lip, and some other organs. Because of the characteristic features of this period a special prelarval period is distinguished in the development of sturgeon as in that of teleosteans (Rass 1941, 1946). This period begins with the emergence of embryos from the membranes and ends when prelarvae begin active feeding. However, until now some authors continue to refer to the prelarvae as free embryos or just as larvae, thereby ignoring fundamental differences of this developmental period both from the embryogenesis and the larval period in the strict sense of the word.

Morphogenesis and differentiation are very intense during the prelarval period. This leads not only to formation of characteristic developmental patterns, but also to the appearance of characteristic structural defects under unfavorable developmental conditions. Improvement of methods for rearing prelarvae requires that these defects be distinguished from those which appear during embryonic development. Furthermore, environmental factors which can lead to malformations have to be found, i.e., have teratogenic effect on prelarvae or even lead to their death. These questions are very urgent at present, since rearing of prelarvae until the transition to active feeding is one of the least developed steps in sturgeon culture. It is at this time, particularly at the end of the prelarval development, that marked losses are common.

Lack of proper understanding of the prelarval period of development which essentially is a period of incomplete organogenesis when the rudiments of numerous organs are still very sensitive to unfavorable conditions, led certain authors to erroneous views. Disturbances of prelarval development were attributed, without any serious proof, to unfavorable environmental effects on

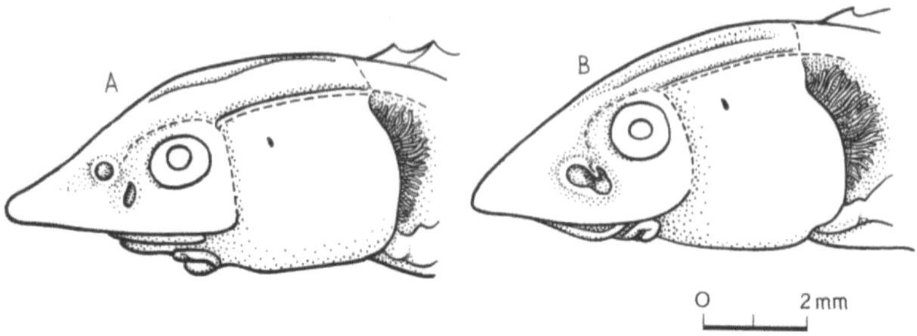

Fig. 65A, B. Abnormal development of the olfactory organ in *A. gueldenstaedti* after rearing prelarvae in the presence of brass mesh. Larvae 15 days after transition to exogenous feeding: **A** Control; **B** the larva developing from the stage of hatching in the presence of a piece of brass mesh (the olfactory opening is not partitioned)

eggs during maturation and on embryos during incubation (Sadov 1941, 1948 a, b, 1950, 1951; Emel'yanov 1951, 1953, 1961; Sadov and Kokhanskaya 1961; Kokhanskaya 1980). This provided a basis for arguing about the correctness of using progressive methods of modern sturgeon breeding like pituitary injections and demucilaging of eggs (see Dettlaff and Ginsburg 1954).

Discussion of these problems required experimental studies to investigate the possibility that induction of malformations was due to unfavorable environmental effects on prelarvae. Such a study has been performed using defects of the olfactory organ as an example (Schmalhausen 1957, 1962). These defects belong to the category studied by the proponents of the views just mentioned. It has been demonstrated that deviations in the development of this organ (Fig. 65) are associated with unfavorable effects on prelarvae rather than with the incubation conditions. At the same time it has been found that such influences also disturb development of other organs.

Thus, it has been established that in a number of cases defective development of prelarvae can be due to unfavorable effects during their rearing. This called for detailed studies of normal development of the prelarvae of various Acipenserid species and elucidation of the factors underlying structural defects. Also this required analysis of the effect of polluted water and various materials used for the construction of rearing reservoirs. The main results of these studies are presented next.

3.2 Stages of Prelarval Development

A description of stages and the timing of normal development of sturgeon prelarvae is of considerable practical significance, since this allows us to compare prelarvae from different batches under different rearing conditions. Furthermore, it makes it possible to estimate their quality and suitability for further breeding. Such descriptions are also necessary for experimental studies. Without them it is impossible to compare the prelarval development of different Acipenserid species.

However, it should be remembered that even prelarvae obtained from eggs of a single female show a certain heterogeneity in the timing of stages in different individuals. This scattering can be due to differences in the initial properties of eggs or the rearing conditions. Furthermore, in different batches of prelarvae this scattering effect could be different. In prelarvae of similar age that have developed from poor-quality eggs or were reared under unfavorable conditions, there is considerable variation in the development of individual traits. In extreme cases this even leads to malformations. Increase in the range of individual variation in sturgeon prelarvae under unfavorable conditions also follows from the data of Sytina and Timofeev (Sytina 1971a, b, 1979; Sytina and Timofeev 1973). The problem of factors responsible for these deviations requires special analysis in every case.

In order to distinguish physiological variation of traits from pathological changes we have first of all to give an as complete as possible and accurate description of the normal prelarval development.

Prelarval development of sturgeon can be divided into two periods: (1) from hatching and up to the beginning of active respiratory movements and (2) from the beginning of these movements to the transition to active feeding (Schmalhausen 1955c). For the convenience of description, ten prelarval stages have been distinguished, i.e., stages 36–45 (Schmalhausen 1968, 1975), which continue the sequence of stages described in embryonic development (see Chap. 2). Some of these stages correspond to the developmental stages of *A. gueldenstaedti* prelarvae described by Dragomirov (1957).

The development of all studied Acipenserid species (*H. huso*, *A. gueldenstaedti*, *A. stellatus*, *A. ruthenus* and *P. kaufmanni*) is very similar. As a result, traits common for all these species were used for describing individual prelarval stages. Therefore, the description of these stages presented for *A. gueldenstaedti* (Schmalhausen 1975) can be used for other species as well. The corresponding stages of *H. huso* prelarval development have been described (Schmalhausen 1968).

Differences between prelarvae of different species at the hatching stage are relatively moderate (Alyavdina 1951b; Dragomirov 1953a, 1961b); they become more prominent by the stage of transition to active feeding. These differences will be dealt with in a special section (Sect. 3.6).

Each developmental stage was characterized according to morphological traits distinguishable upon external examination of the prelarvae as well as certain functional parameters. In this approach the time interval between consecutive stages was rather long, because we wanted at least one new trait to be associated with each stage, rather then just quantitative changes of traits that appeared during previous stages.

The data about changes in the inner structure of prelarvae have been obtained by dissecting and conventional methods of histological analysis of the material. Prelarvae have been studied in *H. huso*, *Acipenser gueldenstaedti colchicus* from the Don, *A. gueldenstaedti* from the Volga, *A. stellatus*, and *A. ruthenus*.

Beginning of each stage was defined as a time of sample fixation, in which prelarvae first showed traits characteristic for this stage.

Materials for the description of *A. gueldenstaedti colchicus* prelarval development were collected on the Don, at the Rogozhkino sturgeon hatchery. Prelarvae were collected at the mass hatching stage from Yushchenko incubators. They were then reared in glass jars in the laboratory at optimal temperatures. Prelarvae of *A. gueldenstaedti colchicus* which have provided material for drawings of normal stages were kept in jars with a thin layer of water (about 2 cm); water was changed daily. The mean water temperature during the rearing of this batch of prelarvae was 18.6 °C.

We shall now turn to the description of consecutive stages of prelarval development.

Stage 36 (Dettlaff and Ginsburg 1954) corresponds to mass hatching of prelarvae; mouth and gill clefts have not yet been formed, the hatching gland is still distinguishable (Fig. 66 and Plates IX and XI, see page 253 and 255).

Prelarvae are 9 to 10.5 mm length (here and below these sizes are given after measurement of the prelarvae fixed in the Bouin fluid). They are rather dark, since a large amount of melanin is still present in epidermal cells. The head is small relative to the trunk and somewhat bent towards the massive ventral region (yolk sac) which is of an elongated ovoid shape. The mouth opening is still absent. The hatching gland is still visible on the lower head surface in front of the mouth invagination. Barbel rudiments are not apparent. Trunk and tail are bordered by the fin fold. The upper and lower blades of the tail fin fold are equal in width, in other words, the tail at this stage is still protocercal. The cloacal rudiment is located between the preanal and postanal parts of the fin

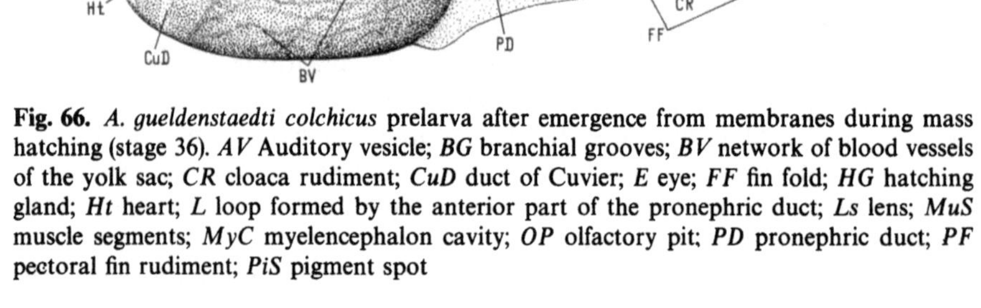

Fig. 66. *A. gueldenstaedti colchicus* prelarva after emergence from membranes during mass hatching (stage 36). *AV* Auditory vesicle; *BG* branchial grooves; *BV* network of blood vessels of the yolk sac; *CR* cloaca rudiment; *CuD* duct of Cuvier; *E* eye; *FF* fin fold; *HG* hatching gland; *Ht* heart; *L* loop formed by the anterior part of the pronephric duct; *Ls* lens; *MuS* muscle segments; *MyC* myelencephalon cavity; *OP* olfactory pit; *PD* pronephric duct; *PF* pectoral fin rudiment; *PiS* pigment spot

fold; these parts are separated from one another by a small groove. Pronephric excretory ducts open into the cloaca. The intestine is still closed. The preanal fin fold begins at the posterior surface of the yolk sac; this anterior part of the fold is distinctly widened but still does not form a process (a keel), which is characteristic of subsequent stages. The dorsal fin fold becomes narrower towards the head and disappears at the level of anterior muscle segments. Generally there are no rudiments of pectoral fins but, when present, they look like hardly noticeable skin thickenings located immediately behind pronephros and the upper part of Cuvier ducts. No rudiments of ventral fins are present.

By this stage the segmentation of the trunk muscles is already completed. There are about 40 muscle segments up to the cloaca rudiment. The lower parts of the anterior segments begin to grow downwards. Segmentation of the tail muscles is incomplete (posterior part of the mesoderm is not yet segmented). Posterior end of the notochord is slightly bent upwards. There are paired rudiments of maxillar, hyoid, and first branchial arches. Gill clefts are not present yet. There are grooves at the place of the first two pairs of gill clefts. Respiration is accomplished through the body surface and a network of vessels located in the posterior part of the yolk sac. These vessels are supplied with blood from the caudal and subintestinal veins, from segmental vessels of the trunk and the developing parts of posterior cardinal veins. In the anterior part they fuse into paired vitelline veins opening into the heart near the Cuvier ducts. Blood has an yellowish color.

Sense organs are weakly developed. Rudiments of olfactory organs are rounded with one external opening. Eyes are not pigmented, with the exception of a small spot adjacent to the region of provisory retinal photoreceptors (Baburina 1972). According to Baburina, prelarvae of *A. gueldenstaedti colchicus* which have just emerged from membranes do not have any photoresponse; they retain this behavior throughout the prelarval period. Auditory vesicles (rudiments of the labyrinths) are located on both sides of myelencephalon and are clearly distinguishable because the covering tissues are transparent. The differentiation of parts of the labyrinth has not yet started. The seismosensory (lateral line) system of the head is represented by paired rudiments of supraorbital, suborbital and temporal lines as well as by paired rudiments of the hyomandibular organ complex (Disler 1949; Dragomirov 1961a; Nikol'skaya 1985). Suborbital rudiments begin under the auditory vesicles, extend beyond eyes and end at the level of the lower edge of eyes. Shorter supraorbital rudiments are located over the eyes.

Stage 37 is the stage at which the mouth begins to open as well as the first pair of gill clefts; there are distinct rudiments of pectoral fins (Plate IX).

Prelarvae are 10.5 to 11.5 mm long. The embryonic pigment is still occasionally retained in the epidermis. The head begins to straighten. The mouth breaks through in its middle part. On the lower side of the head, in front of the mouth there are four rounded tubercles, i.e., rudiments of barbels located near the anterior mouth border. The shape of the yolk sac changes: it becomes more elongated and its ventral surface becomes less convex. Short invaginations are seen through the epidermis on both sides of the yolk sac. They are formed in the wall of the digestive tract rudiment and subdivide its wider part into two regions. These invaginations are oriented in an oblique direction (downward and forward) and coincide with the anterior border of the yolk sac vessel network. The distance between this network and the ducts of Cuvier is increased. This distance continues to increase also later throughout the period of yolk respiration. Rudiments of pectoral fins look like distinct folds. The tail grows faster as compared with the posterior part of the trunk and becomes increasingly longer in relative terms at this and subsequent stages. The anterior end of the preanal part of the fin fold begins to form a keel. The lower part of posterior muscle segments begins to grow forward and downwards. This process is more distinct in the anterior segments. The first gill cleft is open. The auditory vesicles subdivide into several regions. Short rudiments of the lateral line of the seismosensory system of the trunk become distinguishable.

Stage 38 is the stage of appearance of gill filament rudiments on the opercular fold and the first branchial arch (Plates IX and XI).

The length of prelarvae is 11.5 to 12 mm. Rare, small, still rather pale melanocytes can be seen on the upper surface of the head through the epidermis. Similar melanocytes are found on the dorsal side of the first trunk muscle segments as well as on muscle segments and the fin fold in the region of the future dorsal fin. The ventral body region is flattened laterally and ventrally. The invagination subdividing the rudiment of the digestive system to the stomach and intestinal regions reaches the middle of the lateral surface of the prelarva. The tail is slightly bent upwards. The height of muscle segments increases. Their dorsal processes in the region of the dorsal fin produce well-distinguishable rudiments of muscle buds. Ventral processes of anterior trunk muscle segments still do not reach the lower margin of the opercular fold. In the region of the pectoral fin they spread down to its base. Hardly distinguishable rudiments of muscle buds are seen at the level of the future anal fin, they begin from ventral processes of muscle segments. The opercular fold and the first branchial arch carry one row each of short rudiments of gill leaflets. Branchial respiration of prelarvae begins, following the change of direction of blood flow in gill vessels. Olfactory pits become slightly elongated. Pigmentation of the iris begins. Supraorbital line of the seismosensory system increases in length and ends above the olfactory pit. The suborbital line passes outside the lateral barbel and terminates at the anterior margin of its base. The lateral line of the trunk reaches the 4th–6th muscle segments.

Stage 39 is the stage when the invagination of the wall of the digestive system rudiment divides it into two regions: stomach and intestine (Plate IX).

The length of the prelarvae is 12 to 13.2 mm. Their pigmentation is markedly stronger. Melanocytes are darker and spread all over the surface of body and tail muscle segments. Small folds are formed at the corners of the mouth; they are rudiments of the lower lip. The rudiment of the upper lip is subdivided into two parts. Teeth rudiments appear. The wider part of the digestive system rudiment is subdivided into two regions: stomach and intestine. The vascular network of the yolk sac is reduced and the left vitelline vein becomes empty. A short connective tissue cord becomes visible at the right side of the prelarva. It is directed upwards from the intestine and the anterior end of the dorsal rudiment of the pancreas. The cord crosses the rearmost part of the stomach surface and goes further under the ventral processes of muscle segments towards dorsal mesenterium. Ventral processes of the anterior muscles segments almost reach the lower operculum edge. Pectoral fins increase in size and are somewhat displaced downwards. The fin fold is already broadened at the site, where the dorsal fin will be formed, and starts extending in the region of the anal fin and the lower lobe of the tail fin. The tail fin separates from the anal and dorsal fins by shallow grooves. Skin thickenings appear at the sites corresponding to future ventral fins.

The second gill cleft is open. Rudiments of the filaments of the opercular hemibranch (sitting on the opercular fold) and of the first row of the first gill are slightly elongated. They are still short in the second row of the first gill. The first row of filaments of the second gill is laid down. Olfactory pits are trapeziform. Pigmentation of the iris is more noted. Anterior tips of the suborbital lines of the seismosensory system move forward to the level of olfactory organs and bend towards one another; they represent the rudiments of the premaxillary lines. Rudiments of the trunk lateral lines reach the level of the posterior edge of pectoral fins. Rudiments of the accessory row of the lateral line appear.

Stage 40 is the stage when rudiments of ventral fins appear and mandibular movements begin (Plate IX and XI).

Prelarvae are about 13 mm long. Their pigmentation increases. Melanocytes form loose aggregates on both sides of the mesencephalon and anterior part of myelencephalon, on the dorsal side of anterior muscle segments and on the boundary between the axial and ventral muscles (near pectoral fins), on the dorsal fin and along the lateral side of the posterior part of the trunk and the tail. Barbel rudiments become more elongated. The median recess of the upper lip begins to smooth out. The connective tissue cord supporting the intestine is displaced forward but is still located on the posterior part of the stomach. Dorsal rudiment of the pancreas stretches to this level. It is parallel to the intestine.

Bases of pectoral fins still do not reach the middle of the lateral stomach surface. Muscle buds differentiate in them. There is the first evidence for the reduction of the anterior keel-like process of the preanal fin fold, but this process is still quite noticeable. Rudiments of the ventral fins look like narrow longitudinal folds. The lower lobe of the tail fin continues to expand. Ventral processes of muscle segments overgrow the lateral stomach surface halfway and descend below the base of the pectoral fins.

The third gill cleft is still incomplete. The filaments of the opercular hemibranch and the first row of the first gill reach the level of the duct of Cuvier.

Rudiments of filaments of the second row of the first gill are elongated. Rudiments of filaments of the first row of the second gill are well discernible. First respiratory mandibular movements begin at this stage.

Olfactory pits become even more elongated. Two lobes begin to grow from their edges across the olfactory opening; the upper lobe is more developed. The iris is well distinguishable because of its light-brown color. Incompletely separated semicircular canals of the labyrinth are seen through translucent skin. Anterior ends of the right and left premaxillar lines of the seismosensory system on the lower head surface fuse and produce the anterior transverse commissure. It is located at the level of the olfactory organ rudiments (Plate XI, 40). The lateral lines of the trunk seismosensory system do not yet reach the posterior boundary of the stomach; paired rudiments of the accessory row terminate above the pectoral fins.

Stage 41 is the stage when the lobes dividing the olfactory opening make contact, and rhythmic respiratory movements begin (Plate IX).

Prelarvae are 13 to 14 mm long. First taste buds appear on the tips of barbels. The upper lip is constricted in its median part. Short lateral lobes of the lower lip are distinct. The liver margin adjacent to the intestine is subdivided into two parts. Rudiment of the gall bladder is seen in the right part.

The connective tissue cord extending from intestines to dorsal mesenterium has moved forward and is now located across the stomach, midway between its end and the posterior margin of the pectoral fin. Bases of the pectoral fins are located obliquely with respect to the longitudinal body axis and begin to constrict. The keel-like process of the preanal fin fold is markedly reduced. Posterior part of the ventral fins becomes wider. The tail bends smoothly upwards. Ventral processes of the muscle segments have overgrown more than half of the lateral body surface. The gill cleft between the second and third gills is broken through. The second row of filament rudiments has formed on the second gill and they are beginning to form in the first row of the third branchial arch.

Movements of visceral apparatus become stronger and more rhythmic. Lobes subdividing the external olfactory opening are in contact but have not yet fused. The iris is darker. The anterior end of supraorbital rudiment of the head seismosensory system extends over the upper olfactory lobe. Premaxillary canals are distinctly connected on the lower surface of the rostrum; they produce the anterior transverse commissure. As the rostrum straightens, this commissure moves forward and almost reaches the level of the anterior margin of olfactory organs. Developing pit lines (neuroepithelial follicles) of the opercular group are seen on the operculum. The lateral line reaches the level of the spiral intestine. The accessory row of lateral system organs extends beyond the posterior border of pectoral fins. Paired rudiments of the dorsal row appear.

Stage 42 is the stage when lobes partitioning off the olfactory opening fuse and the rudiment of the pyloric appendage appear (Plates X and XI).

Prelarvae are 14 to 15 mm long. Melanocytes appear on the lower surface of operculum and the anterior part of the ventral region. The rostrum continues to straighten. Taste buds appear on the lips. The liver is divided into two lobes. A small tubercle, the rudiment of pyloric appendage can be seen through ventrally, it is separating from the anterior end of the intermediate gut, on its left side. The connective tissue cord located above the stomach on its right side reaches the posterior margin of the pectoral fin. Here, the anterior end of the dorsal rudiment of pancreas is located as well. The pectoral fin has moved down to the middle of the lateral surface of the prelarva and forward. Its anterior edge touches the filaments of the opercular hemibranch and the first gill. Keel-shaped process of the preanal fin fold continues to decrease. Ventral fins change their shape and a lobe forms in their posterior part which does not yet reach the edge of the preanal fin fold. Muscle buds differentiate at the base of the ventral fins. The caudal part of the anal fin broadens. Ventral processes of muscle segments overgrow the entire lateral body surface. Ducts of Cuvier located in front of the pectoral fins become indistinguishable from outside (they are covered by a muscle layer and gill filaments). Rudiments of the first row filaments are well discernible on the third branchial arch. Movements of the visceral apparatus are rhythmic.

Lobes which partition off external olfactory openings fuse. The anterior transverse commissure of the seismosensory system is located in front of the olfactory organs. Rudiments of neuroepithelial follicles become visible on the lower side of the rostrum. The lateral line of the trunk reaches the anterior level of the ventral fin. The accessory row of lateral line organs extends to the end of the stomach. The rudiment of the dorsal row of seismosensory system organs crosses anterior muscles segments obliquely backwards and starts bending to the back. At this stage prelarvae settle to the bottom of the tank.

Stage 43 is the stage when rudiments of secondary filaments appear in the first gill and the ventral fins reach the margin of the preanal fin fold (Plate X, see page 254).

Prelarvae are 15 to 16 mm long. The rostrum acquires a horizontal position. The pyloric rudiment starts subdividing into lobes. The anus opens. The connective tissue cord fastening the intermediate gut to the dorsal mesenterium is no more seen from the outside since it is covered by a muscle layer and is displaced under the pectoral fin. Lobes of the pectoral fins broaden.

Ventral fins reach the margine of preanal fin fold. The keel of its anterior part is fully reduced. Rudiments of secondary filaments appear in the first gill; they are less prominent in the second row than in the first one. Rudiments of the second row filaments appear in the third gill. The fourth gill cleft breaks through. Operculum covers the anterior margin of pectoral fin.

The iris is dark. The anterior transverse commissure of the seismosensory system is displaced to the anterior end of the rostrum but is still visible from the ventral side. Lateral line of the trunk seismosensory system reaches the level of the posterior margin of the ventral fin. The accessory row ends above the spiral gut; the dorsal row at this stages is already bent and its posterior section is parallel to the lateral line. The end of this row is located above the pectoral fin.

Stage 44 is the stage when the edges of ventral fins extend beyond the preanal fin fold and a common rudiment of dorsal scutes appears (Plate X).

Prelarvae are 16.2 to 17 mm long. Pigmentation of prelarvae is much more intense. Melanocytes migrate downwards and cover almost the whole flank of prelarvae. Dorsally over the pectoral fins aggregations of melanocytes appear as spots. Rudiments of bony scales appear on the upper head surface. The relative size of the head increases because of a marked elongation of the rostrum. Now bases of barbels move forward and the tips of barbels do not reach the anterior border of the mouth. Shorter middle barbels terminate at a greater distance from the mouth than the longer lateral ones.

The pyloric appendage continues to subdivide into lobes; most often three lobes can be distinguished. Massive ejection of pigment plugs takes place. In the dorsal fin fold a narrow mesenchyme strand appears which is seen through the epidermis: this is the common rudiment of dorsal scutes. The height of the dorsal fin increases markedly. The anterior edge of the preanal fin fold tapers sharply. Lobes of the ventral fins descend below the margin of the preanal fin fold. The tail end becomes thinner and starts bending downwards. Segmental trunk muscles overgrow a part of the ventral surface of prelarvae.

Rudiments of secondary filaments appear in both rows of filaments of the second gill. The anterior transverse commissure of the seismosensory line has moved dorsally and it is not seen from the ventral side. The lateral line of the trunk seismosensory system extends beyond the posterior level of the ventral fin, and the accessory row does not reach the level of its anterior border. The end of the dorsal row rudiment reaches the level of the posterior edge of pectoral fin.

Stage 45 is the stage of transition to active feeding. Prelarvae are capable of grasping mouth movements, have separate rudiments of dorsal scutes (in *A. gueldenstaedti* and *A. stellatus*) (Plates X and XI, see page 254 and 255).

Prelarvae are 17–18 mm long. This is the stage of prelarvae transition to active feeding. After this stage they are called larvae. Prelarvae are intensely pigmented. The rostrum becomes even more elongated and now the middle barbels are at a considerable distance from the mouth. The lateral barbels almost reach the anterior edge of the mouth. Mandibular and subpharyngeal teeth begin to appear. The pectoral fins are displaced to the ventral side, the preanal fin fold still exists, but now its width is less than the width of the anal fin and the dorsal part of the fin fold. Mesenchyme rudiments of individual dorsal scutes can be distinguished in the fin fold. Dorsal and anal fins are not yet completely separated from the tail fin. The tip of the tail becomes thinner and bends downwards. The ventral lobe of the tail fin which had started to widen by stage 39 is now far wider than its dorsal lobe. The pigment plugs are ejected in most prelarvae. Segmental muscles of the right and left flanks of prelarvae have not yet fused on the ventral trunk surface.

The spiracle has broken through in some prelarvae. Formation of the spiracular hemibranch begins. The first two gills are pectinate. Rudiments of secondary leaflets appear in the first row of the third gill. Rudiments of the first-row filaments have been formed on the fourth branchial arch. The iris is completely pigmented. At this stage subject vision appears and oculomotor muscles begin to function (Baburina 1972). The majority of neuroepithelial follicles on the ventral side of the head are now open. The end of the lateral line extends beyond the level of the middle part of dorsal fin. The accessory row of the seismosensory system almost reaches the level of the anterior margin of the ventral fin, whereas the dorsal row extends beyond the level of the pectoral fin posterior margin.

3.3 Timing of Prelarval Development

In order to determine the timing of various stages of prelarval development characterized above, we have used a batch of *A. gueldenstaedti colchicus* prelarvae developing in the laboratory at temperatures from 16.8 to 19.9 °C (the mean temperature over the whole period of development was 18.6 °C).

In order to obtain the dimensionless characteristic of the timing of successive developmental stages, we have calculated the mean temperature for every interval between fixations (approximately 24 h long), determined the value of τ_0 for this temperature from the curve in Fig. 42A and divided the time of development by the value of τ_0. In addition we have calculated the expected duration of every interval between fixations at 18.0 °C. These data are shown in Table 8.

In order to use these data to predict the timing of these stages at other temperatures, we have to know whether duration of prelarval development and τ_0 value will change proportionally at different temperatures. With this in mind, prelarvae obtained from the eggs of five *H. huso* females were reared from the stage of mass hatching at various temperatures until the transition to active feeding. We have determined the timing of stages 40 (beginning of respiratory movements) and 45 (transition to active feeding) and also calculated the time interval between these stages and expressed it in terms of τ_0 values. These results are shown in Table 9.

It has been found that the relative duration of these intervals is very similar, whereas the absolute duration expressed in hours differed markedly at different temperatures. We concluded that the duration of development changes in proportion to temperature not only during maturation and embryonic development but during the prelarval development as well. Therefore, from the duration of interval between the stage of mass hatching and various stages of prelarval development in τ_0 values we can calculate the timing of various stages at other temperatures (within the spawning range) using the curve relating τ_0 to temperature for *A. gueldenstaedti* embryos (Fig. 42A).

3.4 Prelarval Development During the Period Between Hatching and up to the Beginning of Rhythmic Respiratory Movements

We have already characterized individual stages of prelarval development and presented data on the timing of these stages. Now we would like to follow

Table 8. Chronology of prelarval development of *A. gueldenstaedti colchicus*

Stage number	Time after hatching			Specific features of the stage
	days at 18 °C	hours	τ_o	
36	—		—	Prelarvae at the stage of mass hatching. Mouth opening and gill clefts absent. Hatching gland still visible
37		23 h	28	Barbel rudiments appear. Mouth opening has broken through. Yolk sac begins to subdivide into the stomach and intestinal parts. Rudiments of pectoral fins are distinct as small folds of epidermis. Rudiments of gill filaments absent. Rudiment of the lateral line of the body seismosensory system appears
38	2 days	3 h	63	First melanocytes appear. Endodermal fold separating stomach and intestine incomplete. First muscle buds are formed near dorsal and anal fins. Rudiments of gill filaments appear on operculum and first branchial arch. Lateral line of the seismosensory system reaches the level of the duct of Cuvier
39	3 days	3 h	93	Stomach separated from intestine. Dorsal and anal fins segregated. Lateral line of seismosensory system reaches the level of the posterior margine of the pectroral fin; accessory row of the seismosensory system appears
40	4 days	4 h	123	Rudiment of the ventral fin is seen as a narrow fold of epidermis. Ventral processes of muscle segments in the region of pectoral fin descend below its base. First irregular mandibular movements can be seen. Lateral line of seismosensory system does not reach the level of the stomach end, the accessory row ends above pectoral fin
41	5 days	4 h	153	Margins of the lobes partitioning off the olfactory opening make contact, but do not yet fuse. Mandibular movements frequent. Lateral line of seismosensory system ends above spiral gut and its accessory row goes beyond the level of the posterior boundary of the pectoral fin; short rudiment of dorsal row appears
42	6 days	3 h	181	Rudiment of the pyloric appendage appears. Lobes of olfactory organ fused. Lateral line of seismosensory system reaches the level of the anterior margin of ventral fin; dorsal row starts bending

43	7 days	5 h	213	The snout assumes horizontal position. Ventral fins reach the margin of the preanal fold. Rudiments of secondary filaments appear in the first gill. Lateral line of seismosensory system reaches the level of the posterior margin of ventral fin, accessory row ends above spiral gut; dorsal row is bent and starts growing in parallel to lateral line
44	8 days	2 h	238	Barbel bases are displaced forward and their tips do not reach the anterior mouth boundary. Mass ejection of pigment plugs. Lobes of ventral fins descend below the pre-anal fold margin. Mesenchyme band (common rudiment of dorsal scutes) appears in the dorsal fin fold. Anterior transverse commissure of seismosensory system is displaced dorsally and is not seen from the ventral side; lateral line goes beyond the level of the posterior margin of ventral fin, accessory row does not reach the anterior boundary of ventral fin
45	9 days	1 h	266	Stage of transition to active (exogenous) feeding. Separate scute rudiments are seen in the dorsal fin fold. Lateral line of seismosensory system goes beyond the level of the middle part of dorsal fin, accessory row almost reaches the level of the anterior margin of ventral fin; dorsal row goes beyond the level of the posterior margin of pectoral fin

Table 9. Duration of development of *Huso huso* prelarvae from the stage of mass hatching to the beginning of respiratory movements (stage 40) and transition to active feeding (stage 45)

Female number	Mean temperature during rearing of prelarvae, °C	Duration of development					
		Up to stage 40			Up to stage 45		
		days,	hours	τ_o	days	hours	τ_o
1	18.8	—		—	8 days	0 h	221.5
2	16.0	4 days	18 h	100.5	10	0	225.2
3	16.7	5	6	106.8	9	6	222.0
4	17.3	4	18	105.2	9	18	237.0
5	18.5	4	12	106.0	8	0	217.0

changes of prelarvae during two qualitatively different periods of their development using data on their internal structure.

The main characteristic feature of the first of these periods, i.e., from hatching and up to the beginning of rhythmic respiratory movements, is the preparation to a definitive way of respiration. At this time considerable changes in the gill area take place (Severtsov 1948; Kryzhanovskii 1934; Schmalhausen 1952, 1955a–c; Timofeev 1971).

The visceral apparatus is still rudimentary in prelarvae throughout the analyzed period. Its muscles are laid down before its skeleton. They develop at first in the anterior gill arches than in the hypobranchial area (hypoglossal musculature). A few anterior pairs of muscle segments provide the material, from which a part of muscle elements of hypoglossal muscles are formed; other muscles develop from mesenchyme of the visceral region.

Cartilaginous elements of the maxillar arch appear at stage 38 simultaneously in maxilla and mandible; the chondrification of the hyoid arch begins at stage 40. Teeth rudiments are laid down on mandible and soon thereafter on maxilla. Gills are still absent at the beginning of this period and respiration continues with the aid of a network of the yolk sac blood vessels, which develops and starts functioning long before the embryos emerge from the membranes. At the time of hatching no morphological changes take place which would indicate reorganization of the respiratory system; however, the respiration rate drastically increases (Korzhuev 1941). This appears to be due to motility of prelarvae and also to the higher oxygen level in water as compared with the perivitelline fluid surrounding the embryo.

At the beginning of this period the bulk of venous blood from caudal and subintestinal veins, segmental trunk vessels and nascent parts of the posterior cardinal veins enters the network of the yolk sac blood vessels, where it is saturated with oxygen. Venous blood from the head and anterior part of the trunk goes into the ducts of Cuvier and from them directly into the heart. Thus at this stage the vascular network of the prelarvae is filled with mixed blood.

After hatching of prelarvae blood vessels of the visceral apparatus are rearranged. A few hours after hatching differentiation of aortic arches into afferent and efferent vessels begins in the first two pairs of branchial septa. Then vascular loops are formed, which later grow into the rudiments of gill filaments. This is associated with the change in the direction of blood flow in the commissure connecting the mandibular and hyoid aortic arches (Fig. 67). As a result the mandibular vessel below the commissure degenerates. The head, earlier supplied by mixed blood from the heart through ventral aorta begins to receive oxygen-saturated blood from vessels of gill leaflets and prelarvae start passive gill respiration. Gradually, the process of gill blood vessel formation spreads to other branchial arches and, eventually, definitive gill circulation is established (Schmalhausen 1952, 1955b).

As gill circulation develops, the significance of the vascular network of the yolk sac for respiration of the prelarvae diminishes drastically. This is due to the alteration of direction ol blood flow in the posterior cardinal vein when more

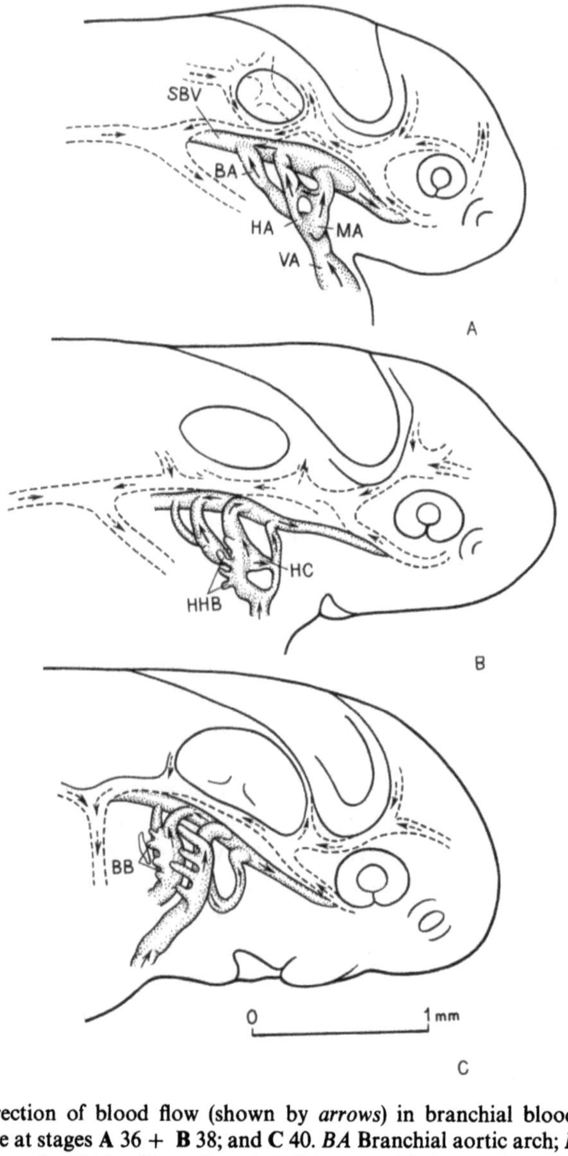

Fig. 67A–C. Direction of blood flow (shown by *arrows*) in branchial blood vessels of *A. stellatus* prelarvae at stages **A** 36 + **B** 38; and **C** 40. *BA* Branchial aortic arch; *BB* rudiment of branchial blood vessels; *HA* hyoid aortic arch; *HC* hyomandibular comissure; *HHB* rudiment of hyoid hemibranch blood vessel; *MA* mandibular aortic arch; *SBV* superbranchial blood vessel; *VA* ventral aorta

and more venous blood flows directly to ducts of Cuvier. Correspondingly, the lower proportion of blood flows to the vascular network of the yolk sac which begins to atrophy.

Change of the blood flow direction in the posterior cardinal veins is gradual and blood from an ever-decreasing number of segmental vessels is supplied to the subintestinal vein. Eventually blood flow into this vein stops. All blood from the caudal vein flows directly to the posterior cardinal veins and from them into ducts of Cuvier bypassing the subintestinal vein. The subintestinal vein degenerates. Thus, the greater part of larval venous blood goes to the heart and gills receive blood which is not saturated with oxygen (Schmalhausen 1955c).

At the beginning of this period, gill rudiments are small and immobile, whereas the mouth is open. By the end of the period, at stage 40, first infrequent mandibular movements are observed and mouth becomes capable of closing.

All these changes of the visceral apparatus and circulatory system create conditions for the transition of prelarvae to active gill respiration. This transition takes place at the beginning of the next period of prelarval development.

We shall now consider changes taking place in the prelarval digestive system (Schmalhausen 1968). At the hatching stage, the digestive system consists of the digestive tube and rudiments of two digestive glands, liver and pancreas. The digestive tube is subdivided into two parts, a bulky anterior part (yolk sac) and a narrow posterior part. It is closed: there is still no mouth and the pharyngeal cavity is separated from the cavity of the digestive tube by a massive septum. Even after the formation of the mouth opening at stage 37, communication of the digestive tube with the environment medium is not yet established. The narrow posterior part of the digestive tube still lacks a lumen.

Separation of the yolk sac into two regions takes place during this period. The anterior region corresponds to stomach and the posterior one to the intermediate gut. Separation takes place by invagination of an oblique fold of the digestive tube wall. The anterior wall of this fold becomes the lower wall of the stomach, whereas its posterior wall becomes the upper wall of the intermediate gut. During separation the network of vessels moves to the region of the intermediate gut and a part of it invaginates. Later some of these vessels disappear, whereas the other ones become vessels of the intestine.

Segregation and formation of the spiral gut with its spiral valve takes place. The intestinal mucous membrane acquires a folded structure and undergoes differentiation: on the surface of cylindrical cells facing the lumen the striated absorbing border (brush border) is formed; ciliated and bottle-shaped cells appear in the spiral gut between these cells. The pancreas during this period consists of three separate rudiments and the dorsal rudiment begins to function. Liver becomes bilobate. The gall duct is formed; at the region adjacent to the liver it is dilated, thereby giving rise to the gall bladder.

Intracellular digestion of yolk in sturgeon begins very early, even during embryonic development. Thus, at the stage of late neurula, cells of the nervous system and sense organ rudiments are almost completely devoid of yolk. As yolk is consumed, pigment granules contained in it are ejected from the cells. Accumulation of granules of the embryonic pigment in the digestive tube lumen produces the melanin plug.

The bulk of yolk is consumed after hatching. Along with the intracellular yolk digestion, which continues after the emergence of embryos from the membranes, yolk also disintegrates in the cavity of the digestive tube. This cavity is filled with endodermal cells overloaded with yolk; these cells have not become an integral part of the wall of the digestive system rudiment. Yolk digestion proceeds with the aid of substances secreted by the cells of the digestive tube wall. Accumulation and secretion of this material takes place even before any gastric glands have been formed and before the brush border of cylindrical cells of the intestine has differentiated. On histological preparations cleavage of yolk platelets can be traced at the sites of their contact with the digestive tube wall, this leads to the formation of a homogeneous layer of yolk. Some authors (Vernidub et al. 1971) have also observed this phenomenon and used the term "parietal digestion" proposed by Ugolev (1963). It should be pointed out, however, that these changes begin in prelarvae even before the formation of the brush border in endodermal cells. According to Ugolev, parietal digestion is associated with differentiation of the brush border and takes place in those regions of the digestive tract where partially hydrolyzed products are present. In other words, it should be preceded by extracellular digestion. According to Korzhuev and Sharkova (1967) proteolytic enzymes appear in *A. gueldenstaedti* prelarvae during the first day after their hatching, i.e., extracellular digestion is already present at this stage. Enzymes present in yolk platelets themselves appear to take part in the cleavage of the yolk phosphoproteins (comprising its main bulk), as in other vertebrates including amphibians (Harris 1946; Gross 1952; Deuchar 1958).

Sense organs in the just-hatched prelarvae are also poorly differentiated. The rudiment of the olfactory organ at this stage represents a rounded sac in which the lower part is formed by a thickened internal epithelial layer (surface layer of epithelium is already atrophied); both epithelial layers are present in the marginal zone. The bottom of the olfactory sac is smooth, and sensory and supporting cells are differentiated in the olfactory epithelium. The olfactory sac is connected with the forebrain by a short branch of the olfactory nerve. The cavity of the sac (olfactory pit) communicates with the environment through a circular primary olfactory opening. As length of the head increases, the olfactory rudiment elongates and its external opening becomes oval. By the end of the period a short rudiment of the first longitudinal fold appears, which partitions off the olfactory sac bottom.

Two lobes appear in the marginal zone of the olfactory organ on the dorsal and ventral side and grow towards one another–the onset of the partitioning off of the primary olfactory opening (Schmalhausen 1962).

Definitive receptor cells differentiate in the eye and retinal layers are formed. Concomitantly, transient photoreceptor cells degenerate in the region of the pigment spot. At the end of the period all major retinal layers can already be distinguished (external nuclear, external plexiform, inner nuclear, inner plexiform, the layer of ganglionic cells and the layer of nerve fibers; see Fig. 72A). Iris pigmentation begins and guanine crystals appear in the eye (Baburina, 1972).

Lines of the seismosensory system are laid down, as in all fish, as thickenings of the inner layer of the epidermis. These lines grow together with the rudiments of corresponding nerves. The main lines on the sturgeon head are pairs of supraorbital, suborbital, and temporal lines, anterior and occipital transverse commissures and several paired lines of the hyomandibular complex. In addition, the seismosensory system of the head includes individual receptor groups and single receptors, as well as sensory pits (neuroepithelial follicles) of secondary accessory organs of the lateral system (Dragomirov 1961a).

Rudiments of all lines of the head mentioned above appear during the first period of prelarval development, as well as the paired rudiments of the lateral line and the accessory row of the seismosensory trunk system. Rudiments of neuroepithelial follicles cannot be detected from outside at this period.

Auditory vesicles differentiate and by the end of the first period incompletely formed semicircular canals are already present.

Pigmentation of the prelarvae trunk surface begins.

3.5 Development of Prelarvae from the Beginning of Rhythmic Respiratory Movements and up to Transition to Active Feeding

Several changes associated with the transition of prelarvae to the definitive mode of feeding take place during the second period of prelarval development. These changes concern both the digestive system and the visceral apparatus. Transformations of the visceral apparatus create conditions for seizure of prey, and sense organ differentiation allows the prelarvae to search for food. Respiratory and circulatory systems also change resulting in a better supply of prelarval organs with oxygen.

The yolk sac at this period loses its initial significance for prelarval respiration. The relative significance of the opercular hemibranch also diminishes. Rhythmic respiratory movements of the visceral apparatus are now synchronous with those of the operculum. Rudiments of secondary gill filaments appear on the leaflets of the first two gills and these become pectinate. Gill filaments develop on other gill arches. After the rhythmic respiratory movements have been established prelarvae start moving near the bottom of the tank.

The pump creating negative pressure for the branchial apparatus is poorly developed in sturgeon (Woskoboinikoff 1928, 1932; Tatarko 1936; Balabai 1939). Respiratory movements of sturgeon make use of the positive pressure pump principle. Its performance is based on devices for compression of the visceral cavity and pumping of water from under the opercula outside. Correspondingly, muscles compressing the visceral cavity and oral valves preventing the backflow of water through the mouth play a major role in the work of the positive pressure or "compression" pump. Sturgeon do not possess oral valves of the type found in teleosteans. Their role is played by a cartilage plate tightly pressed to the bottom of the skull. Lower ends of the hyoid and branchial arches are attached to this plate. Respiration based on the principle of a positive pressure pump cannot be accomplished with mouth closed. The operculum in sturgeon is significantly reduced, as compared with teleosteans, and its development proceeds at a slower rate than that of gill filaments. The delayed development of the operculum is most distinct in *A. stellatus*. Correspondingly, gill filaments remain uncovered for a long time, particularly in the upper part of the branchial arches. This facilitates continuous replacement of water near their surface. Thus, respiration is not disturbed during seizure of the food, although rhythmic movements of the visceral apparatus are suspended.

A characteristic feature of the visceral apparatus of sturgeon fish is the protrusible mouth, which has developed in connection with feeding on benthic organisms. Unlike teleosteans, where the secondary maxilla (developed from covering bones) is able to protrude, in the sturgeon it is the primary mandibular arch which moves forward. It consists of the palatoquadrate and Meckel's cartilages. The mandibular arch is not directly connected with the skull, as in teleosteans, but hangs on the hyomandibular. The mandibular arch is connected with the hyomandibular through a separate cartilaginous element. The hyomandibular is thinned at the site where it is connected with the auditory capsule and can move not only laterally but also in the forward/backward direction. This is necessary for the protraction of jaws.

A joint between maxilla and mandible and the hyoid arch develops during the second period of prelarval life. Also at this time the hyomandibular connects with the auditory capsule and the bottom part of the branchial apparatus chondrifies. Muscles and connective tissue cords attach to the corresponding parts of the skeleton.

Muscles and skeleton of the posterior branchial arches develop. Rudiments of subpharyngeal teeth appear in the first branchial arch. At last, teeth appear on the mandible and the visceral apparatus becomes capable of performing its second main function, i.e., seizure of food.

Capacity of mouth protraction in prelarvae of various Acipenserid species is different: it is highest in *A. stellatus* somewhat lower in *A. gueldenstaedti* and still lower in *H. huso* and *P. kaufmanni.*

Further differentiation of the digestive system consists of the preparation to the definitive digestion (Schmalhausen 1968). Rudiments of tubular glands appear in the cardiac part of the stomach as buds consisting of few cells. Soon lumen appears in these buds and alveoli are formed, which open into the stomach lumen. Rudiments of the glands quickly increase in number and size. The pyloric part differentiates in the stomach. The rudiment of pyloric appendages is segregated. By the end of the period this rudiment already consists of several lobes.

Fat is gradually accumulated from stage 41 on, in cylindrical cells of the intestine with a brush border. This fat is produced as a result of yolk digestion and the corresponding cells greatly increase in size. Fat also accumulates in the liver. Fat accumulation continues throughout this period up to the transition of prelarvae to active feeding. After this stage fat begins to gradually disappear, first from the intestinal cells and later from the liver cells as well. Fat stores support energy expenditure of the fry during searches for food under conditions of insufficient food supply.

Ciliate cells are found in all regions of the digestive tract. The ciliate epithelium covers esophagus and the hindgut. In the stomach, particularly in its pyloric part, ciliate cells undergo the mucous transformation: mucus is accumulated in them, which then is shed into the gastric lumen together with cilia. Typical bottle-shaped cells are found in large numbers in esophagus and in smaller numbers in the midgut and hindgut.

Localization of ciliate and bottle-shaped cells predominantly in the esophagus and hindgut is associated with the function of these regions of the digestive system, i.e., food transport and defecation. Correspondingly, these cells differentiate later than the cylindrical brush border cells, which participate in yolk digestion.

By the end of the second period the melanin plug is gradually moved to the hindgut from the spiral gut lumen. Communication is established between the hindgut and the environment and the anal opening is eventually broken through. This is followed by the ejection of pigment plugs.

Ejection of the pigment plugs is often used to determine the time of transition of the prelarvae to active feeding. Indeed, under normal conditions the ejection of pigment plugs almost corresponds in time with the transition of prelarvae to active feeding and can provide a good guildeline. It should be remembered, however, that under unfavorable developmental conditions the ejection of pigment plugs can take place prematurely.

Late differentiation of the esophagus and the hindgut as well as late resorption of the septum between esophagus and pharynx (lasting almost to the end of yolk nutrition) confirms the conclusion of Dragomirov (1953b, 1961b) and Gerbil'skii (1957) that it is useless to feed prelarvae with microplankton and microbenthos before the end of this period. Matveev's proposal (1953) that prelarvae must be fed after the beginning of movements of the mandible is not justified. Transition of the larvae to active seizure of food can take place only after stage 45 has been reached when their digestive system is ready to absorb this food.

The functioning of all parts of the digestive system begins before yolk is completely resorbed in walls of different parts of the digestive system.

Structure of the visceral apparatus and digestive system by the time of transition of the prelarvae to active feeding allows prelarvae to swallow rather big prey. This is most relevant for *H. huso* prelarvae: under laboratory conditions they are capable of swallowing small larvae of Chironomids, prelarvae of the pike perch and even newly hatched prelarvae of *A. gueldenstaedti* and *A. stellatus.*

A number of changes take place during this period in the sense organs. Lobes which partition off the external olfactory opening come closer and eventually fuse. The suture at the fusion site is soon resorbed. As a result, the incoming and outgoing openings of the olfactory organ are formed, i.e., the anterior and posterior nostrils. A rosette of folds develops at the bottom of the olfactory sac. Ciliate epithelium, similar to that found in the marginal zone of the olfactory pit, is formed at apices of these folds. Sensory olfactory epithelium covers the remaining surface of the olfactory organ.

The lower margin of the posterior nostril is located below the level of the corresponding margin of the anterior nostril. The margin of the septum between the nostrils which borders the anterior nostril is thickened and rounded, whereas that bordering the posterior nostril is thinned and raised as a screen. The water flows through this olfactory organ both passively when the prelarva moves and actively with the aid of cilia.

By the beginning of active feeding the whole iris of the eye is intensely pigmented; the ocular fundus is completely pigmented as well. The main layers of retina and definitive photoreceptor cells are developed all over the ocular fundus. Eye muscles already function and the eye acquires its definitive structure (Baburina 1972).

Differentiation of the seismosensory system continues (Disler 1949, 1960; Dragomirov 1954, 1961a, 1968, 1971). Paired rudiments of the third (dorsal) line appear in the trunk. Development of neuroepithelial follicles takes place throughout this period. The sensory pits of the opercular group differentiate on the operculum at stage 41. At stage 42, other sensory pits of the head differentiate, including those on the lower side of the rostrum. The pores of most follicles open outside at stage 45. Taste buds develop on barbels and lips of prelarvae.

Thus, by the end of this period, organs of food seizure and digestion, and the sense organs of the prelarvae are ready to begin active feeding. At this period the larvae still retain a certain amount of yolk and a large store of fat.

Some authors distinguish one more period of prelarval development, i.e., the period of mixed feeding from the beginning of active feeding and up to complete resorption of yolk. However, sturgeon prelarvae are capable of long starvation (Bogdanova 1967, 1969, 1972b) and can begin active feeding after resorption of their endogenous stores (see, for example, Chusovitina 1963). Thus, mixed feeding is not an obligatory stage in their development.

It should be pointed out that morphogenetic processes are not completed with the end of prelarval development and the transition of the fry to active feeding. Several new traits appear (dorsal, paired lateral and ventral rows of scutes, spikes on the head and on the pectoral girdle, tail thread in Amu Darja *P. kaufmanni*). Body proportions also undergo some change. Drawings of a larva and juvenile of *P. kaufmanni* (Plate XIX) and an adult immature individual of the same species (Plate XX) illustrate the significance of changes which take place.

3.6 Differences Between Prelarvae of Different Sturgeon Species (*A. gueldenstaedti, H. huso, A. stellatus, A. ruthenus,* and *P. kaufmanni*)

The structure of prelarvae of all studied Acipenserid species at the mass hatching stage (stage 36) is rather similar. However, there are some traits which are different and this allows to identify the species to which they belong (Alyavdina 1951b; Dragomirov 1953a).

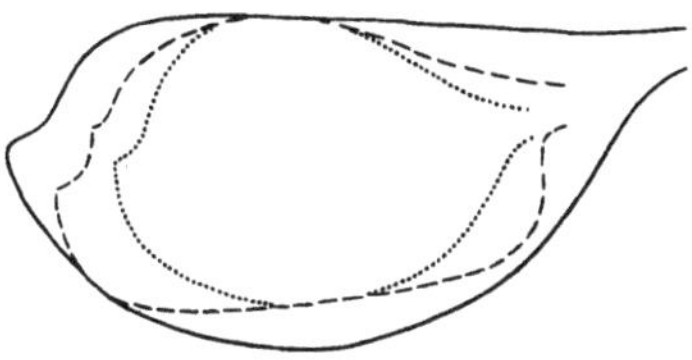

Fig. 68. Outlines of the yolk sac in prelarvae of different Acipenserid species at stage 36; lateral view. *Solid line, H. huso; broken line, A. gueldenstaedti; dotted line A. stellatus*

First, prelarvae differ in their size. Prelarvae of *H. huso* are the largest, the size is somewhat less in the *A. gueldenstaedti*, still less in *A. stellatus* and *A. ruthenus*, and prelarvae of *P. kaufmanni* are the smallest. Prelarvae of *A. gueldenstaedti colchicus* differ from prelarvae of other species by darker pigmentation due to a higher level of the embryonic pigment in the epithelium. There are also differences in the size and shape of the yolk sac (Fig. 68). The yolk sac is biggest and most elongated in *H. huso*. It is also elongated in *A. gueldenstaedti*, whereas in *A. stellatus*, *A. ruthenus* and *P. kaufmanni* the yolk sac is rounded. Its size decreases in the sequence *H. huso-A. gueldenstaedti-A. stellatus -A. ruthenus-P. kaufmanni*.

Rudiments of eyes and olfactory organs at this stage have practically similar diameter. The relative size of eye rudiments in prelarvae of *A. gueldenstaedti*, *A. ruthenus*, and *A. stellatus* are similar, it is somewhat less in *H. huso* and *P. kaufmanni*. The pigment spot on the ocular fundus in the region of transient retinal photoreceptors in *A. gueldenstaedti* and *A. ruthenus* is darker and has a sharper contour as compared with *H. huso* and *A. stellatus* (Baburina 1972). The spot is absent in *P. kaufmanni* (Dragomirov and Schmalhausen 1952).

Prelarvae of all the above species are still unable to swim for a long time in a certain direction after hatching because of the large amount of yolk in the yolk sac and the absence of pectoral fins (Kryzhanovskii 1949). These prelarvae perform alternate movements: they move upwards by active movements of the posterior trunk region and tail and, then, passively settle to the bottom.

The photobehavior (attitude to light) is established in all species at this stage. This reaction is, however, different in various species and appears to be associated with the differentiation of transient retinal photoreceptors (Baburina 1972). The most distinct positive photobehavior is characteristic of *A. stellatus* prelarvae. It is less distinct in *H. huso* and *A. ruthenus* and is quite weak in *A. gueldenstaedti*. Photokinetic response has not been studied in *P. kaufmanni*.

As development proceeds, differences between prelarvae of various species increase and new traits appear which distinguish them from one another. Formation of these differences can be traced in Plates IX–XVIII, see page 253–262.

By the beginning of active feeding (stage 45), prelarvae of most species can be identified without any difficulties. They acquire a number of species-specific traits such as the intensity of pigmentation, length of the head and rostrum, length and position of barbels, width of the mouth, size and position of fins, and some others.

Thus, in prelarvae of *A. gueldenstaedti colchicus* the number of melanocytes increases with development, and by the beginning of active feeding prelarvae are already dark. The dorsal side of the trunk and the tail have the most intense pigmentation. Prelarvae of other species have a lighter color, particularly in *H. huso* and *P. kaufmanni.*

The relative size of the eye in *A gueldenstaedti* prelarvae is somewhat less as compared with *A. ruthenus* but still remains bigger than in *A. stellatus* and *H. huso.* Eyes of *P. kaufmanni* are the smallest. Measurement of eyes in adults *H. huso* and *Acipenser* species (Baburina 1972) and our data about the size of eyes in the adult *P. kaufmanni* demonstrate that these trends are also retained at later stages.

By the beginning of active feeding the relative size of the eyes decreases as compared with the size of the olfactory organ. The change in this ratio is lowest in *A. gueldenstaedti,* whereas in *H. huso, A. ruthenus* and *P. kaufmanni* the longitudinal diameter of the olfactory organ is considerably greater than the diameter of the eye. Also there are some species-specific differences in the eye structure. Thus, in prelarvae of different species the eyes differ in the structure of retina. Specifically, prelarvae of *A. stellatus,* in contrast to many other Acipenserid species, have only one type of photoreceptor, i. e., cones (Baburina 1972). In addition to *A. stellatus,* only prelarvae of *P. kaufmanni* contain a region in the retina (in the middle part of the ocular fundus) where cones are the only type of photoreceptors (Nikol'skaya 1989).

Among the studied species the development of the eye in *P. kaufmanni* has the biggest deviations from the standard pattern. In *P. kaufmanni* the eyes are greatly reduced in size. They are very small and often covered by skin folds from the dorsal side. Even at the stage of transition to active feeding, when the eyes in other sturgeon species are already completely formed, the ocular fundus and iris are only very weakly pigmented in *P. kaufmanni.* At this time the *P. kaufmanni* eye is similar in its structure to the eye of prelarvae of other sturgeon species but at earlier stages of development. By the beginning of active feeding, the lens in *P. kaufmanni* prelarvae almost completely fills the inner eye cavity, whereas in other species it is occupied by the vitreous body and the lens is located in the pupil.

In the species with positive photokinetic behavior at the hatching stage this response weakens after stage 38. As a result, prelarvae settle to the ground. This change of behavior in the *A. gueldenstaedti* prelarvae with their very weak photokinetic response at stage 42 is much more distinct than in other species.

Species-specific differences in the structure of the eye are intimately associated with different ecological conditions of development of prelarvae of various species (Baburina 1972). Such dependence has also been observed in the development of the lateral system organs (Dragomirov 1961a, b).

Barbels located on the lower side of the rostrum are the longest in *H. huso* and *P. kaufmanni.* Lateral barbels are longer than the middle ones in all species. Bases of lateral barbels are located at the same level with those of middle barbels in *A. stellatus, A. ruthenus* and *P. kaufmanni.* In *A. gueldenstaedti* (Alyavdina

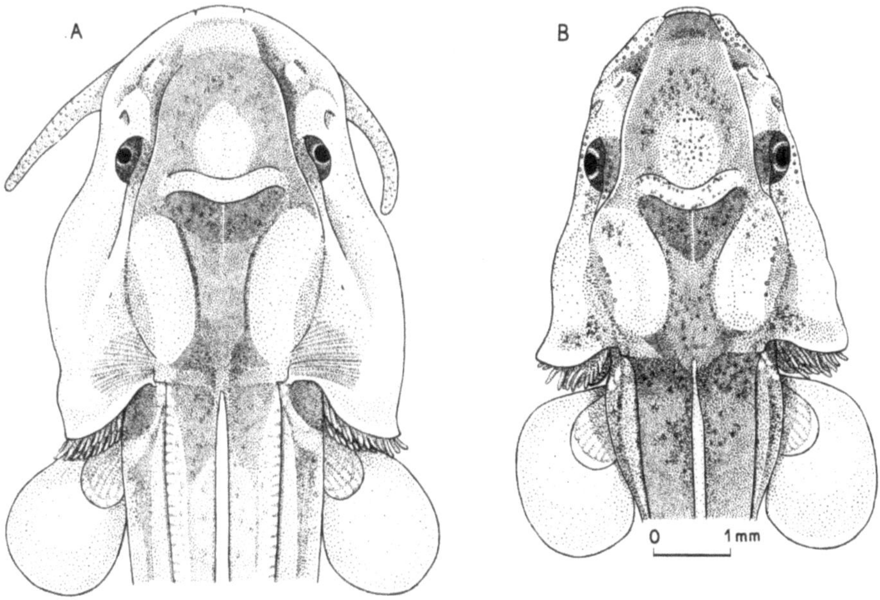

Fig. 69. Head structure of prelarvae in **A** *H. huso* and **B** *A. gueldenstaedti* at the stage of transition to active feeding; view from above

1951b) and *H. huso*, however, they are located behind the bases of the middle barbels. By the stage of transition to active feeding in *A. gueldenstaedti*, *A. stellatus* and *A. ruthenus*, the barbels are displaced closer to the end of the rostrum and their tips no longer reach the anterior margin of mouth (Plate XI, Fig. 70). Bases of barbels in *H. huso* and *P. kaufmanni* are located closer to the anterior margin of mouth at almost half the distance between the end of the rostrum and mouth. The barbel tips protrude beyond the anterior margin of mouth (Plate XVIII, Fig. 70). *H. huso* and *P. kaufmanni* barbels have large rudiments of taste buds.

The rostrum of *A. gueldenstaedti* prelarvae is relatively longer than in *H. huso* (Fig. 69), but is shorter than in *A. ruthenus* and *A. stellatus*. Sensory pits at the lower side of the rostrum in *A. gueldenstaedti*, *A. stellatus*, and *A. ruthenus* are large and at the stage of transition to active feeding the most of these pits are open. The number of sensory pits in these species is far greater at this stage than in *H. huso* and, particularly, in *P. kaufmanni*. Moreover, they are more differentiated (Dragomirov and Schmalhausen 1952). The dorsal row of seismosensory organs is absent in *P. kaufmanni* prelarvae.

The relative size of the mouth is biggest in *H. huso*, followed by *P. kaufmanni*, *A. gueldenstaedti*, and *A. ruthenus*. During *H. huso* development, the mandible forms more and more prominent protrusion directed forward (Fig. 70A′). In prelarvae of all species compared as in the majority of sturgeon the lower lip is interrupted over a great distance (Fig. 70). The interruption is much less in *A. ruthenus* only.

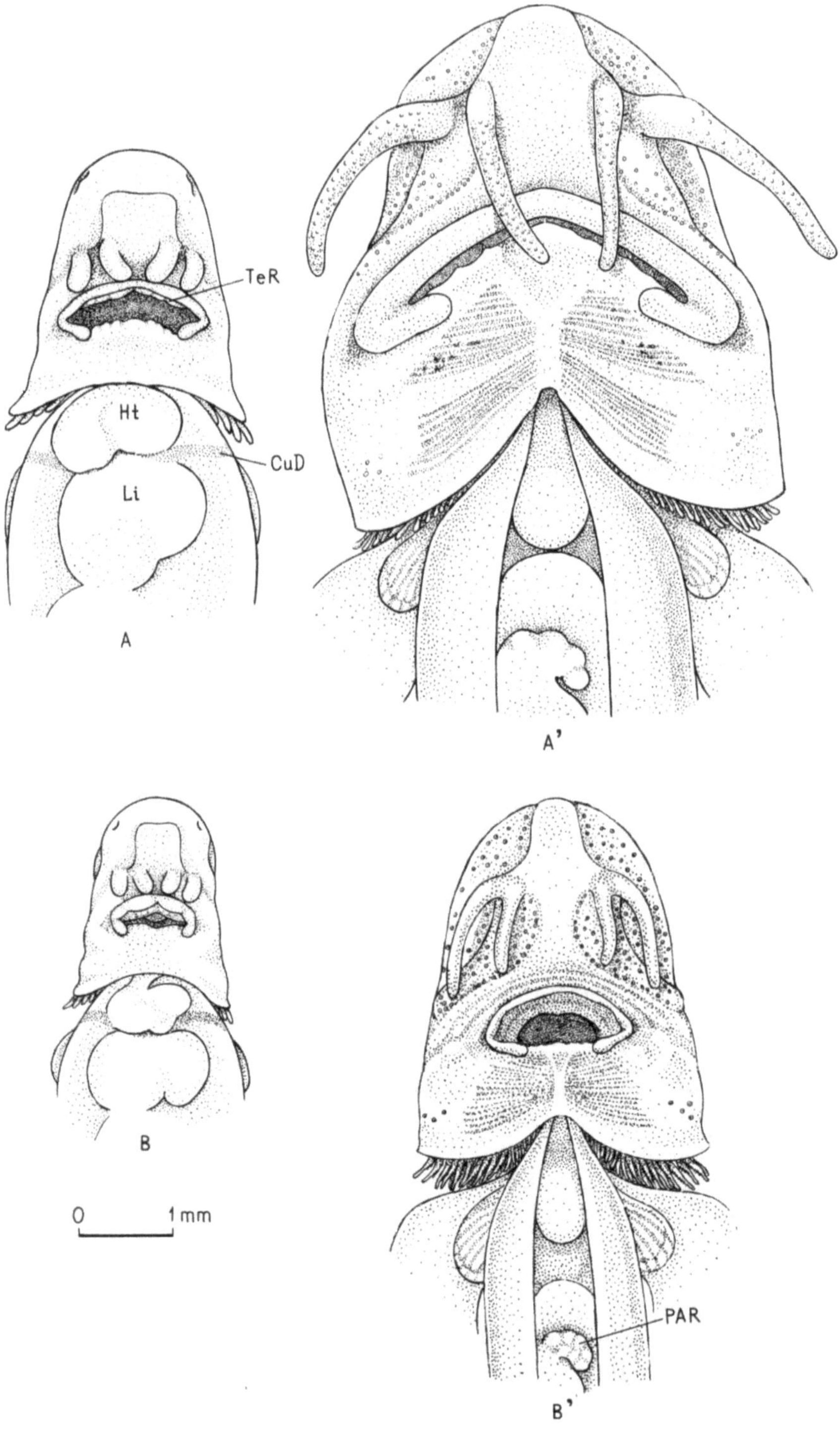
TeR
Ht
CuD
Li
A
A'
B
0
1 mm
PAR
B'

Teeth are largest in *H. huso* and their rudiments become visible earlier than in other species. In *P. kaufmanni* teeth are very small.

Capacity to protrude the mouth at the stage of transition to active feeding is most pronounced in representatives of the genus *Acipenser* and is far less noted in *H. huso* and *P. kaufmanni.*

The rudimentary operculum in prelarvae of *A. gueldenstaedti* and *A. stellatus* is shorter than in *H. huso, A. ruthenus,* and *P. kaufmanni,* and by the time of transition to active feeding it does not yet reach the pectoral girdle. Gill leaflets in these species are covered by the operculum over a smaller distance. This specifically refers to filaments of the opercular hemibranch and filaments located on the dorsal side of the first branchial arch. In prelarvae of other species operculum reaches the pectoral girdle already at stage 42. In all species studied the branchiostegal membranes remain fused with the isthmus. Only in prelarvae of *H. huso* do they fuse with one another and at stage 43 form a free fold, which is characteristic of this species. The head in *H. huso* prelarvae markedly increases in height and width by stage 45, particularly in the branchial area.

Pectoral fins have the biggest relative size in *P. kaufmanni.* All other species produce the following sequence in terms of this trait: *A. gueldenstaedti-A. ruthenus-H. huso.* The anterior margin of the dorsal fin in prelarvae of all species is located at the level of cloaca, apart from in *H. huso* when it is in front. Because of this the anterodorsal distance in *H. huso* is shorter than in other species. In addition, the relative length of the dorsal fin base in *H. huso* prelarvae is greater than in *Acipenser* species and *P. kaufmanni.* The dorsal fin fold in the *A gueldenstaedti* is wide, and individual rudiments of scutes in this region appear by the beginning of active feeding. This fold is narrower in *H. huso* and *P. kaufmanni,* and the scutes are present there as a common rudiment or are hardly noticeable.

The spiracle in *P. kaufmanni* is broken through later than in *A. gueldenstaedti, A. ruthenus,* or *H. huso,* i.e., after the beginning of active feeding. Then it is overgrown and its absence in the adult fish is a taxonomic trait which distinguishes the genus *Pseudoscaphirhynchus* from other Acipenserid genera.

In addition to qualitative differences, prelarvae of various species also have differences which can be characterized only by measurements concerning the proportions between various body regions. They are different in various species and the absolute size of various parts of the trunk and organs cannot be used as a criterion to identify the species of prelarvae: with different size of prelarvae and irregular growth of their body parts these values are difficult to compare. Description of fish in ichthyology commonly uses the relative or percentage ratios of body parts or various organs. It is difficult to apply this method for

Fig. 70. A, A′ head structure of prelarvae in *H. huso*; and **B, B′** *A. stellatus* at stages 40 (**A, B**) and 45 (**A′, B′**); view from below. *CuD* duct of Cuvier; *Ht* heart; *Li* liver; *PAR* rudiment of the pyloric appendage; *TeR* teeth rudiments

Table 10. Relative size of certain body parts and organs of prelarvae in *P. kaufmanni*, *A. ruthenus* and *H. huso* at stage 38

Species	*Pseudoscaphirhynchus kaufmanni*		*Acipenser ruthenus*		*Huso huso*	
Source[a]	1	2	1	2	1	2
Length of prelarva, mm	8.0	7.25	9.6–10.2	8.8	14.1–14.6	13.6
Relative sizes with the length of prelarva taken as 1.0:						
Length of the motional region behind the yolk sac	0.63	0.63	0.63–0.65	0.61	0.61–0.62	0.60
Length of preanal fin fold	0.23	0.22	0.27–0.28	0.27	0.26–0.28	0.25
Length from cloaca to tail tip	0.40	0.38	0.35–0.38	0.38	0.34–0.35	0.38
Width of a tail rudiment	0.11	0.12	0.13	0.13	0.14–0.16	0.16
Length of the pectoral fin base	0.08	0.07	0.06–0.07	0.07	0.06	0.06
Longitudinal eye diameter	0.02	0.02	0.03	0.03	0.03	0.03

[a] 1, According to Dragomirov (1954); 2, according to Schmalhausen (present work)

prelarvae—first, the parameters common in ichthyology cannot always be applied to prelarvae because of their continuing development and absence of various reference points for measurements. On the other hand, it should be also remembered that these ratios change out of proportion during development.

Disproportionate growth of rudiments and body parts characteristic of the embryonic period continues during the prelarval period and appears as the altered proportion between the sizes of various parts of the body and organs in prelarvae. Thus, as prelarvae continue to grow, there is a quick increase in the size of the body from the cloaca to the tip of tail, whereas the distance from the gill cleft up to the cloaca changes slowly. Correspondingly, the relative size of the first parameter increases, whereas that of the second one decreases (Plates IX, X, XII–XVII). The relative distance from the tip of rostrum to the barbel bases decreases due to the increased distance between barbel bases and mouth (Plate XI, Fig. 70).

Changes in body proportions in prelarvae of various Acipenserid species (genera) often proceed in the same direction, i.e., towards either an increase or a decrease in the relative size of a certain body part or organ, but these changes are different in their extent. Certain proportions show small changes, but in terms of several parameters, some considerable species-specific differences can be noted.

The material at our disposal is insufficient for any statistical analysis of measurements. This is particularly true of *P. kaufmanni* for which only individual prelarvae have been available at various developmental stages. However, during yolk feeding the population is far less variable than after the transition of prelarvae to active feeding. Therefore, when using prelarvae that were developing under optimal conditions, we can compare parameters obtained by measurement of just a few individuals. This comparison is of considerable interest because such data are scarce in the literature.

Table 10 shows results of measurements of the stage 38 prelarvae by two different authors in different years. These data have been obtained with three Acipenserid species belonging to various genera and they explicitly demonstrate the high stability of body proportions and their similarity in various representatives of the Acipenseridae. At the same time, even at stage 38, we could see certain species-specific differences, for example, in the size of eyes, length of the motional part of the prelarva and of the preanal fin fold.

Comparison of 24 parameters for all consecutive stages of prelarval development in *H. huso*, *A. gueldenstaedti*, *A. ruthenus*, and *P. kaufmanni* allowed us to pool these parameters into the following three groups.

1. Stable proportions showing small variation depending on the stage and similar in different species include the length of the base of the ventral fins and the minimal body height.

2. Proportion changing depending on the stage and similar in different species. This group of parameters includes the relative antero-anal and antero-ventral distances, distance from the gill cleft to cloaca and from the tip of the

rostrum to barbels, the maximal height of body and head, width of the tail, which diminish as the development proceeds, as well as the relative distance from cloaca to the tip of tail, length of the head and rostrum, height of the dorsal and pectoral fins, postorbital distance, length of the lateral barbel and distance from barbels to the margin of the mouth which increase. These changes are fast during the prelarval development, but by the end of it their rate diminishes and they can even alter their direction.

3. Ratios changing with the developmental stages and different in different species include the relative size of the eye, width of the mouth, and distance between the margins of lower lip at the place of its interruption.

Differences in the relative sizes of various body parts between prelarvae of different genera and certain species of the *Acipenseridae* can be observed even with the naked eye (Plates IX–XVIII). However, in closely related species, like *A. gueldenstaedti* and *A. stellatus*, plastic features (that are easily distinguished in adult individuals) such as position of barbel bases relative to the rostrum tip and the upper lip, relative length of the head and the rostrum cannot be used as criteria for the identification of the species: differences between *A. gueldenstaedti* and *A. stellatus* prelarvae and larvae in these features are not statistically significant (Ginzburg 1939). Muscle buds in fins at the stage of transition to active feeding are still incomplete and, therefore, their number does not allow us to identify the species of prelarvae. According to the same author, the safe feature which distinguishes *A. gueldenstaedti* from *A. stellatus* at any age is the ratio between the mouth width and the distance between the lateral barbel base and the opposite margin of the lower head surface along the line passing through the barbel bases. In *A. gueldenstaedti* prelarvae the mouth is wider than this distance, whereas in *A. stellatus*, as a rule, it is similar. Differences between *A. gueldenstaedti* and *A. stellatus* for this criterion are highly significant.

The above differences in the structure and proportions of various body parts in prelarvae of different Acipenserid species are not so great as to make impossible their common characterization. Therefore, classification of developmental stages of *H. huso* prelarvae proposed by us previously (Schmalhausen 1968) is applicable for all studied species. This classification is based on traits appearing simultaneously which characterize every stage and underly the overall organization of a prelarva without, however, being relevant to less significant species-specific differences.

Several objections have been raised against this method for determining the developmental stages of sturgeon prelarval development (Sytina and Timofeev 1973). According to these authors, the ratio between certain traits and the degree of their expression varies in different individuals. They explain this phenomenon by natural variation and weak correlations between individual traits at prelarval developmental stages. Therefore, they have proposed determining the stage of development on the basis of individual distinct morphological criteria, rather than the sum total of diagnostic traits. However, our observations made over many years have demonstrated that when prelarvae develop under optimal conditions, the transition from one stage to another is usually synchronous and, therefore, easy and accurate determination of the stage, from the sum total of traits which characterize the stage, is possible. Thus, the presence of a complex of traits characteristic of a given stage can be a criterion for typical prelarval development. When conditions are unfavorable, development is no longer synchronous in terms of appearance of various traits and this "desynchronization" can be so great that it becomes impossible to determine the stages in such prelarvae and compare them with the developmental stages of normal prelarvae.

In addition to species-specific differences in the structure of the sturgeon prelarvae, there are also differences in relative duration (τ_n/τ_0) and the rate of their development. Reliable judgements about this requires comparison of relaive timing of consecutive developmental stages in prelarvae of several species. The data available to us are rather limited and allow only some preliminary conclusions. We present these comparison of the relative duration of development in *A. gueldenstaedti* and *H. huso* prelarvae from the stage of mass hatching (stage 36) and up to stages 39–45 in the number of τ_0.

	39	40	41	42	43	44	45
A. gueldenstaedti	93	123	153	181	213	238	266
Huso huso	87	106	131	155	180	205	225

It follows from these data that *H. huso* attains a given developmental stage faster than *A. gueldenstaedti* if time is expressed in τ_0. The difference in the relative duration of a similar developmental period between *H. huso* and *A. gueldenstaedti* (as well as *A. stellatus*) appears for the first time at the end of embryonic development (Igumnova 1979).

In addition, the duration of *A. stellatus* prelarval development was determined for the Don and Volga rivers. In two batches of *A. stellatus* prelarvae from the Volga that were developing at mean temperatures of 17.2 and 20.3 °C the relative duration of the period from stage 37 to stage 40 was practically identical: 96.0 and 94.5 τ_0, respectively. The relative duration of the period from stage 37 to stage 45 in prelarvae of the Volga *A. stellatus* that were developing at a mean temperature of 17.2° was 308 τ_0. Prelarvae of the Don *A. stellatus* were developing at the mean temperature of 21.3 °C. Duration of the corresponding period in them was 296 τ_0.

The relative duration of development of *A. gueldenstaedti* prelarvae from stage 37 to stage 40 equals 95 τ_0 and from stage 37 to stage 45–238 τ (see Table 8). This means that synchronous development of *A. stellatus* and *A. gueldenstaedti*, established earlier for the embryonic period (Dettlaff and Dettlaff 1961) proceeds also during the first part of the prelarval period before the onset of respiratory movements. Later the relative duration of periods to identical developmental stages in *A. stellatus* is greater than in *A. gueldenstaedti.*

In *H. huso* prelarvae from two batches that were developing at mean temperatures of 15.2 and 17.6 °C the period from stage 37 to stage 40 occupies 69 to 75 τ_0 vs. 95 τ_0 in *A. gueldenstaedti* and 94.5–96 τ_0 in *A. stellatus*. The period from stage 37 to stage 45 (at mean temperatures of 15.6 and 18.3 °C in *H. huso* lasts 186–203 τ_0 instead of 238 τ_0 in *A. gueldenstaedti* and 296–308 τ_0 in *A. stellatus*. Although the relative duration of similar developmental periods for *H. huso* in two batches somewhat differes; in both cases it takes fewer τ_0 as compared with *A. stellatus*.

Data for prelarvae of five *H. huso* females (see Table 9) on the duration of the period from stage 36 to stage 45 also demonstrate that relative duration of the whole prelarval development in *H. huso* is less than that of the period from stage 37 to stage 45 in *A. gueldenstaedti* and *A. stellatus*.

Differences in the relative duration of similar periods of embryonic and prelarval development can be characterized by the criterion of the relative rate of development (RRD), i.e., the ratio τ_n/τ_0 for one species to τ_n/τ_0 for another species (Dettlaff 1986). The results of such comparison for embryos and prelarvae of several sturgeon species are shown in Table 11.

During the first part of the embryonic period all studied species develop at a similar relative rate, i.e., RRD = 1. During the second period of embryonic development *H. huso* and *A. ruthenus* begin to develop faster than *A. gueldenstaedti* and *A. stellatus* (RRD = 1.15). During the prelarval development the relative rate of *H. huso* development continuously increases, i.e., RRD rises. From stage 40 a difference appears also between the rate of development in *A. gueldenstaedti* and *A. stellatus*: *A. gueldenstaedti* starts to develop faster than *A. stellatus* (RRD = 1.24).

Differences in the relative rate of development found between various species of sturgeon lead to the hypothesis that these differences could contribute to the divergence of these species in the course of evolution.

Table 11. The relative rate of development of embryos and prelarvae of Acipenserid fish. (Dettlaff 1986)

Species[a]	Criterion of the relative rate of development (RRD)					
	In embryos for periods between fertilization and stages				In prelarvae for periods from stage 37 to stages	
	13	18	26	35	40	45
As:Ag	1	1	1	1	1	1.24
As:Ar	1	1	1	1.15	—	—
As:Hh	1	1	1	1.15	1.36	1.52–1.60
Ag:Ar	1	1	1	1.15	—	—
Ag:Hh	1	1	1	1.15	1.36	1.28
Ar:Hh				1	—	—

[a] Ag, *Acipenser gueldenstaedti*; Ar, *A. ruthenus*; As, *A. stellatus*; Hh, *Huso huso*

3.7 Defects of Prelarval Development

Industrial waste and other river pollutants like crude oil, aromatic compounds, pesticides and heavy metal salts adversely affect fish fry (Petru 1965; Luk'yanenko 1965, 1967b; Metelev et al. 1971; Mironov 1973; Danil'chenko 1977). Phenols play an important role because of their wide presence and harmful effect on water basins (Flerov 1973).

The Effect of Phenol

The data on the action of phenol on fish development are rather limited (for teleosteans, Volodin et al. 1965, 1966; Volodin 1973; for sturgeon, Schmalhausen 1962, 1971, 1972, 1973).

Prelarvae of sturgeon are relatively insensitive to the general toxic effect of phenol, although as prelarvae continue to develop, phenol toxicity for them increases quickly. A concentration of phenol equal to 40 mg/l (sublethal for the prelarvae) kills the larvae within a few minutes. The toxic effect involves three phases, which have been described as neurotoxic symptoms for adult sturgeon (Flerov 1965; Luk'yanenko 1967a). It appears that phenol acts on larvae in a different way than on prelarvae and specifically as a poison which selectively affects the central nervous system.

Relative resistance of prelarvae to the toxic effect of phenol allows them to continue development in water contaminated with phenol. However, their development under such conditions is abnormal. The presence of phenol, as of

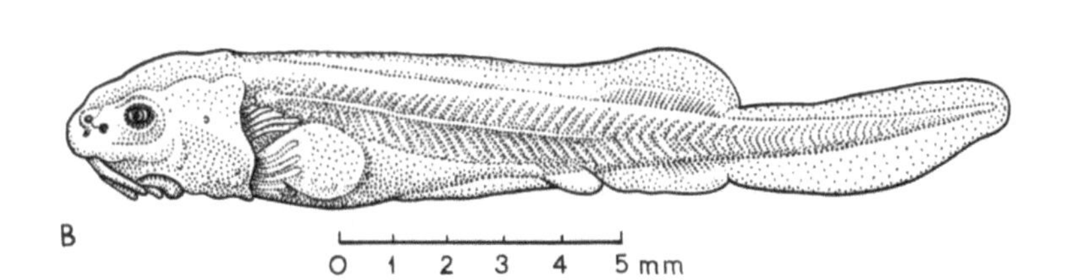

Fig. 71A, B. The effect of phenol on the development of *A. gueldenstaedti* prelarvae (Schmalhausen 1971). Prelarvae at the stage of transition to active feeding reared under normal conditions (**A**) or in the presence of phenol (40 mg/l) after hatching (**B**)

other studied teratogenic substances, impairs certain metabolic processes necessary not only for the normal prelarval life, but also for the normal morphogenesis.

Phenol in concentrations of 10 and 40 mg/ml markedly inhibits prelarval pigmentation (Fig. 71). It has been demonstrated (Schmalhausen 1973) that phenol reversibly inhibits pigmentation of the skin and formation of the pigment in eyes by inhibiting melanin synthesis in melanoblasts and damaging the melanocytes already formed. If prelarvae devoid of the pigment were transferred into pure water, the melanocytes appeared in the skin after 1 or 2 days and pigmentation of eyes intensified. This observation suggests that inhibition of melanin synthesis by phenol is reversible. However, depigmented prelarvae still do not reach the extent of pigmentation characteristic of the control prelarvae of the same age.

Phenol added at a concentration of 40 mg/ml, in addition to inhibiting pigment formation in the eyes of experimental prelarvae, leads to lysis of many retinal cells (Fig. 72B). In some cases disturbances of circulation in the eye accompanied by hemorrhages have been observed (Schmalhausen 1973). It appears that the harmful effect of phenol is mediated by inhibition of oxidative processes, as can be deduced from the properties of phenols (Waters 1966) and changes of some physiological characteristics (Solmann 1949; Andreev et al. 1979). Retina, with its high respiration rate, should be particularly vulnerable to the inhibition of oxidative processes (Piri and van Geiningen 1968).

In addition to the inhibition of melanin synthesis, phenol also inhibits lipid metabolism of prelarvae. Soon after placing prelarvae into water containing phenol (10–40 mg/l) they show increased motor activity requiring high lipid consumption to replenish energy losses. Later, phenol leads to a delayed yolk resorption and a part of the fat remains in the protein-bound state. Lipid absorption in the intestine is also inhibited. As a result, prelarvae are almost completely depleted of fat stores necessary for them to survive possible starvation during downstream migration. The liver of such prelarvae is small and very compact (Fig. 73). Liver cells contain very few lipid inclusions. These inclusions are also scarce in midgut walls (Schmalhausen 1972). This appears to be associated with the ability of phenol, as well as of numerous other organic compounds, to inhibit the effect of several enzymes of lipid metabolism (Haldane 1930).

Pyloric appendages (Fig. 73) and the folds of intestinal mucosa in such prelarvae are underdeveloped and resorption of the septum partitioning off the pharyngeal cavity and esophagus is delayed. In contrast, the ejection of pigment plugs is premature.

Defects of the visceral cartilages have also been observed in prelarvae that developed in the presence of phenol. Mandibular cartilage undergoes the maximal damage which results in a characteristic mouth malformation (Fig. 74A′). This defect is clearly seen even with the naked eye. Mandible in such larvae falls inside the oral cavity and, as a result, the surface of the palate becomes visible from outside.

Study of the visceral apparatus excised from experimental and control prelarvae at the stage of transition to active feeding has shown that in some

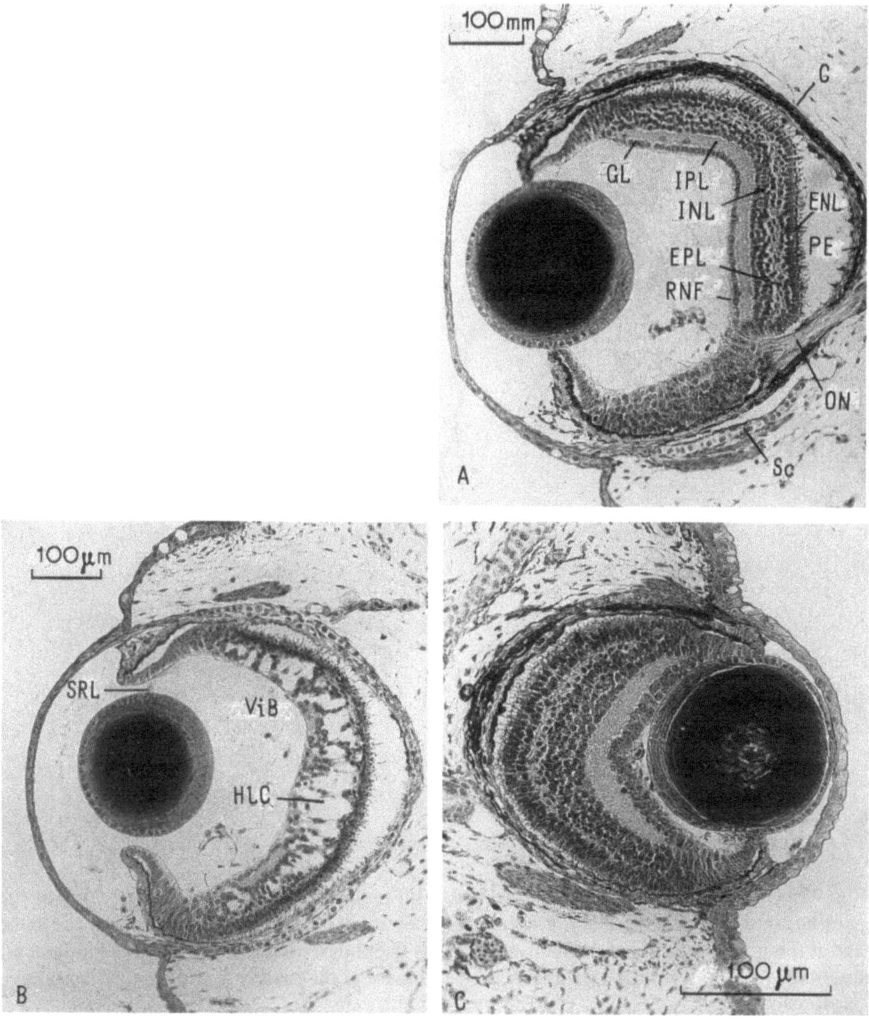

Fig. 72A–C. Structure of the eye in *A. gueldenstaedti* prelarvae developing from the stage of hatching in the presence of phenol (40 mg/l) (**B**) or a piece of brass mesh (**C**); **A** control of the same age at stage 45; (**A, B** Schmalhausen 1973). *C* Choroid; *ENL* external nuclear layer; *EPL* external plexiform layer; *GL* layer of ganglial cells; *HLC* hollow cavity in place of lysed cells; *INL* internal nuclear layer; *IPL* internal plexiform layer; *ON* optic nerve; *PE* pigment epithelium; *RNF* retinal nerve fibers; *Sc* sclera; *SRL* slit between retina and lens; *ViB* vitreous body

experimental prelarvae all cartilaginous elements are underdeveloped. Different cartilages are underdeveloped to a different degree and, as a result, the proportion between them is altered. Most drastic is the inhibition of development of the first two branchial arches, mandibular and hyoid (Fig. 74B′). Disturbed correlation between elements of these arches prevents the formation of the protrusible mouth so characteristic of *A. gueldenstaedti* and *A. stellatus*. Prelarvae with such

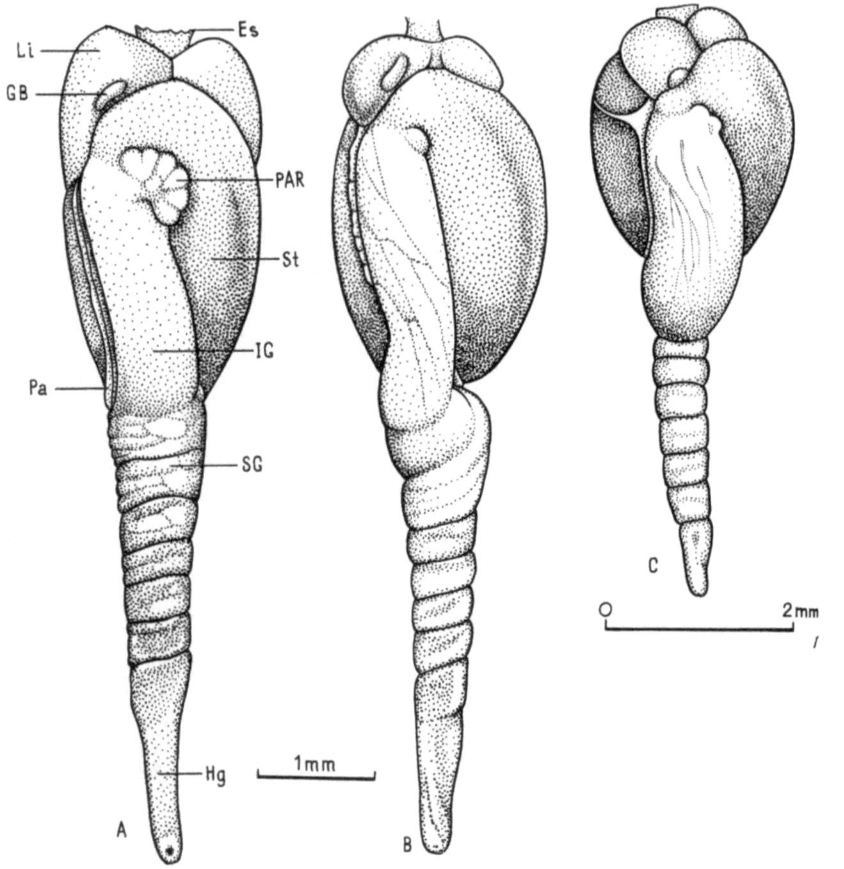

Fig. 73A–C. Digestive system of *A. gueldenstaedti* prelarvae developing after hatching in the presence of phenol (40 mg/l) (**B**) or a piece of brass mesh (**C**). **A** Control prelarva of the same age at the stage 45 (**A, B** Schmalhausen 1972). *Es* Esophagus; *GB* gall bladder; *Hg* hindgut; *IG* intermediate gut; *Li* liver; *Pa* pancreas; *PAR* rudiment of the pyloric appendage; *SG* spiral gut; *St* stomach

Fig. 74A–C. The effect of phenol (40 mg/l) on the development of visceral apparatus in *A. gueldenstaedti* prelarvae. **A, B, C** Control prelarvae at stage 45; **A′, B′, C′** prelarvae of the same age that were developing in the presence of phenol from the stage of hatching **A, A′** View of the head *from below*; **B, B′** dissected visceral apparatus, view from below. **C, C′** First gill, lateral view. *BA IV* fourth branchial arch; *Cb* ceratobranchialia of the 1st-3rd branchial arches; *Cp* copula; *Hb* hypobranchialia of the 1st–3rd branchial arches; *HCa* hyoid cartilage; *Hh* hypohyale; *HmCa* hyomandibual cartilage; *Ih* interhyale; *MCa* Mekkel's cartilage; *PCa* palatoquadrate cartilage; *Sy* symplecticum

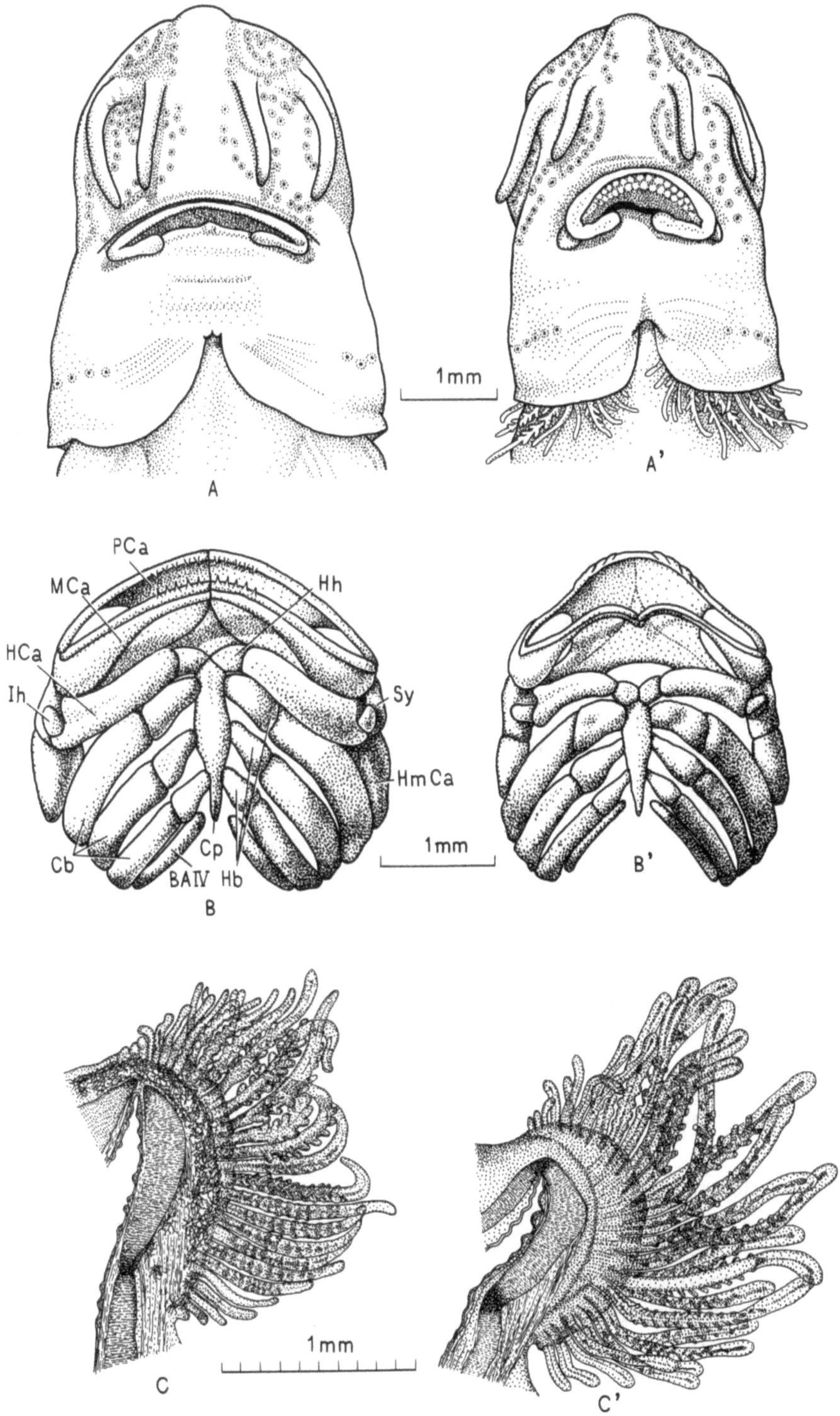
1mm
A
A'
PCa
MCa
Hh
HCa
Ih
Sy
HmCa
Cp
Cb
BAIV
Hb
B
1mm
B'
1mm
C
C'

defects are incapable not only of protraction/retraction but even of simple seizing movements. Therefore such prelarvae die after having depleted their yolk stores.

Similar malformations of the visceral apparatus have been described in the trout reared at the hatchery (Teichmann 1957).

It is known that carbohydrate metabolism disturbances form a basis of various chondrodistrophies. Such abnormalities have also been observed experimentally in several vertebrates (Ancel 1945; Landauer 1947, etc.).

Degenerative changes have also been observed in skeletal striated muscles. Orientation of muscle fibers can be disturbed and vacuoles appear in fibers between myofibrils; also cavities can be formed at those places, where muscle tissue underwent degeneration.

Connective tissue and muscles surrounding various organs are also underdeveloped. Comparison of the consecutive stages of gill development has shown that gill filaments of experimental prelarvae continue to grow in length longer than in the control, and differentiation of secondary gill filaments is delayed in them, as compared with the control. As a result, gills of prelarvae that were developing in the presence of phenol are longer than in the control larvae (Fig. 74C and C′). In this case the gills are not covered by the operculum.

When *A. stellatus* prelarvae are kept under the conditions of inadequate water aeration, the length of gill filaments can also increase (Schmalhausen 1955a).

Characteristic defects of *A. gueldenstaedti* and *A. stellatus* prelarvae which developed in the presence of phenol, include as we have seen drastic inhibition of pigmentation, structural defects of visceral apparatus, digestive system, and the eyes. Defects in the structure of the olfactory organ, decrease in the body size, and circulatory disturbances have also been observed.

The behavior of prelarvae changes as well. The first response to the presence of phenol in water appears to be the drastic increase in motor activity. Normal responses of prelarvae to light, water flow, and vibrations are inhibited. The excited state of prelarvae gradually turns into depression. When prelarvae are returned to pure water during the period of excitation, the excitation ends abruptly and sometimes is even followed by the inhibition of motor activity.

The Effect of Heavy Metals

Heavy metal salts are known to be toxic for fish even at very minute concentrations (a few micrograms per litre). Reversibility of poisoning is very low (Luk'yanenko 1967a). Pieces of brass mesh placed into jars with the sturgeon prelarvae reproducibly inhibit their growth and differentiation, and decrease the viability (Schmalhausen 1957, 1962). When the prelarvae are reared in jars in the presence of a piece of brass mesh, the toxic effect of heavy metal ions did not appear until the stage of transition to gill respiration. Only at these stages the mortality appeared, but growth inhibition became visible soon after hatching.

Growth slowing down affected mostly the length of those parts of the body, which grow very quickly during prelarval development. As a result, body proportions in the prelarvae are disturbed, they look curved, the septum between nostrils is underdeveloped, barbels as well as operculum and gill filaments are short, fins underdeveloped (Fig. 75). Dissection of such prelarvae has demonstrated defects of the digestive system. The esophagus and the spiral gut are markedly shortened and the intermediate gut is flattened and extended along the bulky stomach. The rudiment of pyloric appendage is absent or poorly developed (Fig. 73). Examination has shown that the intestinal mucosa is thinner and its folds are less prominent than in the normal state. Stomach glands are absent and cleavage of yolk platelets in the stomach proceeds far more slowly than in control prelarvae of the same age. When the control prelarvae pass to active feeding, the experimental prelarvae still have a bulky stomach filled with yolk. The uncleaved yolk is a mechanical obstacle for the formation of the fold separating the intestine from the stomach and, therefore, formation of the intestine is abnormal (Figs. 73, 75D and 76). Biochemical observations lead

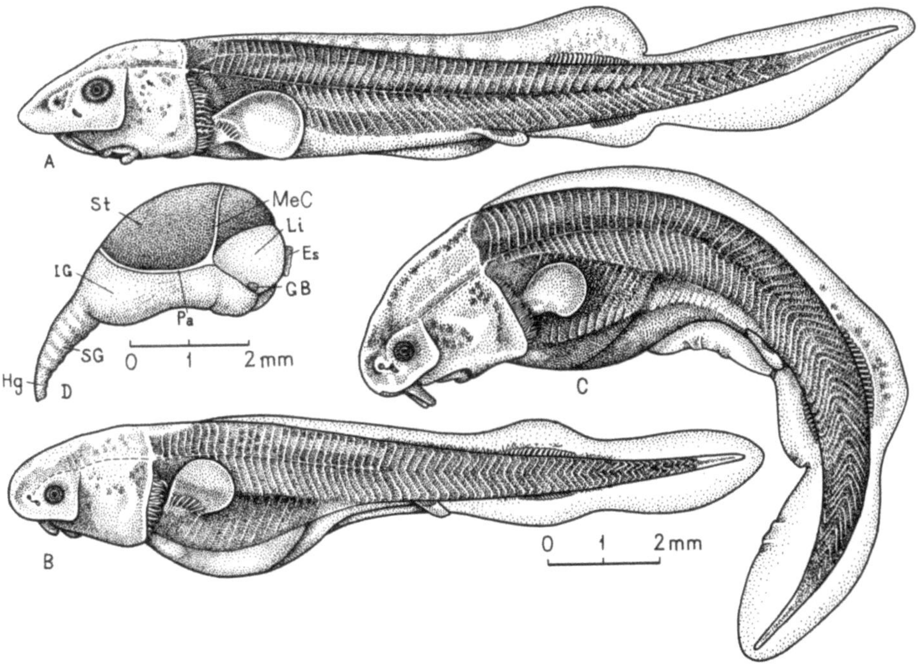

Fig. 75A–D. Development of *A. gueldenstaedti* prelarvae in the presence of a brass mesh. **A** Control prelarva at stage 45. **B** Prelarva of the same age developing from the stage of hatching in the presence of a peice of a brass mesh. **C** Prelarva of the same age developing in Chalikov incubator in the river. **D** Dissected digestive system of the curved prelarva from Chalikov incubator. *MeC* Mesenteric cord. For other designations, see Fig. 73

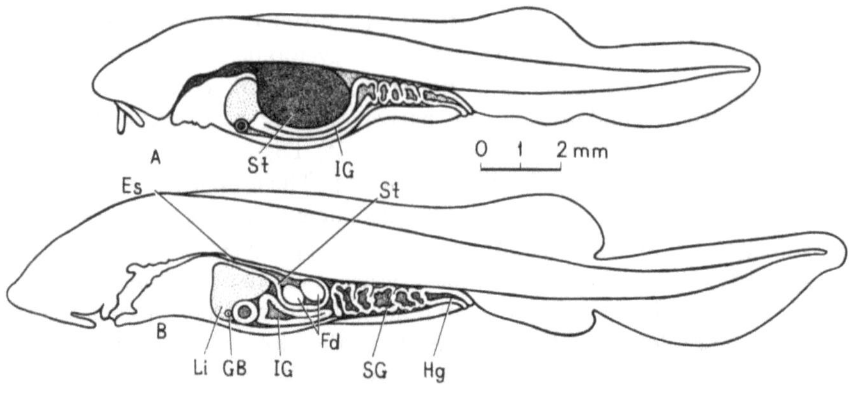

Fig. 76A, B. The structure of *A. gueldenstaedti* prelarvae. **A** Developing from the stage of hatching in the presence of a piece of brass mesh and **B** the control prelarva of the same age shortly after transition to active feeding. *Fd* Food. for other designations, see Fig. 73

to the hypothesis that one factor underlying developmental disturbances induced by heavy metal ions is associated with the inactivation of enzymes participating in the cleavage of yolk (Gross 1954; Flickinger 1956; Hewitt and Nicholas 1963). As a result, the amount of products of this cleavage which may be necessary for growth and differentiation of prelarvae is diminished. Later, yolk platelets are lysed as well as the stomach wall and the entire bulk of the lysed yolk falls out into the body cavity. Such prelarvae are not viable.

Prelarvae that were developing in the presence of brass mesh showed degenerative changes of striated muscles, similar to those that have been described for the prelarvae poisoned with phenol.

Morphogenesis of the eye is also disturbed in such prelarvae. There is no normal thinning of retina. In eyes, having usually a smaller size, the vitreous cavity is incredibly small and the lens is almost adjacent to the retina, which is far thicker than the eye retina of the control prelarvae. The anterior eye chamber is also greatly reduced. Thus, normal proportions between the sizes of the eye and lens are disturbed (Fig. 72).

In many cases the body of the prelarvae contains regions where epidermis undergoes necrosis or where disorganized proliferation of epithelial cells is observed. It remains to be seen whether this proliferation is due to chemical damage of integuments or this is a manifestation of some viral disease affecting weak prelarvae.

Rearing of prelarvae in the presence of a piece of a brass mesh at different developmental stages has demonstrated that the toxicity of heavy metal ions diminishes with age.

Behavior of experimental prelarvae differs from that of the control ones, as well as of those prelarvae that were developing in water containing phenol. Soon after the prelarvae have been placed into a jar containing the brass mesh, they become almost immobile and lie at the bottom for long periods of time without any movement. Their response to light, water current, and vibrations is weak, and thereafter the response ends completely. Inhibition of growth and decreased survival of prelarvae reared in cages with brass mesh has also been observed by other authors (Semenov 1958; Kolodkova and Shevchenko 1976).

The Effect of Substances Present in Fresh Wood

The action of substances extracted from fresh wood also results in defective organogenesis. However, toxicity of these substances is so high, that prelarvae soon die. Numerous whitish tubercles, similar to those found in prelarvae that were developing in the presence of the heavy metal ions (proliferation of epithelial cells), have been seen on the body surface of such prelarvae. Edema was a typical defect of prelarvae which were developing in the presence of fresh wood.

It should be remembered that the action of various toxic substances that occur in rivers greatly depends not only on their concentration but also on environmental factors, such as temperature, oxygen concentration, pH, concentration of CO_2, water hardness, water flow rate and others (Doudoroff and Katz 1953; Lloyd 1961; Luk'yanenko 1967a, b).

For example, there is a feedback between the toxic effect of heavy metal ions on fish and the water hardness. This is explained by decreased solubility of heavy metal salts in the presence of calcium salts, which produce insoluble complexes with them. In addition, calcium ions affect cell permeability and decrease the extent of penetration of the toxic substances into cells (Jones 1938, 1939; Luk'yanenko 1967a, b). Water hardness has almost no effect on the toxicity of most organic compounds (Metelev et al. 1971).

On the other hand, it should be remembered that toxicity of these substances is different for different fish species. It also depends on the fish age, their capacity of adaptation to the harmful environment, and on the possibility of penetration of harmful substances inside the organism, as well as on other factors (Flerov 1965; Luk'yanenko and Flerov 1965; Luk'yanenko 1967a).

The data presented in this section provide evidence that defects of prelarval development are due, most of all, to unfavorable effects during the period of rearing. In the absence of any harmful environmental factors, the prelarval development proceeds normally and tansition to the larval period is not associated with any serious losses. This view is also shared by other authors (Afonich et al. 1971).

Chapter 4 Environmental Conditions of Spawning and Development

4.1 General

4.1.1 The Concept of Thresholds

Animals and plants continuously interact with the environment, which does not remain constant for any long period of time; all poikilothermic animals live and develop under the influence of fluctuating temperature, light, oxygen level, and other factors. However, reproduction and development of any species remains normal only within a certain range of the factors' fluctuations. Outside this range, the reproduction stops, development is disturbed and embryos or prelarvae eventually die.

Usually, there is an upper as well as a lower threshold and the range of optimal and suboptimal conditions is located between these thresholds.

The set of conditions allowing spawning is referred to as the spawning complex. Limits of their fluctuations are usually somewhat more narrow as compared with the threshold conditions.

4.1.2 Spawning Conditions

Most animals actively search for favorable life conditions. This is due to an inborn or instinctive pattern of behavior that has developed during evolution of the species. Similarly, by instinct, parents ensure normal living conditions for the offspring. Such conditions can differ drastically for various fish species. Fish spawn eggs in water of different salinity in oceans and seas, brackish waters or fresh water bodies. Reproduction takes place in different seasons. For example, Cyprinid and Percifurm fish spawn during spring–summer, whereas numerous Salmonid species spawn in the autumn or winter. Eggs of some species develop suspended in water, whereas in other species they develop after being attached to the bottom at the spawning ground or to aquatic plants. Some fish species deposit their eggs into nests, incubate them in mouth or display other forms of parental care (Kryzhanovskii 1949; Soin 1968; Balon 1975).

Acipenserid fish reproduce in rivers and almost all of them spawn in spring and summer. They migrate upstream sometimes for many hundreds of kilometers until they find conditions appropriate for spawning. Females spawn eggs in shallow parts of the river with swift current and dense, usually pebbly ground; here eggs attach to pebbles or sand particles or stay in cracks between stones (Khoroshko 1968).

After reaching the region of spawning grounds the females start spawning only when water temperature is optimal for oocyte maturation and embryogenesis. If, during spawning, the temperature suddenly rises above the optimal one, the spawning stops. Females leave the spawning grounds and go into holes (ditches) (see Dettlaff 1970a). Cases have been described when spawning stopped even at the spawning temperature, but when water level at the spawning ground decreased drastically (Doroshin and Troitskii 1949; Khoroshko et al. 1974). Thus, there are different factors which may play the limiting role in the whole set of spawning conditions.

It is known that females, which are prevented from migration to spawning grounds by dams (like Volgograd and Tsimlyanskii dams) and cannot find appropriate spawning conditions, do not spawn at all. Atresia of follicles has been observed in such females (Faleeva 1965, 1979; Barannikova 1968).

Spawning conditions in various Acipenserid species are similar (although not fully identical) and therefore sometimes they can spawn on the same spawning grounds; in rare cases this leads to interspecific hybrids. Some of these hybrids are viable (Nikolyukin 1952). They are found in the catch and are well known to fishermen and fish breeders. The extent of hybridization under natural conditions, however, is limited because the time of spawning and spawning temperatures of different species overlap only in a narrow range. Recently, however, the flow of rivers is increasingly regulated and this somewhat shifts the time of spawners' migration into such rivers. As a result, the time of spawning of different species changes and there is increased abundance of the fish belonging to different species at the remaining spawning grounds. Therefore, the probability of interspecific hybridization increased. Bester, which now occurs in large numbers in open water bodies is a serious threat for the survival of pure Acipenserid species. Bester is particularly abundant in the Azov sea basin where it fattens and attains maturity by the age of 7 years (Burtsev 1969). Despite active protests of numerous scientists against the introduction of bester into sturgeon water bodies (see Barannikova et al. 1979), this error has already been made. Only the future will see how serious will be its consequences.

4.1.3 Relationships of the Developing Embryo with the Environment at Various Developmental Stages

The relationships of the embryo with the environment change continuously during development. Such changes are particularly prominent in those animals and plants whose embryos develop for long periods of time, including more than one season. In such cases, the complexes of conditions favorable for different developmental stages can be quite dissimilar.

This, however, is not the case for sturgeon since in this group embryonic development is completed within a short period of time (from 2 to 10 days, depending on temperature). Fluctuations of environmental conditions during such a short period of time are random, they do not have any definite trend and the entire development proceeds under practically uniform conditions. This is also true for the development of prelarvae, which continues for 6–11 days depending on the fish species and temperature. The prelarval development proceeds normally in the temperature range which is somewhat broader compared with the spawning temperatures. Under such conditions, temperatures sublethal for maturing oocytes, eggs during fertilization or during the first cleavage divisions, are still favorable for the development of later embryos and prelarvae.

However, the relations of developing embryos and prelarvae with the environment can change even when the environment shows only minor variation: requirements of the embryos as well as their sensitivity to unfavorable effects change during development; a given environmental factor can play different roles at different stages of maturation and development. This can be illustrated by a few examples.

Development of the embryo is associated with a continuous increase in oxygen consumption (Tatarskaya et al. 1958; Korzhuev et al. 1960; Khakimullin and Molodykh 1985). Between fertilization and hatching, oxygen consumption by one embryo increases 20 times in *A. gueldenstaedti*, 27 times in *A. stellatus*, and 50 times in *A. nudiventris*. When these values are normalized by weight, the corresponding coefficients are equal to 15, 25, and 39, respectively (Korzhuev et al. 1960).

Sensitivity of embryos to a given unfavorable effect of the environment also changes during development. Thus, elevated water temperature (30–32 °C) during fertilization and early cleavage kills the egg, whereas at the end of gastrulation it leads to drastic developmental defects (abnormal structure of the head, heart and tail). After the onset of heart beating, however, such temperatures no longer have any serious effect on the embryo structure, whereas during the period before hatching they lead to paralysis (Dettlaff and Ginsburg 1954).

On the other hand, similar effects at different developmental stages can lead to different consequences. For example, swift current at the spawning ground during spawning leads to the dissemination of fertilized eggs. During the embryonic development, this prevents settling of the silt and contributes to better oxygenation conditions. A strong water current during hatching contributes to quicker release of the hatching enzyme and increases mobility of the embryos in the membranes enhancing their more synchronous hatching.

At last, by consuming oxygen from water and excreting carbon dioxide and metabolic products the embryos themselves can modify their environment. For example, if several incubators are supplied by water from a common source and the flow rate and temperature are similar, but are loaded with different amounts of eggs, the actual conditions of embryonic development in such incubators will also be different. Fish breeders know this phenomenon well. Overloading of the incubator with eggs leads to delayed development, increases the percentage of developmental defects and losses due to mortality. Thus, the optimal load of the incubator with eggs is also an important condition for the normal development of the embryos.

The development is normal only if environmental conditions are favorable for the embryo at all developmental stages, including the most sensitive ones. Thus, the range of spawning conditions is defined by the most sensitive stages and, when these conditions are met and egg quality is good, the development of embryos is normal or "typical" at all stages.

The complex of spawning conditions, in addition to the presence of males, includes certain qualities of the ground and water, temperature, oxygen and light conditions, as well as water level in the river and the current rate. Most of these conditions retain their significance when eggs are obtained artificially and incubated at sturgeon hatcheries.

Knowledge of conditions necessary for the reproduction and normal development of different Acipenserid species is critical for correct organization of work at all steps of fish breeding. Most detailed information about such conditions can be obtained from observations at spawning grounds during natural spawning. Unfortunately, such data are scarce and it is improbable that they will be complemented to any significant extent under the present conditions. The more important are observations of such kind made on remaining natural spawning grounds.

Additional information about the range of favorable conditions can be obtained in studies of the effect of various environmental factors at different levels and in different combinations. Such effects on sturgeon maturation and development can be studied in specially designed experiments or under hatchery conditions. In addition to temperature, flow rate, oxygen concentration, and other factors, to which sturgeon have adapted during evolution, there are other factors which do not belong to spawning factors in the strict sense of the word but which exert adverse effects. These factors include water pollution with phenols, salts of heavy metals, and other pollutants. It is important to know the levels of such factors which still do not have a teratogenic effect. It is equally important to know the rate of accumulation of teratogenic substances in the fish.

4.2 Temperature Range

We shall start our discussion with the problem of temperatures favorable for maturation and development of different Acipenserid species. Some ideas about these temperatures can be drawn from their effects on the beginning and end of natural spawning, from laboratory experiments and hatchery experience when we determined temperature limits allowing normal development of embryos and prelarvae. Of course, it is important that other conditions should be maintained as optimal, e.g., flow rate, oxygen level. The upper boundary of optimal temperatures can be also evaluated from the curve relating τ_0 to the temperature (Fig. 42). The temperature at which τ_0 stops diminishing or begins to increase is already damaging for embryos.

Huso huso. Temperature conditions favorable for the maturation of females and incubation of eggs are in the range 7–17 °C (Igumnova 1974, 1975b). At these temperatures, embryos have been successfully obtained at Rogozhkino and Volgograd hatcheries and Kurinskii experimental hatchery. At temperatures exceeding 17–18 °C the ovulation is impaired and the quality of the eggs deteriorates. At still higher temperatures (20–22 °C) increasing scattering of τ_0 values provides evidence for variable temperature damage of eggs obtained from different females.

As concerns the spawning temperatures, only the unpublished data of Chalikov are available (see Gordienko 1953), according to which spawning of *Huso huso* takes place at 8–15 °C.

Acipenser ruthenus. Initially eggs of this species on the Kama River were collected at temperatures from 10 to 13.9 °C (Persov 1957). Later at the Volgograd sturgeon hatchery ovulation was obtained at 6 °C but the ovulated eggs did not develop after insemination. Eggs which have matured at temperatures from 9 to 20 °C developed normally (Igumnova 1985a). Temperatures above 20 °C are already unfavorable, since τ_0 does not diminish in this range.

Spawning of *A. ruthenus* in the Volga and Kama has been observed at temperatures from 10.3–10.4 to 15–16 °C (Arnol'd 1915; Lukin 1947; Ginzburg 1967). In the lower Volga flow it takes place at 8.9–17 °C (Khoroshko, pers. comm.). There is evidence (Shmidtov 1939) that the optimal temperatures for mass spawning of *A. ruthenus* are 13.5–17 °C. The maximum temperature at which spawning could be observed was 20.6 °C (Babaskin 1930) and 20.7 °C (Shmidtov 1939). Spawning stopped when temperature increased to 21 °C. When temperature diminished to 9.4 °C the spawning was also temporarily stopped (Lukin 1947).

Acipenser stellatus. Temperatures favorable for incubation of *A. stellatus* eggs are in the range from 14–15 to 25–26 °C (Gerbil'skii 1949; Nikiforov 1949; Dettlaff and Ginsburg 1954). These temperatures have also been found optimal for *A. stellatus* fry (Derzhavin 1947). At 27–28 °C the embryos of the Volga population of *A. stellatus* are damaged (Sytina et al. 1985). The unfavorable effects at 27–29 °C are also clear in τ_0 values for the Don population of *A. stellatus* (see Fig. 42): when water temperature changes from 27 to 29 °C τ_0 stops diminishing and even has a tendency to increase. Eggs of *A. stellatus* are damaged at 29 and 30 °C.

At 12 °C the mortality of embryos is high and percentage of malformations is also high (Nikiforov 1949). This temperature is considered to be the threshold for embryos of the Volga and Don populations of *A. stellatus*. Maturation of the oocytes is atypical at 12 °C: a considerable proportion of the embryos obtained by fertilization of such oocytes develops with abnormalities (Dettlaff 1970a).

Natural spawning of *A. stellatus* in the lower Volga flow takes place at 16–25 °C (Khoroshko 1968), and in Kuban' at 15–25.4 °C (Kulinchenko 1939;

Doroshin and Troitskii 1949). According to Doinikov (1936), the minimum temperature at which female with ovulating eggs was caught in Kuban' was 13.8 °C and the maximal temperature 27 °C. In the Don females of *A. stellatus* with ovulating eggs appeared at 15 °C and the mass catch of such females was observed at 17–24 °C (Doinikov 1936; Derzhavin 1947). In the Kura river at 15–16 °C "empty" females migrating back after spawning were caught, whereas mass spawning was observed at 20–25 °C or above (Derzhavin 1922). The maximal temperature at which spawning of the Kura population of *A. stellatus* has been observed was 29.6 °C (Derzhavin 1922). However, it is improbable that this temperature is normal for *A. stellatus* spawning.

Thus, temperatures favorable for maturation of females and development of embryos of *A. stellatus* in the Volga, Don, Kura, and Kuban' are similar and equal to 15–25 °C.

Acipenser gueldenstaedti and *A. persicus.* Spawning of the vernal *A. gueldenstaedti* in the Volga has been observed at temperatures varying from 8 to 16 °C (Alyavdina 1951a; Barannikova 1957; Khoroshko 1968). Spawning of the hiemal *A. gueldenstaedti* of summer migration was observed at 9–11 °C, whereas spawning of the hiemal sturgeon during the autumn migration was at 15–13 °C (Barannnikova 1957). Maturation of the hiemal females kept for some time in holding tanks proceeded normally at 18–19 °C after the injection of the pituitary suspension (Molodtsov 1979). According to Molodtsov, embryos developed normally at 18–19 °C and prelarvae were not damaged if the temperature was quickly raised from 19 to 21 °C.

Oocytes of *A. gueldenstaedti colchicus* according to our observations mature and ovulate at 8–9 °C, but a certain proportion of these oocytes is activated and, correspondingly, is not fertilized after insemination. At 10–11 °C there is no egg damage or activation and their fertilization and development is normal. However, temperatures from 12 up to 20–21 °C are most favorable. In this range the duration of various developmental periods changes in proportion to the temperature (see Fig. 64). Above 21 °C and up to 23–24 °C this proportionality is increasingly disturbed and the embryos become damaged at temperatures 25 °C or above. The embryos are most sensitive to the elevated temperature during the period of fertilization and the first cleavage divisions.

The τ_0 determined for the vernal population of *A. gueldenstaedti colchicus* of spring migration hardly diminishes when the temperature is raised above 22.5 °C, and at 24.5 °C it has a tendency to increase (see Fig. 42). Thus, according to this criterion, temperatures exceeding 22 °C are suboptimal for this species.

Embryos of the vernal population of *A. persicus* of spring migration from the Kura develop normally at 10–22 °C (Vernidub 1952). At temperatures above 25 °C the ovulated oocytes do not cleave after insemination and die, whereas if the developing embryos are exposed to this temperature they are impaired (Gerbil'skii et al. 1951). The vernal *A. persicus* of autumn migration in the Kura spawns at 14–23 °C (Vernidub 1952), but spawning stops at 25–26 °C (Derzhavin 1947; Dettlaff 1970a).

Until the 1970s *A. persicus* from the Volga was not distinguished from the

Russian sturgeon *A. gueldenstaedti Brandt* and was referred to as the late vernal or summer-spawning *A. gueldenstaedti*. Later it has been demonstrated that this sturgeon is identical to *A. persicus* in Kura (see Artyukhin 1979). Spawning of *A. persicus* in Volga takes place at 18–23 °C (Alyavdina 1951a; Barannikova 1957).

Thus, temperatures varying from 9–10 °C up to 20–21 °C appear to be favorable for the maturation of oocytes and development of embryos of all biological groups of *A. gueldenstaedti*, its subspecies and *A. persicus*. Embryos of *A. gueldenstaedti colchicus* and *A. persicus* are damaged at 25 °C and spawning of the vernal *A. persicus* of autumn migration stops at 25–26 °C. Comparison of temperature ranges for the reproduction of various biological groups of *A. gueldenstaedti* demonstrates that although in nature their reproduction takes place at different time and at somewhat different temperatures, different groups are not isolated in terms of these parameters. Still, such differentiation of different *A. gueldenstaedti* populations is a biologically important phenomenon since it allows a more extensive and complete use of various river conditions for reproduction; furthermore, the larvae and juveniles can use the available food resources more efficiently.

Although the temperature ranges for the oocyte maturation and embryonic development in various biological groups of *A. gueldenstaedti* are similar, temperature conditions under which batches of good quality eggs can be obtained at the hatcheries differ depending on the temperature at which spawners migrate into rivers for spawning.

Acipenser baeri. Migration of this species in the Lena river for spawning takes place at temperatures of no lower than 8–9 °C, and mass migration begins at 12–14 °C (Sokolov and Malyutin 1977). Water temperature at which the actual spawning takes place is unknown. The females caught at the sites of mass spawning show a good response to the pituitary injection at 14–18 °C; above 20 °C the maturation time increases and, obviously these temperatures are not optimal (Berdichevskii et al. 1979).

In experiments with incubation of *A. baeri* eggs collected in the Lena and transported to Azerbaijan, temperatures from 11 to 15 °C were most favorable for embryonic development (Nikol'skaya and Sytina 1978b). At lower and higher temperatures the loss of embryos during incubation increased considerably and at 23 °C none of the embryos survived. High embryonic mortality outside the narrow temperature range could be due, in these experiments, to the lowered viability of embryos resulting from long transport time at 6–8 °C. The embryos were transported when they were at stages of the first cleavage divisions.

Temperature Range for the Development of Acipenserid Prelarvae

Data on this subject are rather scarce. It has been demonstrated (Bogdanova 1972a,b) that prelarvae of *A. gueldenstaedti*, *A. gueldenstaedti colchicus* and *A. persicus* as well as prelarvae of *A. baeri* from the Lena and lake Baikal can pass to exogenous feeding at 11.0–21.5 °C, i.e., at the interval close to the temperature optimum for the embryos. At lower temperatures (11.0–13.8 °C) exogenous feeding begins later when the size and weight of prelarvae is diminished. However, irrespective of the time period before the beginning of exogenous feeding, the structure of the digestive system in the embryos was similar. A characteristic morphological feature of prelarvae that had developed at lower temperatures was the absence of rudiments of dorsal scutes in the fin fold by the time of their transition to active feeding.

4.3 Oxygen Content of Water and Gas Metabolism in Embryos

Oxygen consumption by Acipenserid embryos as well as by embryos of other poikilotherm animals varies with temperature, provided water is well saturated with air. It has been demonstrated that the total consumption of oxygen by *H. huso* embryos for one τ_0 is minimal at 16 °C and, hence, energy expenditures of the embryo are minimal at this temperature (Ozernyuk 1985)[1]. Below 12 °C or above 20 °C the rate of oxygen consumption increases sharply.

In the presence of oxygen deficiency in water, the first response is the decreased respiration rate of embryos. Further oxygen deficiency is accompanied by various developmental defects (Privol'nev 1947). There are few data about those oxygen levels at which such developmental defects appear in Acipenserid embryos.

It has been noted that decrease of oxygen level in water below 80% of saturation (6 mg/l at 20–22 °C) has an adverse effect on the embryos and prelarvae of *A. gueldenstaedti*, *A. stellatus* and *H. huso* (Vernidub 1951).

The effect of varying oxygen level in water has been studied in more detail using embryos of the Kura and Volga populations of *A. gueldenstaedti* and *A. persicus* (Yurovitskii and Reznichenko 1961, 1963). The incubation was conducted in running water at temperatures 16–20 °C with the necessary concentrations of oxygen set by blowing nitrogen through water. As the oxygen level decreased down to 5–6 mg/ml embryonic development slowed down, hatching took place over a more extended period of time and, in addition, a large number of malformations was observed (mainly, with hypertrophy of the heart and edema of the pericardium). The hatched larvae had lower weight and size as

[1]The temperature range at which energy expenditure in the embryos of poikilothermic animals is minimal is called the "temperature optimum" of development (Alekseeva and Ozernyuk 1987; Zinichev and Zotin 1988; Zotin et al. 1989). We believe that it is more correct to call this range "temperature-energy optimum of development" in order to distinguish between a narrow temperature range with minimal energy expenditures and a far wider range of optimal temperatures in which development proceeds normally for a given animal species. The width of the range of optimal temperatures depends on evolutionary adaptation of poikilothermic animals to reproduction and development under fluctuating temperature conditions.

compared with those that developed under normal oxygen saturation, i.e., 7.5–9.5 mg/l. In water containing 3–4 mg/ml oxygen, development was distinctly atypical and none of the embryos survived.

Absence of water flow affected *A. gueldenstaedti* embryos similarly to oxygen deficiency (Yurovitskii and Reznichenko 1961). In the absence of water flow, the diffusion of oxygen and carbon dioxide does not support the necessary rate of gas exchange, and oxygen deficiency develops despite normal saturation of water with oxygen.

These data are in line with information about the level of oxygen in water at spawning grounds. During natural spawning of *A. stellatus* oxygen levels vary from 9.0 to 6.6 mg/ml in the Kuban' and from 9.3 to 8.4 mg/l in the Volga river (Kulinchenko 1939). During *A. gueldenstaedti* spawning, oxygen level limits in the Volga are 10.1–8.2 mg/ml (Alyavdina 1951a, 1953) and 9.8–7.2 mg/l in the Kura (Derzhavin 1947).

Correspondingly, incubators at hatcheries should be supplied with water about 100% saturated with oxygen; the flow rate should be enough for the exchange of water around the eggs.

Another factor which should be avoided is clumping of eggs. If this takes place, then embryos located in the center of clumps develop more slowly, with various defects and usually do not reach the hatching stage.

One more factor also important for maintaining gas exchange within the normal limits is the concentration of embryos in the incubators. Overloading of the incubator with embryos leads to oxygen deficiency, decreases gas exchange and results, as already mentioned, in delayed development and increased percentage of malformations.

4.4 Light Conditions

There is no information about the effect of different light intensity on embriogenesis. The development is somewhat delayed when eggs are incubated in complete darkness (Kulinchenko 1939). In experiments with *A. gueldenstaedti* during the period from hatching and up to transition to active feeding (Semenov 1957) direct sun light retarded their growth and differentiation; various malformations were common and the survival of prelarvae was markedly decreased. Under daylight of moderate intensity or darkness, development proceeded normally, and the growth rate was higher than under direct sun light.

In nature eggs develop in more or less turbid water and at a marked depths, i. e., under a weak light.

Incubators with eggs in hatcheries should be protected from direct sun light since such exposure can lead to the damage of embryos and appearance of malformations.

As concerns prelarvae, due to their photobehavior response they actively migrate to light conditions optimal for their development. In different Acipenserid species and at different stages of prelarval development the attitude to light is different (see Sect 3.6).

4.5 Quality of Water

Water optimal for the development of sturgeon embryos should be neither acidic nor alkaline. During the spawning season the pH of water on the spawning ground fluctuates around neutral or is weakly alkaline: pH varies from 6.5 (Lukin 1947) and 6.6 (Stroganov 1938; Kulinchenko 1939) up to 7.7 and sometimes up to 8 (Alyavdina 1951a). There is evidence (Babaskin 1930) that at pH below 6.4 and above 7.5–8 *A. ruthenus* embryos are damaged.

Sturgeon embryos and fry are adversely affected by various substances which pollute rivers. Deleterious substances are present in industrial waste, oil-products, pesticides washed from fields and other pollutants. For eggs developing on spawning grounds oil is particularly dangerous which settles to the bottom of the river together with particles of river silt. The harmful effect of crude oil is largely determined by toxicity of naphthenic acids which dissolve in water. It has been demonstrated in experiments with *A. stellatus* eggs that naphthenic acid at concentrations above 20 mg/l exert a definitely lethal effect (Derzhavin and Digurova 1951).

According to other data the deleterious effect of naphthenic acids on eggs of *H. huso*, *A. nudiventris*, *A. gueldenstaedti* and *A. stellatus* appears at far lower levels: concentrations varying from 0.5 to 1 mg/l lead to the developmental arrest of embryos at the late gastrula–neurula stages, while concentrations above 1 mg/l result in the instant death of embryos during the period from fertilization to hatching (Rustamova 1974).

At relatively high temperatures (within the spawning range) the teratogenic action of various deleterious substances present in water increases drastically.

Recently, water pollution in rivers and water bodies on which sturgeon hatcheries are located, became particularly pronounced (Kozlov 1990; Vlasenko 1990). It has been proposed that hatcheries for commercial sturgeon breeding should use purified water and operate on closed water supply cycles (Androsov et al. 1990). For such hatcheries optimal combinations of various environmental factors will have to be determined.

4.6 The Range of the Ecological Optimum

Embryonic development of poikilothermic animals is determined by both intrinsic factors such as their genotype and physiological condition and environmental factors. Optimal combinations of external factors and internal (intrinsic) factors will be referred to as the range of the ecological optimum (Shelukhin et al. 1990).

It has been traditional in ecological studies to define limits for the optimal values of a certain environmental factor by varying its intensity and keeping other conditions optimal. Such studies, however, are insufficient to determine the range of the ecological optimum, since major environmental factors may be present in different combinations with their effects virtually inseparable (see Dettlaff and Dettlaff 1982). Therefore, it is important to know more about the consequences of joint action of various factors and in different combinations.

Environmental factors affect the development of sturgeon in more than one way. Thus, increase in temperature within the spawning range leads, on the one hand, to a decrease in the level of dissolved oxygen and, on the other, stimulates metabolism and oxygen consumption by the embryo (see Dettlaff and Ginsburg 1954; Zotin and Ozernyuk 1966).

Various combinations of environmental factors can move the limits of optimal values of each of them. This is illustrated, for example, by our experiments on rearing *A. stellatus* prelarvae in hard water (from an artesian well) without intense aeration (Schmalhausen 1983). At 17 °C prelarval development was normal. At 20–21 °C from stage 40 onwards it was drastically affected, although this temperature is within the optimal range, provided other conditions are favorable. In the control prelarvae hatched from the eggs of the same female in the settled Moscow tap water developed normally at 25–26 °C under intense aeration.

Another example of a change in optimal conditions when individual environmental factors are combined in different ways is provided by the effect of various combinations of water temperature and salinity on physiological condition of sturgeon fry (Shelukhin et al. 1990). These observations made with Caspian water have demonstrated that in summer at 20–24 °C salinity of 4–7 p.p.m. is most favorable, whereas at lower temperatures during autumn–winter months (but above 6 °C) higher salinity is optimal.

Determination of the range of the ecological optimum can be very important for the improvement of fish culture. However, studies of this kind are at an early phase, since they can be conducted only when biology of the species is known in great detail and such knowledge requires extensive experimental studies. The number of necessary observations can be reduced, if we define criteria that can allow us to judge the effect of joint action of several main environmental factors.

One such criterion is the relative duration of development τ_n/τ_o, which in the range of optimal temperatures is practically constant (see Sect. 2. 8. 3). Change in τ_n/τ_o ratio within this range can serve as an index of deviation from the optimum for a given set of environmental factors as concerns their effect on development.

The method of parabolas developed by Zotin and his associates (Yaroslavtseva et al. 1991) is another approach for the assessment of joint action of various environmental factors on the development of poikilotherms (specifically of fish). It has been reported that oxygen consumption by the embryo calculated per τ_o depends on temperature and this dependence can be described as a parabola (for sturgeon, see Ozernyuk 1985). As the embryos develop, the shape of parabola and its position on the temperature scale show regular changes (Alekseeva and Ozernyuk 1987; Ozernyuk 1988; Zotin et al. 1989). Dependence of the duration of development on salinity is also parabolic (Yaroslavtseva et al. 1991). If the shape of such parabolas and their position with respect to coordinates differ form the reference, this would provide evidence for the deviation of a certain set of external conditions from the optimum.

Elucidation of the range of the ecological optimum will be all the more important, if commercial breeding of sturgeon is to take place under conditions different from the natural conditions on spawning grounds.

Chapter 5 Control of Incubation Conditions

5.1 General

Before 1950 sturgeon eggs were incubated at hatcheries in the river in Seth Green incubators. The results depended on the rate of water flow in the river, height of flood, surge winds and other factors. Care of the embryos was extremely labor consuming. In 1950–1952 P. S. Yushchenko (Rogozhkino station of Azdongosrybvod) constructed the first on-shore incubator of mesh-drum type for the incubation of demucilaged sturgeon eggs. In 1953 he developed an incubator with a moving frame, which received wide recognition. Such incubators (Yushchenko 1957) are still widely used at sturgeon hatcheries.

Each incubator consists of two metal boxes; the bottom of the inner box is covered by a brass mesh. Four incubators are mounted on one table. The most important element of the incubator is a mobile frame with paddles located under the mesh wire bottom of the inner box. Metal strips of zig-zagged shape are soldered to the paddles in such a manner that when the frame moves, a vertical water current is created. Eggs caught by this water current move upwards and then gradually settle to the bottom of the incubator; the movement of eggs in water provides for good gas exchange. Paddle movement is achieved in the following way. A swing bucket is fastened to one end of the lever under the table. Water from the incubators flows into this bucket. When the bucket is filled with water it goes down, position of the lever is changed and the frame together with attached paddles moves to one side. When the bucket is full of water, it swings and the counterbalance returns the lever to the initial position and the frame moves to the other side. Periodicity of the frame movement is controlled by the rate of water flow into the swinging bucket. In a modification of this apparatus the swinging bucket is replaced by a turbine.

Recently, new incubators ("Osetr" type) developed at the Azov Research Institute of Fishery have been introduced at sturgeon hatcheries (developed by Fedchenko and Gorbacheva). In these devices a box with mesh bottom containing the incubated eggs is fixed in an oblique position and moves periodically in the vertical plane (a bucket fixed to a hinge which turns over after it is filled with water ensures movement in the system). Movement of the box in the vertical plane creates mixing of eggs over the whole surface of the mesh bottom. In addition, dead embryos having lower specific weight as compared with the living ones are concentrated at the exit from the box and can be easily removed. The hatched prelarvae swim out through a hole in the box wall and therefore are not to be collected manually as in the Yushchenko incubator.

In addition to these types, the water-jet apparatus of Kazanskii and certain other devices have been used for the incubation of demucilaged eggs.

Sturgeon eggs are not incubated in the attached state at the hatcheries, although, for certain specialized tasks this method can be convenient. A tray incubator for sturgeon eggs has been developed by Sadov and Kokhanskaya (Sadov and Kokhanskaya 1961, 1963; Kokhanskaya, 1980).

Incubation conditions in various apparatus should be selected on the basis of conditions known to be optimal for sturgeon development (see Chap 4). However, the available data are incomplete and in some cases require further consideration. As concerns the number of eggs per incubator and the amount of water, which should flow through the apparatus per unit of time, such problems have to be solved individually for each type of incubator under specific conditions.

Finding optimal conditions requires skills in the evaluation of incubation conditions from the characteristics of the development. Fluctuation of environmental conditions during incubation should not extend beyond the " spawning range" and reach its threshold values. Then with good-quality eggs embryos develop normally and there is no need to create specialized conditions for different stages. But if the development takes place under conditions close to the threshold ones, special attention should be paid to meeting the necessary conditions during the first half of the incubation period and shortly before hatching. It is safe, however, to avoid such conditions in general and not to begin incubation before spawning temperatures are reached.

In obtaining eggs, insemination, demucilaging and incubation it is desirable not to pool eggs from several females, since if the quality of eggs is different, the yield of prelarvae from better-quality eggs under these conditions will be lowered because of *Sprolegnia* growth.

Furthermore, in determining the number of eggs of *A. stellatus*, *A. gueldenstaedti*, and *H. huso* which can be placed in one incubator, correction should be made for the number of embryos per 1 g of eggs, since oxygen consumption and the rate of metabolism depend not only on the total weight of eggs but also on the number of embryos.

Recently, it has been repeatedly proposed that incubation lines with controlled temperature should be established, because variation in water temperature during the incubation may be responsible for the loss of embryos and developmental defects (see, for example, Romanov 1979). However, it has been demonstrated in special experiments performed with *A. stellatus* and *H. huso* embryos (Igumnova 1985b) that a singificant lowering of the incubation temperature within the spawning range (by 3.4–6.6 °C and 5.4 °C respectively) with subsequent return to higher temperatures does not impair or kill embryos. This was true for all stages examined from fertilization up to organogenesis. When the temperature was decreased below the limit of the spawning range (down to 6.7 °C and 3.9–4.2 °C, respectively) with subsequent return of embryos to spawning temperature, the development was not damaged if temperature treatments were applied during gastrulation and organogenesis. Such cooling had deleterious effect, i.e., produced developmental arrest and killed embryos, only during the period of fertilization and of the first two cleavage divisions.

Normal prelarvae having high viability developed from *H. huso* embryos subjected to drastic cooling (Igumnova and Dubinin 1987).

It should be pointed out here that temperature changes measured over a 10 year period in the incubation section of the Volgograd sturgeon hatchery never exceeded 3–4 °C per day and were never accompanied by increased egg losses. Therefore, there was no harmful effect of such temperature changes during the incubation period. Constant temperature regime is not a biologically indispensable condition for the normal development of sturgeon embryos. However, this does not rule out the need for thermoregulation in the incubation section, which may prolong the breeding season and increase throughput, particularly at hatcheries equipped with thermoregulated production lines for keeping spawners.

In addition, increasing pollution of rivers necessitates sturgeon breeding in purified water that can be recycled. Then incubation of eggs and rearing of prelarvae will be under the complete control of fish breeder. Probably it would be appropriate to conduct the process under the conditions most economical from the point of view of conservation of energy reserves of the embryos and prelarvae (energy optimum of development; see Sect. 4.3).

Several criteria can be recommended for finding best incubation conditions and for judgements as to how good is a given set of conditions for development. These criteria include duration of development, synchrony of hatching, number and structure of malformations, extent and sources of the loss during incubation and viability of the juveniles.

5.2 Duration of Development as a Criterion for the Evaluation of Incubation Conditions

Curves relating the duration of various developmental periods to temperature have been plotted for the embryos of *A. gueldenstaedti*, *A. stellatus*, *H. huso* and *A. ruthenus* (see Fig. 81). The abscissa shows mean temperature over a given period of development (in degrees); the ordinate shows time after insemination up to a certain stage (in hours). Curve I shows the timing of the beginning of gastrulation (stage 13, Plate III); curve II shows the timing of gastrulation completion (stage 18, Plate III); curve III shows the timing of fusion of the lateral plates at the place of heart rudiment formation and beginning of formation of the rudiment of posterior trunk and tail regions of the embryo (stage 26, Plate V); curve IV corresponds to the beginning of hatching at different temperatures (stage 35, Plate VIII).

The curves have been plotted taking into account the timing of the corresponding developmental stages in advanced embryos, i. e., in 5–10% of embryos in the sample that first reach a given stage. Embryos of *A. gueldenstaedti colchicus*, *A. stellatus*, and *H. huso* were incubated in Yushchenko incubators at the Rogozhkino sturgeon hatchery on the Don. These results were supplemented by the data for *A. gueldenstaedti*, *A. stellatus* and *H. huso* of the Volga river. Vast material on the duration of development of *Huso huso* from the Volga has been obtained by Igumnova (1975b); she has also plotted curves for the Volga sterlet (Igumnova 1985a).

A. gueldenstaedti, *A. stellatus*, and *H. huso* from different rivers (provided egg quality is good and incubation conditions are close to optimal) attain similar stages at the same time at similar temperatures. When conditions are less than favorable, for example, when the flow of water is insufficient or when the incubator is overloaded, the rate of development slows down and the period of hatching is delayed and prolonged. This is easily observed if different amounts of eggs are placed into adjacent incubators supplied with water from a common source and the rate of development and duration of hatching are monitored. The experiment can be modified: the number of eggs can be similar (and low) but the flow rate through one incubator can be normal whereas the flow rate through the second incubator is slowed down. In both cases the development will proceed more slowly in the incubator with less than optimal conditions and the hatching will start later and will be less synchronous.

In experiments conducted for developing the guidelines as concerns the number of eggs which can be loaded into a Yushchenko incubator the following trend has been observed. Demucilaged eggs were placed into several incubators; when the number of eggs was low and water circulation intense, hatching of embryos began 95 h after insemination, after 110 h hatching of all normal embryos was completed. In the adjacent incubator at a similar incubation temperature but with a higher number of eggs and insufficient circulation of water the first larvae appeared only within 129 h, i.e., with a 34-h delay.

Thus, if the hatching is markedly delayed as compared with the control embryos developing under favorable conditions (or as compared with the time shown in Fig. 81), more attention should be paid to the incubation conditions, specifically to conditions of gas exchange (to the amount of eggs incubated in the incubator and the flow rate).

Duration of the hatching period, i.e., time during which mass hatching is complete and all normal embryos come out of membranes, in *A. gueldenstaedti* and *A. stellatus* usually amounts to 15–20% and in *H. huso* (Igumnova 1979) to 15–28% of the total duration of embryonic development (between fertilization and hatching of the first larvae; see Fig. 81). However, the duration of this period is rather variable since it depends not only on temperature, gas exchange and other incubation conditions, but also on some other specific factors (current, impacts), necessary for the synchronous release of the hatching enzyme and for the emergence of embryos from the membranes. Generally, hatching is delayed and occurs over a longer time under conditions which are less than optimal.

Figure 81 illustrates the timing of various stages at different mean temperatures which may be of interest for fish breeders. Thus, curves I (stage 13) and II (stage 18) limit the period of gastrulation, when unfertilized eggs die and, therefore, it is convenient to record losses due to the presence of unfertilized eggs. Curves II (stage 18), III (stage 26) and IV (stage 35) show the timing of stages convenient for the assessment of the structure of embryos (i.e., for seeing how typical their structure is; see Appendix G).

Curve IV shows time to hatching of the first embryos and the duration of incubation at different mean temperatures. These data are necessary for the calculation of the incubator throughput; they also provide an opportunity to predict the timing of hatching and allow assessment of incubation conditions. Drastic delay of development points out to poor incubation conditions or low quality of eggs.

In order to use such curves we have to calculate mean water temperature for a given incubation period (see Appendix E); of course, knowledge of insemination time is also needed.

5.3 Normal Embryonic and Prelarval Development

Abnormal development of embryos can be due not only to poor or unfavorable incubation conditions but also to poor quality of eggs or incorrect biotechnics of insemination, when a large percentage of eggs is fertilized by more than one spermatozoon.

"Weak" eggs with a low turgor are easily damaged and produce numerous malformed embryos. External incubation conditions which may be close to threshold ones and still good for normal eggs can damage "weak" egg batches. Thus, the range of favorable conditions is narrower for weak eggs as compared with the good-quality ones. Weak or poor-quality eggs should not be used in experiments aimed at finding optimal incubation conditions.

As concerns malformations resulting from the penetration of supernumerary spermatozoa into the egg, at later stages embryos with such malformations cannot be practically distinguished from the embryos deformed due to other mechanisms. In order to exclude this source of malformation in assessment of the incubation conditions, percentage of polyspermic embryos has to be determined at the stage of four blastomeres for a given batch of eggs (see Appendix G) and this value has to be subtracted from the fertilization percentage.

If incubation of good-quality eggs inseminated by a semi-dry method with the correct sperm dosage still gives a considerable rate of developmental defects, this points to non-optimal incubation conditions.

Assessment of development can be performed at different stages, however, it is most convenient to take samples for this purpose at the end of gastrulation or beginning of neurulation (Plate III, 18; Plate IV, 19, 20), at the stage of heart rudiment formation (Plate V, 26, 27) and beginning of hatching (Plate VIII, 35, A, B). If conditions are good and gastrulation is synchronous, then in the majority of embryos the yolk plugs are approximately similar in size and the neural plate appears when the yolk plug has already disappeared and blastopore edges have closed. If, however, there is a big difference in the size of yolk plugs between different embryos and the neural plate forms in the presence of yolk plugs having different size (see Fig. 56), this means that incubation conditions were non-optimal: water temperature was too low, the incubator was overloaded with eggs, or the water flow was too slow. Such disturbances of development can occur even when water supply is interrupted for a short period of time during gastrulation.

Developmental abnormalities are easily detected at stages 26–27 (Plate V), stages of heart rudiment formation and separation of the rudiment of the posterior trunk and tail regions. At these stages embryos with the underdeveloped anterior body part (see Sects 2.6.3, 2.7.2) are easily identified. Such malformations point out to poor environmental conditions during maturation of spawners rather than during development.

Samples taken at these stages can be conveniently used for the assessment of egg quality, since defects of this type are very rare in good-quality eggs.

By the hatching stage some embryos with developmental defects have already perished and it is difficult to distinguish them from the dead unfertilized eggs. However, different types of malformations are conveniently distinguished at these stages: embryos with the edema of the pericardium, heart defects, bent embryos and others. Less prominent disturbances of embryo structure at the stage of hatching which become clear-cut only later, as well as disturbances due to incubation conditions but appearing at the larval stages, cannot be quantified at this time. The assessment of structure and viability of prelarvae during their growth (particularly, during transition to active feeding) can help to detect such developmental defects and in this way provide an additional criterion for the assessment of the conditions of maturation of breeders and egg incubation. However, in fact this criterion must be used with great caution, since conditions under which prelarvae live after hatching markedly affect their growth and development (see Sect. 3.7).

5.4 Losses During Incubation, Their Magnitude and Sources

The main source of egg losses during incubation is unfertilized eggs (activated and non-activated) which die during the period of gastrulation of the normal embryos. This loss depends on the quality of eggs and sperm but does not depend on incubation conditions. In contrast, death of developing embryos in a batch of good-quality eggs points to poor developmental conditions. In order to make any conclusions about incubation conditions from the magnitude of embryo losses we have to subtract losses due to the presence of unfertilized eggs (as well as the percentage of polyspermic eggs if this is a considerable fraction) from total losses.

Generally, it is very difficult to make accurate estimates of losses during the incubation period at the hatcheries: from the middle of incubation period degenerating unfertilized eggs are infected with *Saprolegnia*, and must be removed since actively spreading *Saprolegnia* can also infect living embryos. Therefore, counting dead embryos in a batch of eggs at the end of incubation gives underestimation of embryo losses. When chemical methods are used to control *Saprolegnia* (treatment with Malachite Green or other substances), such assessments can reflect true losses. However, even in such cases the development of *Saprolegnia* is not fully suppressed and, therefore, it is also necessary to collect and remove the infected eggs.

Another source of error should also be remembered: when embryos infected with *Saprolegnia* are siphoned off, a certain number of normal living embryos is also lost from the incubator together with the infected material. In any final comparison of the number of prelarvae in the incubator with the number of fertilized eggs placed there such embryos automatically will be counted as unexplained "losses". Of course it can be speculated that such losses may be due to temperature fluctuations or other unfavorable influences. Therefore the best estimates of losses for a given batch of eggs can be obtained by analysis of the sample taken from the incubator before mass development of *Saprolegnia* starts.

In order to analyze factors responsible for egg losses during incubation it would not be sufficient to analyze one such sample. It is best to also use samples that have been used for the control of typical morphology of development (see Appendix G).

Samples taken at the end of gastrulation or beginning of neurulation (stages 18–20) can be used for detecting the main source of losses due to unfertilized eggs (both non-activated and activated). Degenerative changes of such eggs are rather advanced at this time. With good-quality eggs and under favorable developmental conditions the magnitude of losses by the end of gastrulation should be of the same order as the number of unfertilized eggs during cleavage (this is why sometimes fertilization per cent is determined not during cleavage but at the time of gastrulation by counting the number of living embryos). However, in batches of "soft" eggs and under poor incubation conditions, the damaged eggs and embryos which die at the beginning of incubation are added to non-viable (dead) unfertilized eggs.

If during incubation of good eggs the percentage of losses at the end of gastrulation is considerably higher than percentage of unfertilized and polyspermic eggs in a given batch, this leads to the conclusion about poor conditions of development.

The second sample taken at the stage of heart formation (stages 26–27) should demonstrate whether losses due to death of fertilized developing eggs increase with incubation time. Such a conclusion can be made not only from the high percentage of dead eggs as compared with the first sample but also from the presence of damaged embryos, which have not yet decomposed. Their death usually follows from the rupture of yolk sack wall from which white yolk-rich cells leak out; the head of the embryo and axial organs of the trunk region remain distinguishable for a long time.

The third sample taken at the beginning of hatching serves to check, whether embryonic mortality is also associated with the second part of the incubation period. Two to three dozen dead embryos (eggs) have to be collected from such samples and outer membranes removed from them with forceps (usually these membranes are smeared with silt and therefore non-transparent). If this material does not contain any considerable number of dead embryos (with distinguishable head, dorsal region and tail), it can be safely concluded that there is no significant increase in losses.

With good-quality egg batches and under favorable developmental conditions, only a few malformed embryos do not survive the incubation. Under poor conditions the malformed embryos lose their viability earlier, whereas if the conditions deteriorate drastically, normal embryos are lost as well.

Appendix A–H

Appendix A: Quick Method for the Determination of Gonad Maturity in the Sturgeon (Kazanskii et al. 1978)

It is known that when pre-spawning stage IV of gonad maturity is completed (transition from incomplete to completed stage IV), polarization in oocytes of the older generation increases. The GV is displaced towards the animal pole and passes from the region of course-granular yolk to the region with fine-granular yolk inclusions. It has been demonstrated that only when oocytes reach this stage do they acquire capacity to mature in response to a single injection of the pituitary suspension. The technique for determining the degree of oocyte polarization with the purpose of selecting spawners for breeding was proposed by Trusov (1964b, 1967) and improved by Kazanskii et al. (1978).

The degree of oocyte polarization is determined using material collected from the caudal part of the ovary. The hollow probe is introduced at an angle of about 30°. Follicles taken out with the probe are boiled in water for 2 min in a test tube and then are cut with a sharp safety razor blade exactly along the axis passing through the animal and vegetal poles (Dettlaff et al. 1965). Then the following distances are measured under a dissecting microscope: (1) from the upper boundary of the GV to the oocyte membranes in the region of the animal pole and (2) from the animal pole to the vegetal pole (Fig. 77). The index of oocyte polarization is obtained with the formula: $l = A/B$. Such measurements should be made for several oocytes of each female.

Lower values of l correspond to higher polarization of oocytes and, therefore, to a more advanced stage IV of gonad maturity. It has been determined experimentally that the normal response of *A. gueldenstaedti* follicles to pituitary injection is observed, if l is below 0.07. The highest polarization of oocytes usually corresponds to $l = 0.03$–0.025.

Rough estimates can be made on the basis of another criterion, i.e., the ratio of A to the diameter of the GV. Oocytes in which A is equal to or less than 0.5 of the GV vertical diameter are usually ready to respond to a pituitary injection. Since, however, the shape of the GV varies, this criterion is less reliable.

Migration of GV into the region of fine-granular yolk may be used as an additional criterion for the assessment of the degree of polarization.

Authors of this method successfully used gradual injections with females in which l was above these values.

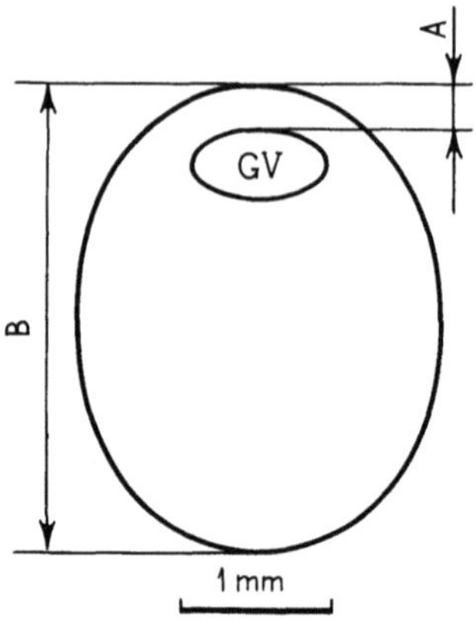

Fig. 77. Schematic representation of a cut through the oocyte taken from the gonad of IV stage of maturity. *GV* Germinal vesicle; for other designations, see text.

Appendix B: Duration of Oocyte Maturation Time in Vitro as a Criterion for Selecting Sturgeon Spawners for Breeding (B. F. Goncharov)

It has been demonstrated in experiments with *A. stellatus* collected in the lower Volga flow, that oocytes of different females have different duration of maturation in vitro in response to progesterone. It has also been demonstrated that when oocyte maturation lasts longer than 18 τ_0 (see Sect. 1.5.1) the quality of eggs collected after pituitary injection is usually poor. This observation provided a basis for the technique, making use of the gonad condition for vital identification of fish that are unsuitable for hatchery purposes (Goncharov 1976).

Assessment of the spawners by this method is performed in the following way.

A stock solution of progesterone is prepared in ethyl alcohol. Concentration of this solution is calculated as described: 0.01 ml of the solution should contain the amount of hormone necessary to give concentration of 5 μg/ml in the solution that will contain the follicles. For example, 10 mg of progesterone are dissolved in 2 ml of 96% ethyl alcohol; 0.01 ml of this solution contains 50 μg of the hormone and then, if 10 ml of Ringer solution is added to the dish, the necessary concentration 5 μg/ml will be obtained. Stock solution of progesterone in ethyl alcohol can be stored at room temperature throughout the breeding season. It should be remembered, however, that the vessel with progesterone solution should be tightly closed.

The necessary amount of the stock solution of progesterone is added with a micropipette onto the bottom of the dry dish and ethyl alcohol allowed to evaporate. Then Ringer solution modified for sturgeon is added to the dish (see Appendix C).

Removal of follicles from the female, incubation conditions, and recording of the results are described in Appendix C.

After 18 τ_0 follicles are fixed by boiling, cut along the animal-vegetal axis and the presence or absence of the GV in oocytes is noted; then the percentage of matured oocytes is determined. Females, in which 90 to 100% of oocytes have no GV by that time, can be used for breeding with high chances of success. Since the condition of oocytes in females is mostly uniform, it is enough if 15–20 oocytes are collected. The time of sampling 18 τ_0, as already mentioned, was determined for *A. stellatus*. It requires some adjustments for other Acipenserid species.

It should be remembered that this method for selecting spawners for breeding is not universal. It does not allow us to detect those females, in which poor quality of eggs may be due to degenerative changes. Therefore the effectiveness of this technique will be greater if the number of females with latent degenerative changes of oocytes is low.

Appendix C: Comparative Quantitative Determination of the Gonadotropic Activity in Sturgeon Pituitaries (Goncharov 1972)

Comparative quantitative estimation of the gonadotropic hormones in preparations of sturgeon pituitaries is based on their ability to induce oocyte maturation in certain media.

Modified Ringer solution for cold-blooded animals is used for the incubation of sturgeon follicles. This solution contains 6.5 g NaCl, 250 mg KCl, 300 mg $CaCl_2$, and 2 g of $NaHCO_3$ per liter of distilled water (Goncharov 1978). The addition of antibiotics, i.e., penicillin (500 000 units) and streptomycin (0.25 g) per liter of the solution is desirable but not absolutely necessary.

For testing, acetone-dried pituitaries are ground into a fine powder and the necessary amount of the powder is weighed. The measured volume of Ringer solution is added simultaneously to all preparations tested. The suspension of the pituitary acetone-dried powder is left at room temperature (20–22 °C) for 1 h to extract gonadotropins. Then sequential dilutions of the initial suspension are prepared in such a way that they span the whole concentration range, in which oocytes undergo maturation from 0 up to 100%.

Follicles can be conveniently removed from sturgeon ovaries with a hollow metal probe through a puncture in the female body wall. Under such conditions follicles are detached from one another. In order to wash follicles from the probe it is put into a wide test tube filled with the Ringer solution and the probe is gently rotated. After the necessary quantity of follicles are collected, they are washed with several portions of Ringer solution until the fluid becomes completely clear. Follicles are then transferred from the test tube into dishes with solutions to be tested using a glass pipette with a rubber bulb. Pipette diameter should not be much larger than the follicle diameter. If this condition is met, follicles are located in the pipette one above another and can be quickly counted. The same number of follicles has to be put into the tested solutions (no more than five or six per 1 ml of the solution).

Follicles have to be put into dishes with the minimal amount of Ringer solution. Any contact between the pipette and surface of the hormone-containing solutions should be avoided by all means. Follicles extracted from the female gradually lose their sensitivity to gonadotropic hormones, therefore they should be put into tested solutions as quickly as possible.

It should be remembered that the follicles have to be equally distributed over the bottom of the dish and completely covered by the solution. Throughout the whole incubation period the temperature should be within the range of spawning temperatures. After 24–36 h of incubation in the tested solutions the follicles are fixed by boiling. Then they are cut by a sharp safety razor blade along the animal-vegetal axis under a dissecting microscope. Forceps are used to handle the follicles and the presence or absence of the GV in oocytes is recorded. If the GV is not seen on the section, then the halves of the oocyte must also be cut to be sure that it is not away from the plane of the cut. Percentage of matured oocytes is calculated for each sample; oocytes in which the germinal vesicles underwent breakdown are classified as mature.

In order to estimate the significance of the differences in the results obtained from various solutions, the results are treated statistically. It is recommended that the method of probit-analysis be used (Belen'kii 1963). This involves determination of the effective dose leading to maturation of 50% oocytes (ED 50) and confidence intervals are selected for a certain significance level. The ratio of ED 50 of the tested preparations characterizes the relative content of gonadotropic activity in every such preparation.

Since the dependence between gonadotropin dose and its effect on the maturation of sturgeon oocytes in vitro is generally rather steep, two or even one significant point (percentage of matured oocytes differing from 0 and 100%) can be obtained by using two-fold dilutions. Therefore, to improve the accuracy of determination it is preferable to use concentrations differing by less than two. It is also clear that the accuracy of the assay will be greater if more follicles are used to determine the effect of each concentration.

This method is appropriate for the comparative evaluation of gonadotropic activity in sturgeon pituitaries. The activity, however, cannot be expressed in standard units, because follicles of different females differ in sensitivity to gonadotropins. However, comparable data can be obtained in experiments with follicles of different females if a "reference" preparation is used throughout the experimental season. This "reference" preparation should be kept under conditions that prevent the loss of activity, i.e., in a sealed container in the refrigerator.

The capacity of sturgeon gonadotropin to induce oocyte maturation in amphibians (frogs and toads) provides an opportunity for using these species as test objects for comparative studies of gonadotropic activity.

Appendix D: Relative Characteristic of Developmental Duration

In this book we have repeatedly used the relative characteristic of developmental duration for various periods of development, since it allows us to calculate the duration of these periods at different temperatures without making additional observations. The duration of periods from the beginning of hormone action and up to certain stages of oocyte maturation or from insemination to a certain stage of embryonic development or from mass hatching up to certain stages of prelarval development is given in this book not only in astronomical time units (days, hours, or minutes) at a certain temperature but also in the number of τ_0. τ_0 is the duration of one mitotic cycle during synchronous cleavage divisions (see Sect. 2.4). Since it has been demonstrated (Dettlaff and Dettlaff 1961; Dettlaff et al. 1987) that the duration of various developmental periods changes proportionally in the range of spawning temperatures, when we know the duration of a certain period in τ_0 at a certain temperature, its duration at a different temperature can be easily calculated. For this purpose we have to use the curve shown in Fig. 42 from which the value of τ_0 at the required temperature can be found. This should be multiplied by the number of τ_0 corresponding to the duration of this period.

We give the following example: stage 13 in *A. gueldenstaedti* embryos begins 19.5 τ_0 after insemination; at 18 °C this time is 16 h 15 min. We may be interested in finding the duration of a period from insemination to that stage at 12 °C. In order to find this time we have to use the curve in Fig. 42A relating the τ_0 value to temperature in the *A. gueldenstaedti* and to find the value of τ_0 at 12 °C. For this purpose, we have to find the point on the curve corresponding to the abscissa of 12 °C and, then, the corresponding value on the ordinate. Specifically, in this case τ_0 at 12 °C is 85 min. When we multiply 85 by 19.5 we find that the duration of this period at 12 °C will be 27 h 37 min.

Appendix E: Determination of Time for Injection of Pituitary Suspension and Inspection of Females After Pituitary Injection for Obtaining Eggs at the Optimal Time at Different Temperatures

E.1: Determination of Time for Injection of Pituitary Suspension to *A. gueldenstaedti*, *A. stellatus*, *A. ruthenus*, and *H. huso* Females for Obtaining Eggs by the Beginning of the Working Day (Dettlaff et al. 1965)

1. Calculate the mean temperature 1 day before the injection.
2. On the horizontal axis of the plot describing the egg maturation for the corresponding species (Fig. 78A–D) find the point corresponding to this

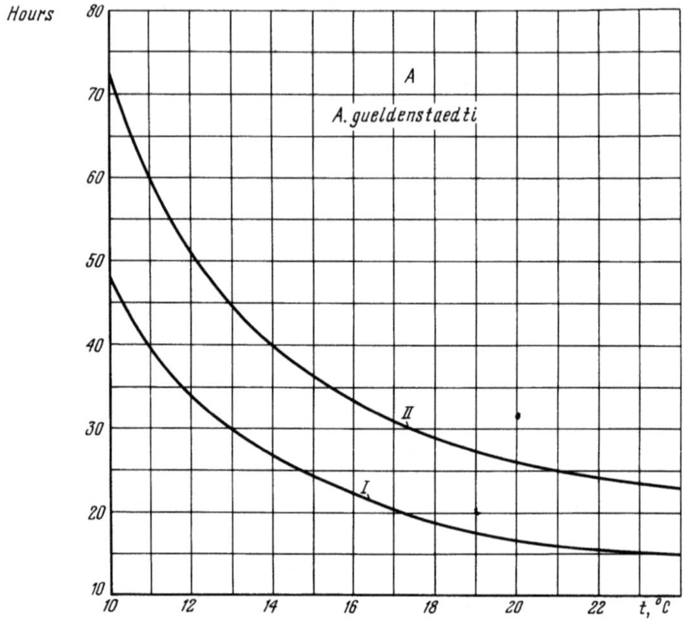
Hours
A
A. gueldenstaedti
II
I
t, °C

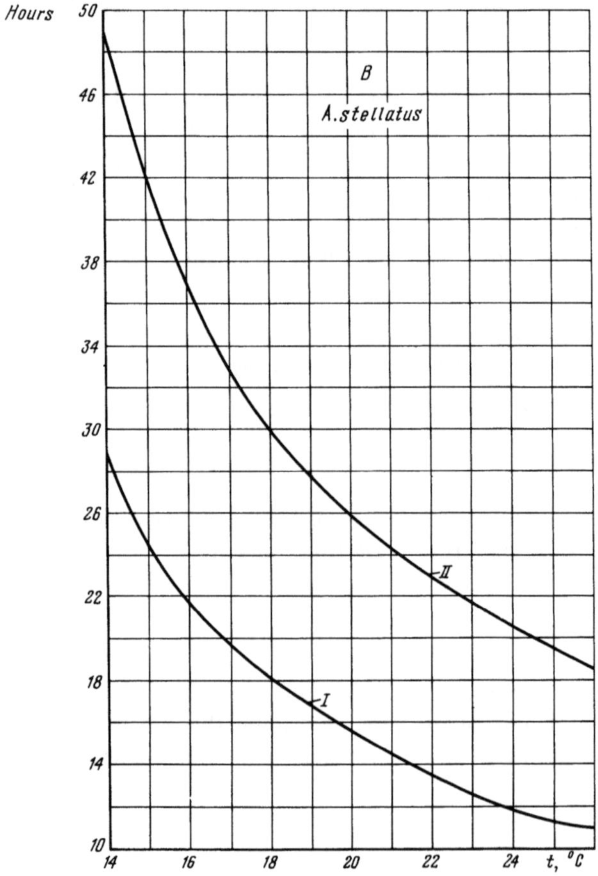
Hours
B
A. stellatus
II
I
t, °C

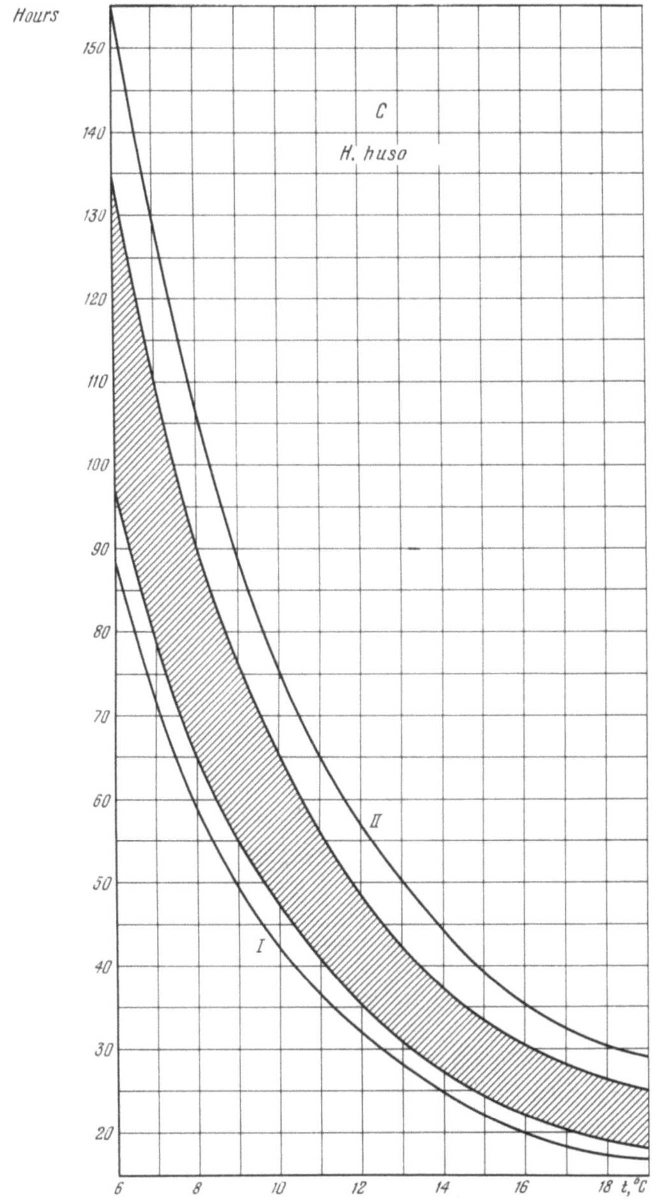

Fig. 78A–D. Duration of maturation of **A** *A. gueldenstaedti*; **B** *A. stellatus*; **C** *H. huso*; **D** *A. ruthenus* females as a function of temperature (**C**, Igumnova 1974; **D**, Igumnova, pers. comm.). *Curve I* maturation time for the first females; *curve II* for the females that did not mature by this time the probability of collecting good quality eggs is low. *Abscissa*, mean temperature during maturation, °C; *ordinate*, time after pituitary injection to maturation (in hours). **C** Hatched band designates the time interval during which most females mature

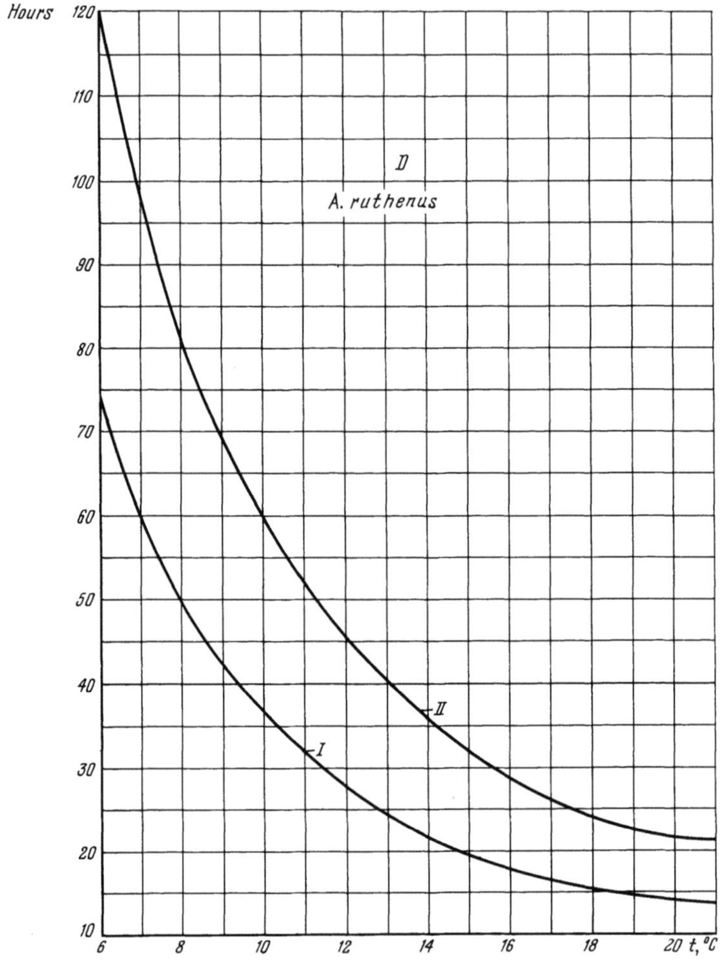

78D

temperature and draw a perpendicular up to intersection with the curve I. Draw a perpendicular from the intersection point to the vertical axis and read the number of hours necessary at a given temperature for the maturation of first females after the injection.

3. Substract the obtained number of hours from the hour when the females should be ready for subsequent work. The resulting value corresponds to the time when females should be injected.

E.2: Determination of Time for Inspection of Females and Collecting Eggs

1. Calculate the mean temperature during maturation period since this temperature can differ from the mean temperature before injection. For this purpose

at 7 p.m. before the day of collecting eggs and at 7 a.m. on the day of collecting eggs calculate the mean temperature for the period beginning from the time of injection.
2. Find on the horizontal axis the point corresponding to the mean temperature during the maturation period and draw a perpendicular up to intersection with the curves.
3. The point of intersection of the perpendicular with curve I will show when the first females become mature. This number of hours should be added to the time of injection and we will obtain time when the inspection of females should be started.
4. Intersection with curve II shows the time after which the probability of obtaining eggs from females, which do not show signs of maturation, is very low.
5. For determining the time of the female dissection within the time period between curves I and II we have to use features of maturation common in fish breeding: soft abdomen, eggs are ejected as a jet, when the female is raised by its tail its abdomen is strongly retracted.

The use of the curves markedly reduces the number of necessary female inspections (they should not be inspected before the time corresponding to curve I). This helps to determine more accurately the proper time for collecting the eggs. In this way losses of eggs at the hatcheries which may be due to their "undermaturation" or "overmaturation" are reduced.

E.3: Example for Calculation of the Maturation Time and Time of Female Inspection

Females of *A. gueldenstaedti* were injected on the 5th of May at 8 a.m. at 14.3 °C. From Fig. 78A the approximate time can be determined during which first females mature at this temperatures. For this purpose a perpendicular is drawn from the horizontal axis point corresponding to 14.3 °C up to the intersection with the lower curve (I). This corresponds to 26 h. Then if the temperature after the injection is exactly 14.3 °C the female must be inspected for the first time 26 h after the injection. However, the temperature usually does not remain constant. If the temperature starts decreasing after the injection, this time will be extended, whereas if the temperature rises, the time will be reduced. Therefore at 20–22 h after the injection the mean temperature over the maturation period must be calculated and the time of female inspection should be revised on the basis of the graph used.

The mean temperature throughout this period should be the arithmetic mean temperature, since at the hatcheries the temperature is usually measured at 7 a.m., 1 p.m. and 7 p.m. However, the duration of intervals between measurements has to be properly adjusted. This must be done in the following way: for measurements made at 7 p.m. and 7 a.m. on the next day one intermediate point should be added (×) corresponding to 1 a.m. In order to obtain temperature at this time we have to add the temperatures at 7 p.m. and 7 a.m. and divide the sum by two. In addition, we have to add the temperature on the first and then on last day of a given measurement period divided by two and to use this temperature instead of temperature of the last point, the first point has to be discarded. These temperature values have to be summed and their sum divided by the number of observations.

Date	Time, h	t, °C
May 5	7	14.3 (do not add)
	13	14.5
	19	15.6
	x	15.1
May 6	7	14.6
	13	14.8
	19	16.0
	x	15.7
May 7	7	15.4
	13	16.0
	19	(16.3) replace by $\frac{14.3 + 16.3}{2} = 15.3$
		153.0: 10 = 15.3 °C

The mean temperature over the period in question equals 15.3 °C. If temperature measurements were conducted at different time intervals, this technique should not be used, particularly if temperatures differ very much. In order to determine the mean temperature under these conditions we have to calculate the sum total of the product degrees × hours and to divide it by the sum total of hours.

An example of the calculation is shown below:

Date	Time, h	t, °C	Product of the mean temperature over a given time interval by the number of hours between measurements (degrees × hours)
May 7	9	14.3	14.9 × 3 = 44.7
	12	15.5	15.6 × 8 = 124.8
	20	15.7	15.0 × 11 = 165.0
May 8	7	14.3	
			Total 22 h 334.5°C × h

The mean temperature is then equal to 334.5: 22 = 15.2 °C

Assume that the mean temperature over time which passed after the injection is 15.2 °C. Now on the graph (Fig. 78A) we find the intersection points of the perpendicular drawn from the horizontal axis point corresponding to 15.2 °C with the curves I and II. The intersection point with curve I corresponds to 24 h and with curve II to 35 h. This means that females must first be inspected after 24 h, whereas keeping of females that show no symptoms of maturation by 35 h should be discontinued.

Appendix F: Artificial Insemination of Sturgeon Eggs (Ginsburg 1963)

1. The following equipment is necessary for performing insemination: dry pans for eggs, dry vessels for the sperm (aluminum or preferably enamelled dippers); enamel-coated bucket, measuring cylinder for 100 ml, microscope, several glass slides, a glass rod, and a pipette.
2. Eggs have to be collected in dry pans individually from each female. It is not recommended to pool eggs from different females, since the quality of eggs cannot always be determined by visual inspection and, as a result, a batch of good-quality eggs can be mixed with that of bad-quality ones giving a low fertilization rate. This leads to the deterioration of the conditions for the incubation of good-quality eggs and the care of eggs is more complicated.

Eggs have to be placed in a cool place and protected from direct sunlight. In order to ensure good results, insemination should be performed without delay, preferably no later than 10 to 20 min after collecting eggs.

3. Sperm has to be collected in dry vessels, preferably individually from each male. Sperm with a milky-white or slightly yellowish color has to be used for insemination.

Under the conditions of male shortage, when sperm of only one or two males can be used for insemination of a given batch of eggs, examination of sperm quality under the microscope is needed. Sometimes even sperm with good visual appearance may be inactive. If eggs are inseminated with a mixture of thick sperm from three to five males, then it is not necessary to verify its quality under the microscope.

Under conditions of acute shortage of males, sperm can be kept for some time on ice as an exception. When sperm is collected for subsequent storage, it is particularly important to prevent its contact with water. The sperm is poured in clean dry glass vessels such as wide test tubes or beakers (individual beaker has to be used for each male). These containers are then tightly sealed and placed on ice. Sperm of good quality can usually be used for insemination over 2 or 3 days. Since sperm quality gradually deteriorates during storage, the capacity of the spermatozoa to undergo activation after the addition of water has to be checked under a microscope before use.

Control of Sperm Quality

A small drop of the sperm should be picked up by a dry glass rod and put onto a glass slide under the microscope. The iris diaphragm of the microscope condenser should be closed in such a way that spermatozoa can be clearly distinguished. A drop of water is added to the sperm with a pipette, the sperm mixed with the rod and immediately viewed under the microscope. When sperm is of good quality, all or almost all spermatozoa show active forward motion.

4. Mixture of the sperm from three to five males is prepared (10 ml of the sperm per kg of eggs) and diluted with water 200 times. In practical terms, for insemination of any amount of eggs below 4 kg, 40 ml of "dry" sperm is diluted in a bucket of water.

If sperm quality is less than optimal (after dilution with water a large proportion of spermatozoa remains immobile), then about 100 ml of the "dry" sperm should be diluted in one bucket of water. If only very liquid, watery sperm is available, the maximum amount should be taken and diluted in the amount of water fully covering the eggs.

Insemination should be performed in the following way: the excess coelomic fluid should be decanted from the pan with eggs. The proper amount of the sperm should be measured with a cylinder, poured into the bucket with water, quickly mixed and immediately added to eggs. Eggs have to be carefully agitated with the diluted sperm for 3–5 min. Thereafter water with the sperm has to be decanted and demucilaging of eggs started.

Appendix G: Sampling for Determination of Fertilization Percentage, Extent of Losses, and Percentage of Abnormal Embryos

Determination of Fertilization Percentage

For determining the fertilization percentage samples should best be collected at the stage of the 2nd cleavage division (4 blastomeres, stage 7 in Fig. 79). The time of sampling can be determined from the curves in Fig. 80, showing temperature dependence of the onset time for the 2nd and 3rd cleavage divisions in eggs of *A. gueldenstaedti* (A), *A. stellatus* (B) and *H. huso.* (C) and *A. ruthenus* (D).

1. Calculate mean temperature over the time period after insemination (see Appendix E).
2. On the horizontal axis in Fig. 80 find the point corresponding to this temperature and draw a perpendicular up to intersection with the curves. Samples have to be collected during the interval limited by curves I and III close to curve II, i.e., at the stage of the completed 2nd division.

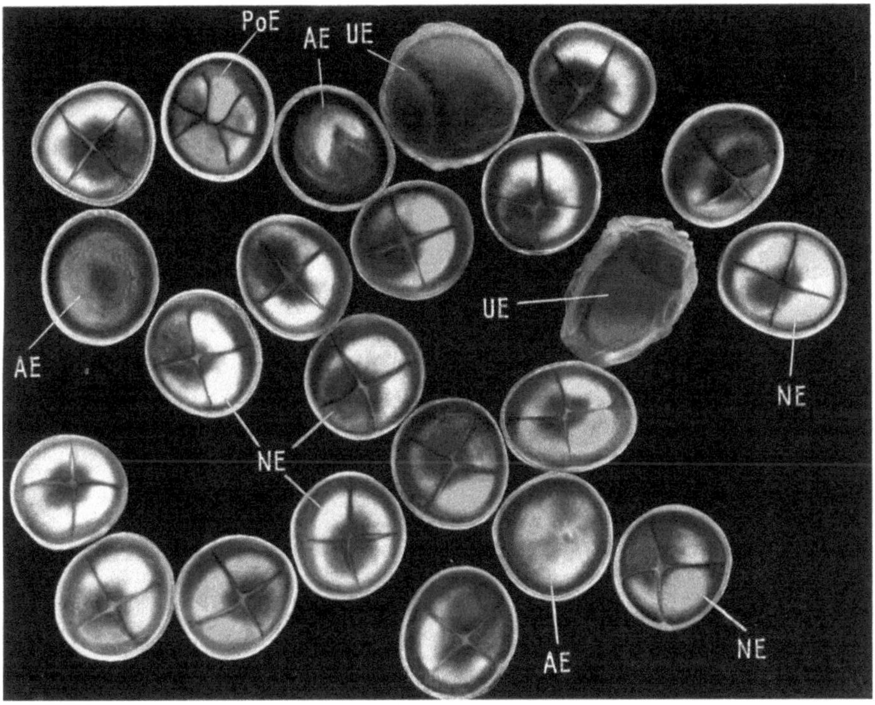

Fig. 79. A sample of *A. gueldenstaedti colchicus* eggs at the stage of second cleavage division (Ginsburg and Dettlaff 1955). *AE* Activated egg; *NE* normal egg; *PoE* polyspermic egg (with the number of blastomeres exceeding the norm for a given stage); *UE* unactivated egg

Before the sample is collected, thoroughly mix eggs in the incubator and take 200 to 300 eggs.

3. First the material should be sorted out by separating cleaving and non-cleaving eggs, then they have to be counted and the fertilization percentage determined. This corresponds to the percentage of cleaving eggs. Activated unfertilized eggs begin to cleave with a delay (see Sect. 2.4) and at the recommended time usually they have no furrows on their surface.
4. Sometimes it is necessary to determine the percentage of polyspermic eggs, which show abnormal development (see Sect. 2.4) and are a source of losses during incubation. With this in view a sample of eggs is fixed with formaldehyde (one part of 40% formaldehyde is mixed with nine parts water).

In a time interval limited by curves I and III in Fig. 80, normal eggs fertilized with one spermatozoon show four blastomeres, whereas polyspermic eggs have six and more (see Fig. 43A′–E′). If the technique of insemination was correct, usually no more than 4–6% of polyspermic eggs are present in a batch of good-quality eggs.

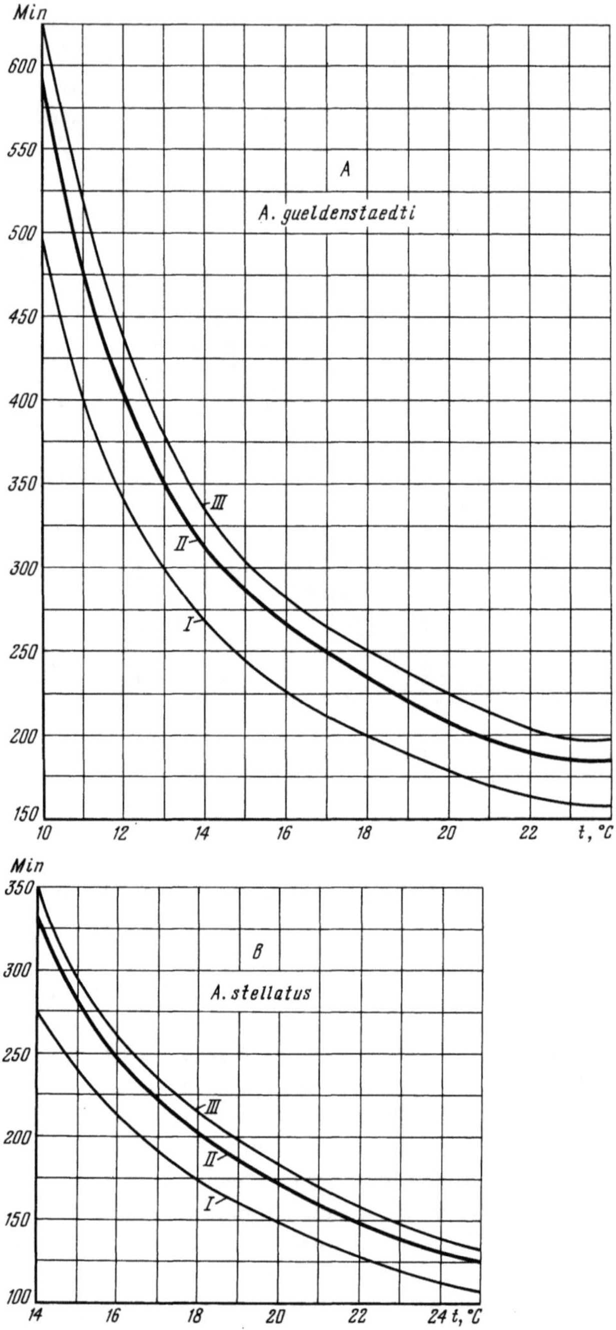

Fig. 80. Time of collecting samples to determine percentage of fertilization in **A** *A. gueldenstaedti*, **B** *A. stellatus*, **C** *H. huso* and **D** *A. ruthenus* depending on the incubation temperature (**A, B** Dettlaff 1965a; **C** Igumnova 1975b; **D** Igumnova pers. comm.). *Curves* show time from insemination to the appearance of the cleavage furrows of the second (*I*) and third (*III*) cleavage division. *II* Best time for collecting a sample. *Ordinate*, Time after insemination, min; *abscissa*, mean temperature, °C

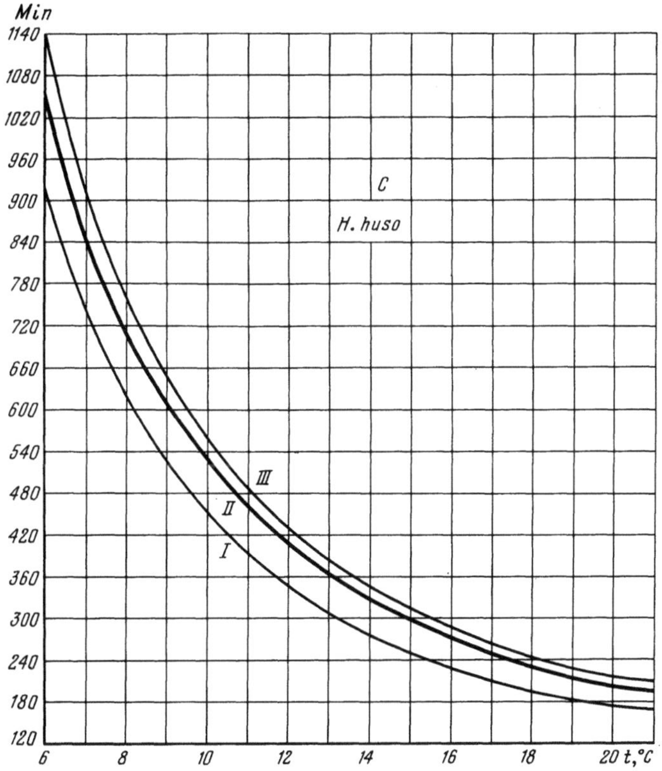

Min
1140
1080
1020
960
900
840
780
720
660
600
540
480
420
360
300
240
180
120
6
8
10
12
14
16
18
20 t,°C
C
H. huso
III
II
I

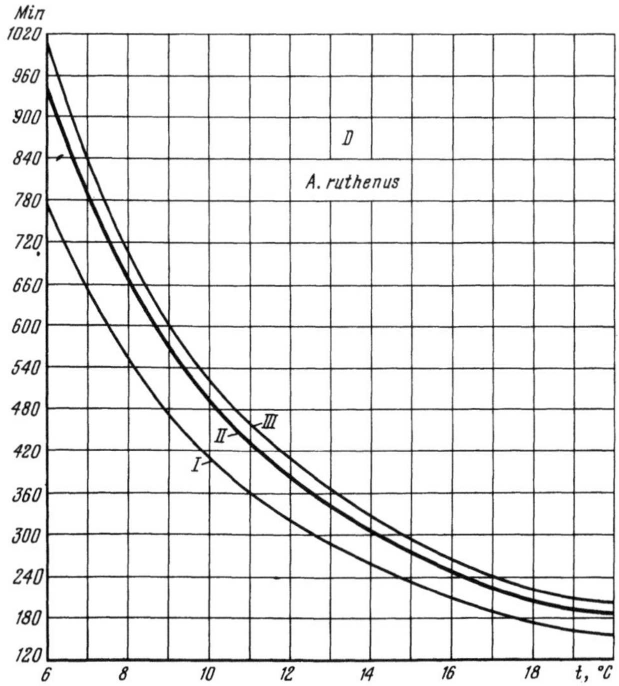

Min
1020
960
900
840
780
720
660
600
540
480
420
360
300
240
180
120
6
8
10
12
14
16
18
t, °C
D
A. ruthenus
III
II
I

Estimation of Losses and Determination of the Percentage of Abnormally Developing Embryos

Usually it is recommended to collect several samples: (1) at the stages of late gastrula and early neurula (stages 18–19, Plates III, IV), (2) at the period of heart rudiment formation (stages 26–27, Plate V) and (3) before the beginning of hatching (stage 35, Plate VIII).

Time for collecting these samples can be determined from curves on Fig. 81 (for samples 1 and 2 from curves II and III, slightly above, and for sample 3 from curve IV). Before sampling eggs have to be thoroughly stirred in the incubators, since dead eggs usually assemble together among the living ones. No fewer than 200 to 300 eggs must be taken.

Sample 1 (Stages 18–19)

Usually all unfertilized eggs die off by the end of gastrulation and typically developing embryos do not have yolk plugs or these plugs are very small.

First, dead eggs have to be separated from the sample. They usually have a marbled appearance and their percentage is to be determined from the total number of eggs in the sample (this value characterizes losses mainly due to unfertilized eggs). Then developing embryos with malformations like the large yolk plug of an irregular shape have to be counted. Their number depends on the quality of eggs and conditions of insemination rather than on incubation conditions.

In the remaining part of the sample we usually inspect how similar and typical is the structure of different embryos. If embryos with yolk plugs of various sizes are present and neurulation starts before the yolk plug has fully submerged inside the embryo, this provides evidence for unfavorable developmental conditions (overload of the incubator with eggs, poor flow conditions or wrong temperature). If the conditions are improved, such deviations may gradually disappear.

Sample 2 (Stages 26–27)

If *Saprolegnia* has not been removed before the onset of these stages in the incubator (usually this is not done), then the percentage of losses can be determined. For this purpose degenerated eggs and dead embryos have to be separated. Particular attention should be given to dead embryos (in these it is often possible to distinguish the head and axial structures of the trunk region among light yolk-rich cells forming a loose mass). Under good incubation conditions, losses in sample 2 are usually equal to or only slightly higher than in the first one.

In living embryos at these stages disturbances in head development can be easily determined (shortening, underdevelopment or complete absence of anterior parts of the body). The number of embryos with defects of this type characterizes the quality of eggs rather than incubation conditions.

Sample 3 (Stage 35)

In order to quantify the number of embryos with drastic structural defects at the end of the incubation period, 300 to 500 embryos have to be collected (after thorough agitation of eggs in the incubator) and put into a glass jar with fresh water. Hatching under these conditions lasts at least 24 h and, therefore, water should be replaced every few hours (beforehand separating the hatched prelarvae and placing them into a different jar). Excess warming up of water should be avoided, since this can lead to paralysis and embryos will have a curved shape. When the majority of prelarvae have already been hatched, they have to be placed into another vessel and the eggs remaining in the jar examined. This material always contains dead unfertilized eggs. It can also contain dead embryos (usually with drastically abnormal structure) as well as living embryos which are still enclosed in membranes. If the hatching is not yet complete, then some of these embryos can have normal structure. Membranes have to be removed from all non-hatched embryos with sharpened forceps and embryos examined.

By counting the total number of living prelarvae and embryos remaining in membranes, the percentage of abnormal embryos is calculated. When the quality of eggs is good and incubation conditions are favorable, the percentage of abnormal embryos is very low.

Appendix H: Duration of Incubation of *A. gueldenstaedti*, *A. stellatus*, *H. huso*, and *A. ruthenus* Embryos at Different Temperatures and Estimation of the Incubation Conditions from the Rate of Development

1. Calculate the mean temperature over the period of incubation.
2. Find the point corresponding to the mean temperature over the incubation period on the horizontal axis (Fig. 81) and draw a perpendicular up to the intersection with curve IV; from the intersection draw a perpendicular to the vertical axis and read the number of hours post-insemination at a given mean temperature for hatching of the first few embryos.
3. Add 15–25% to the time estimate found to obtain time from insemination to the end of mass hatching. Duration of the hatching period varies with different batches of eggs and cannot be predicted accurately. It depends not only on the temperature but also on the quality of eggs, load of the incubator and, particularly, on the water flow rate and agitation intensity of eggs during the period immediately preceding hatching.

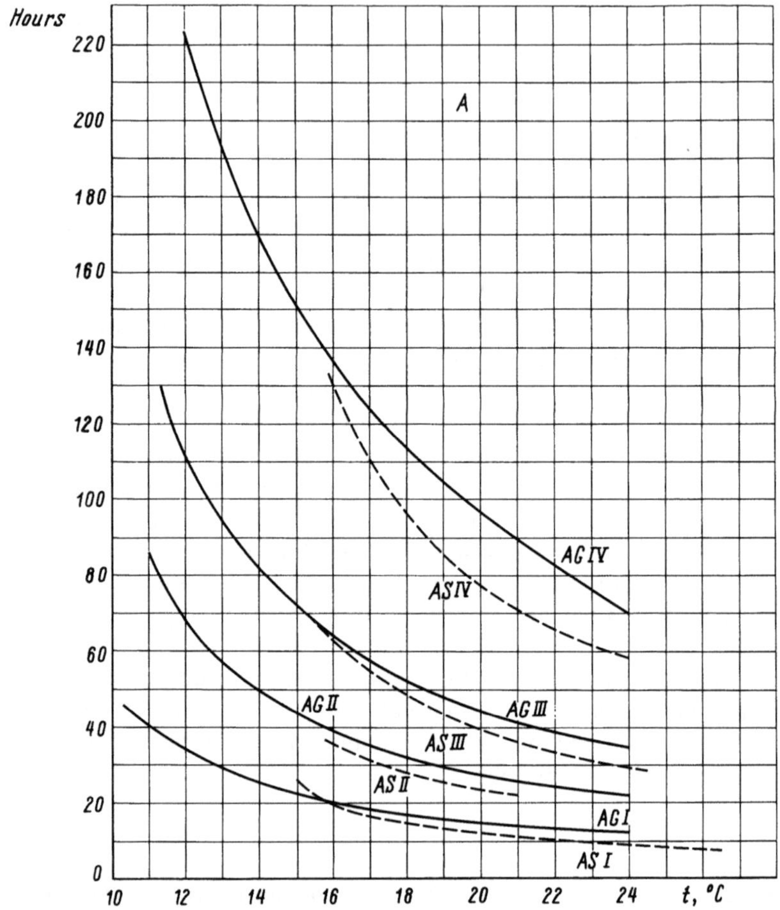

81A

Fig. 81. Duration of development of the embryos of **A** *A. gueldenstaedti* (*AG*), *A. stellatus* (*AS*), **B** *Huso huso* (*HH*), *A. ruthenus* (*AR*), and the time of collecting samples for estimation the normality of development as a function of temperature (**A** Dettlaff 1965b, **B** Igumnova, 1975a, 1985a). *I* Time from insemination to the beginning of gastrulation (stage 13); *II* end of gastrulation (stage 18); *III* stage of fusion of lateral plates before the information of heart rudiments (stage 26); *IV* stage of hatching of individual prelarvae (stage 35). *Ordinate*, time after insemination, hours; *abscissa*, mean temperature, °C

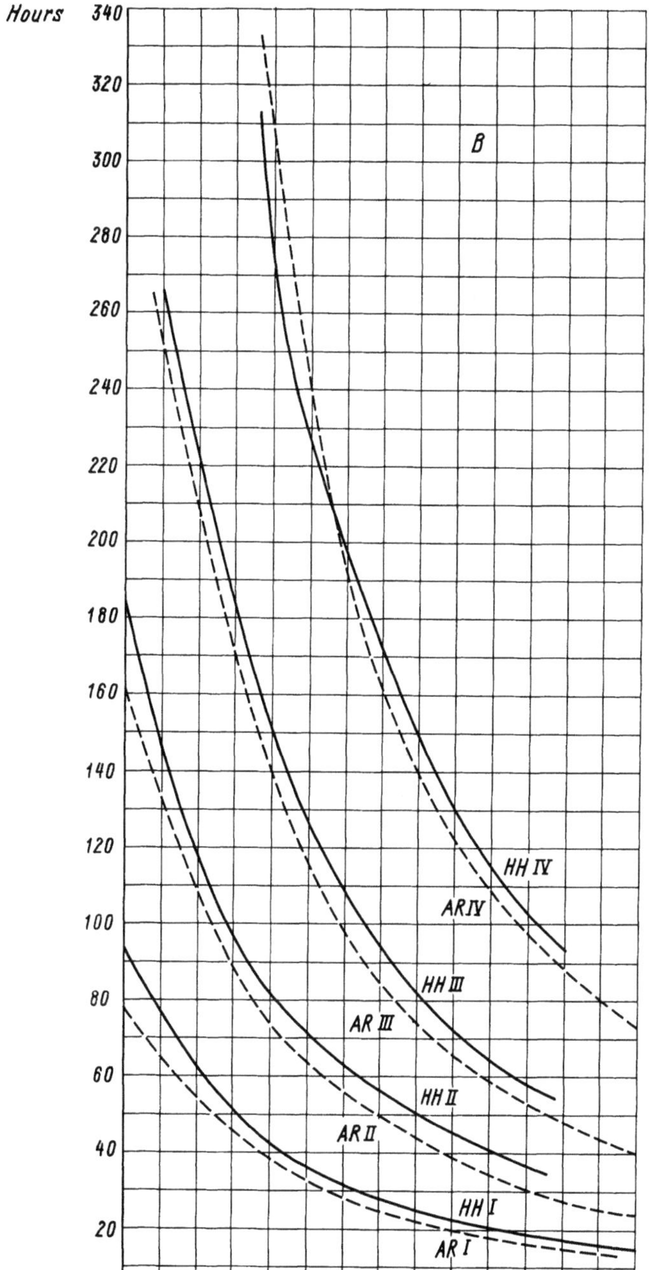
Hours
340
320
300
280
260
240
220
200
180
160
140
120
100
80
60
40
20
0
6
8
10
12
14
16
18
t, °C
B
HH IV
AR IV
HH III
AR III
HH II
AR II
HH I
AR I

81B

These data allow us to predict hatching time in batches of eggs developing at different temperatures and to determine the throughput of the incubator as well as time when prelarvae have to be transferred from incubators into tanks for subsequent rearing.

4. Overloading the incubator with eggs and insufficient water flow create oxygen deficiency. As a result, embryos develop more slowly than under the normal incubation conditions. Comparison of the development duration (in a given batch of eggs) with the normal duration of development at the same temperature (Fig. 81) makes it possible to assess incubation conditions.

Plates I –XX

Plates I–VIII. Normal embryonic development of *A. gueldenstaedti colchicus. Numerals* below the drawings correspond to the numbers of developmental stages. For other abbreviations, refer to the list of abbreviations

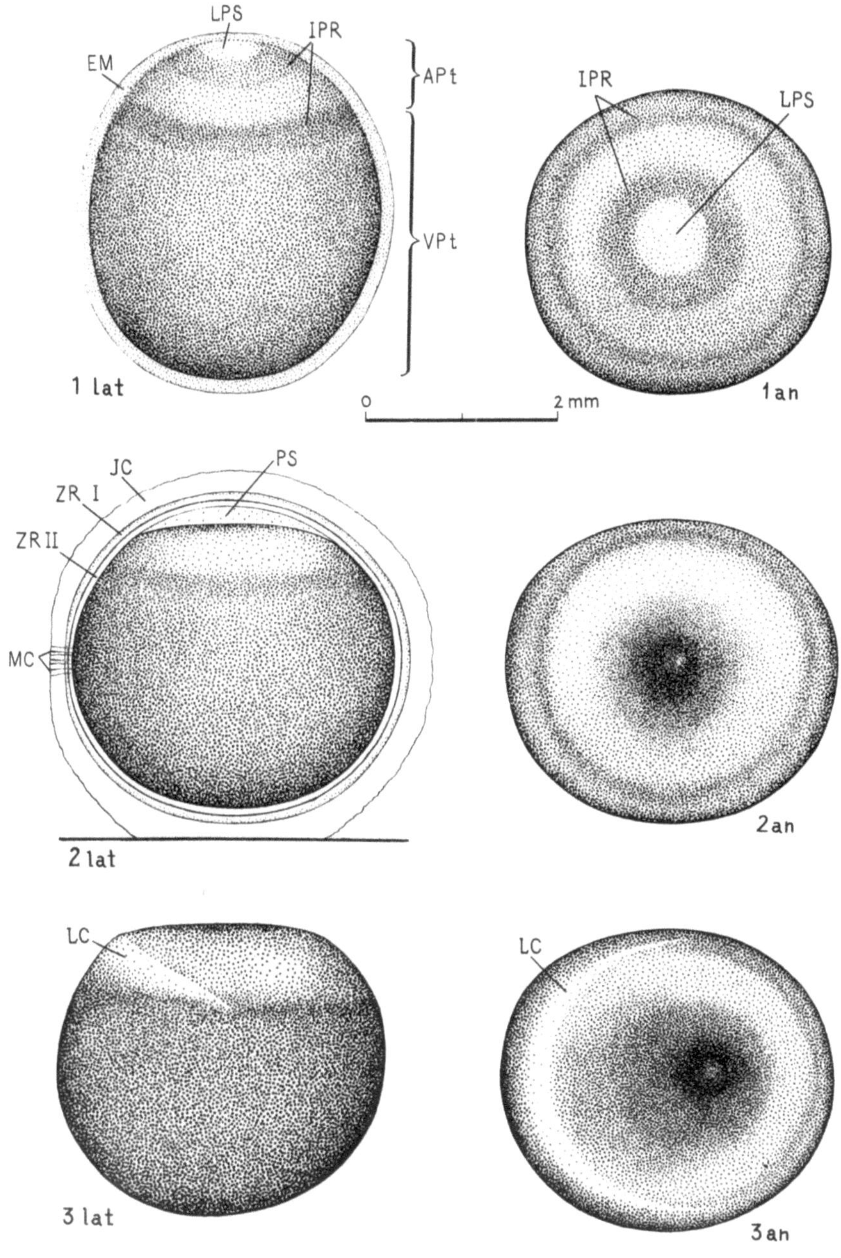

Plate I. An unfertilized egg (stage 1); during the first minutes after insemination the egg has a similar external appearance. Egg after turning and secretion of hydrophilic colloid (stage 2) and at the light crescent stage (stage 3)

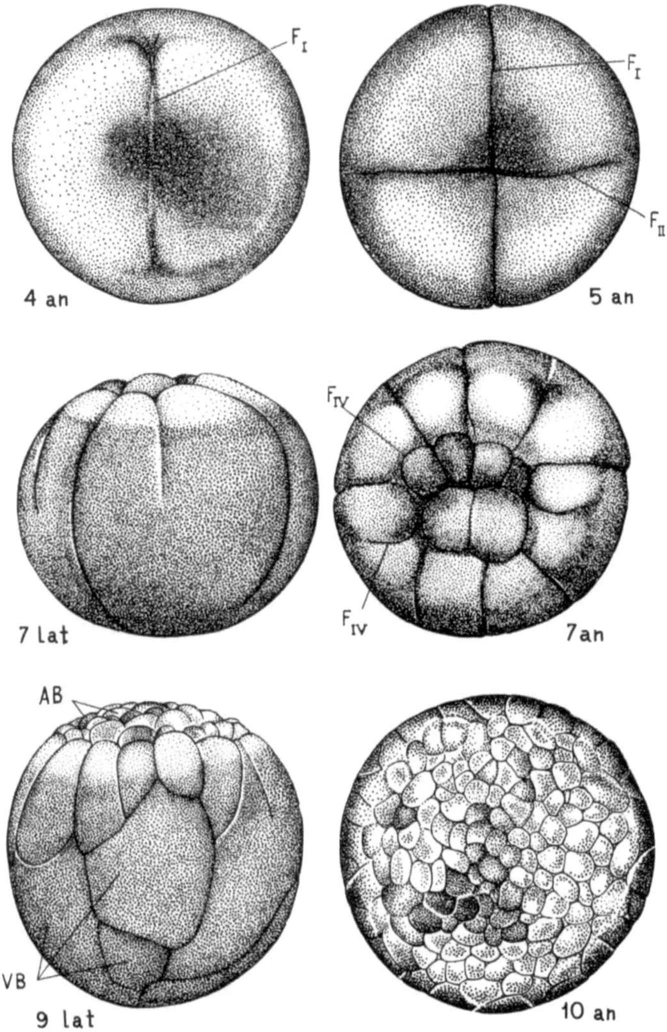
F_I
4 an
F_I
F_II
5 an
7 lat
F_IV
F_IV
7an
AB
VB
9 lat
10 an

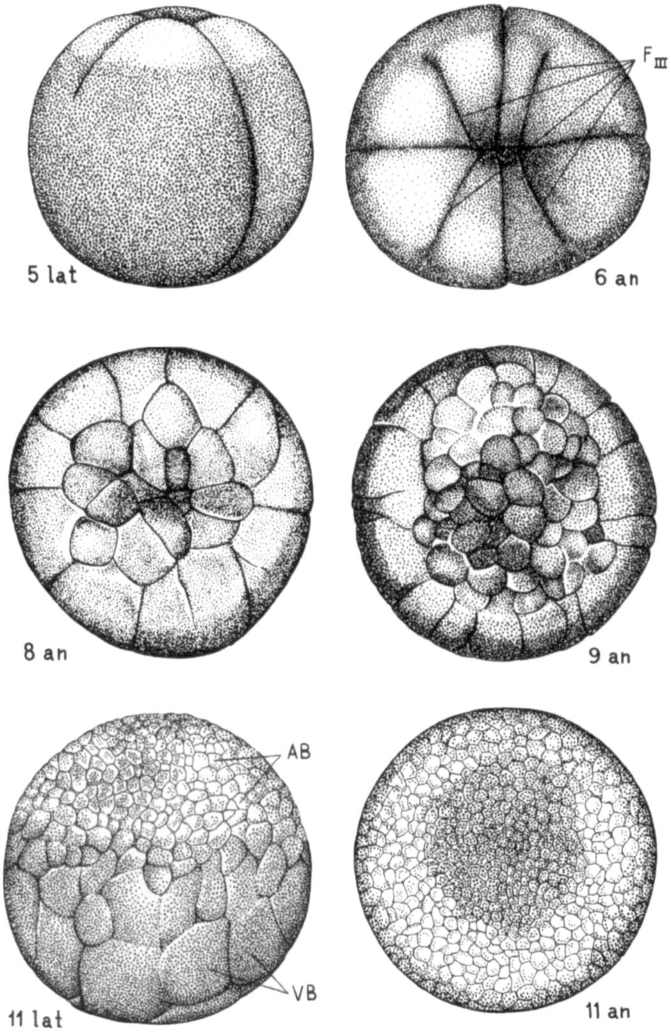

Plate II. Stages of the 1st (stage 4), 2nd (stage 5), 3rd (stage 6), 4th (stage 7), 5th (stage 8), and 7th (stage 9) cleavage divisions; late cleavage (stage 10); stage of early blastula (stage 11)

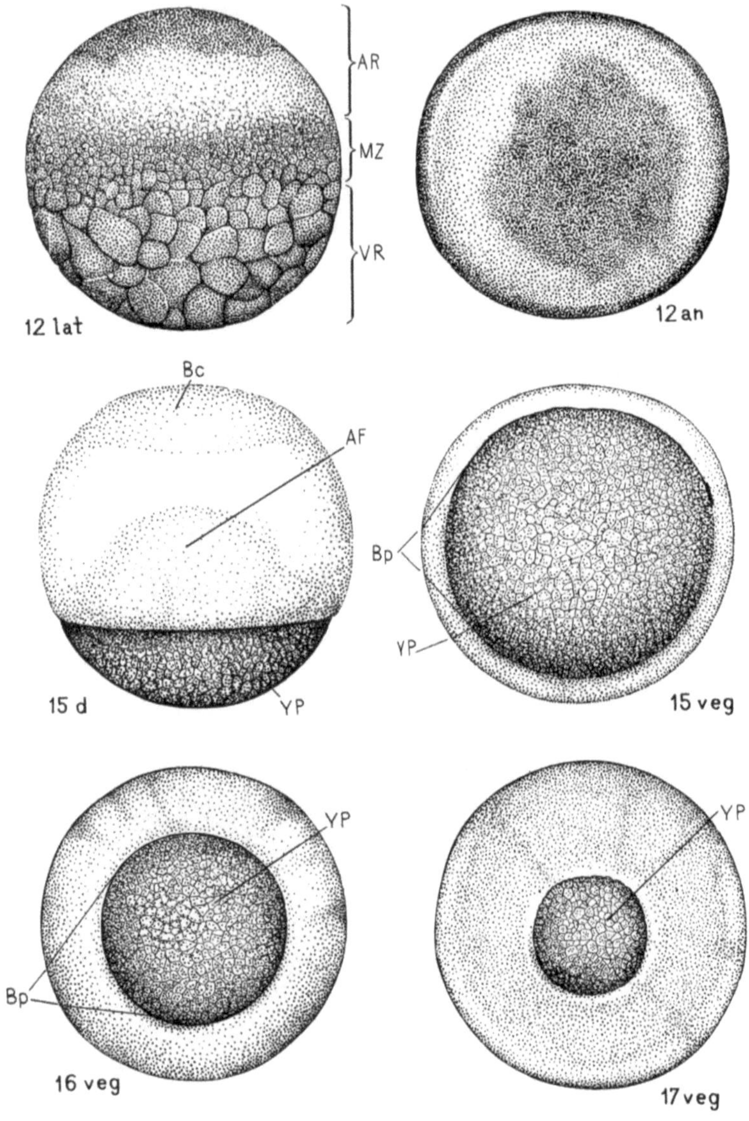
AR
MZ
VR
12 lat
12 an
Bc
AF
15 d
YP
Bp
YP
15 veg
YP
Bp
16 veg
YP
17 veg

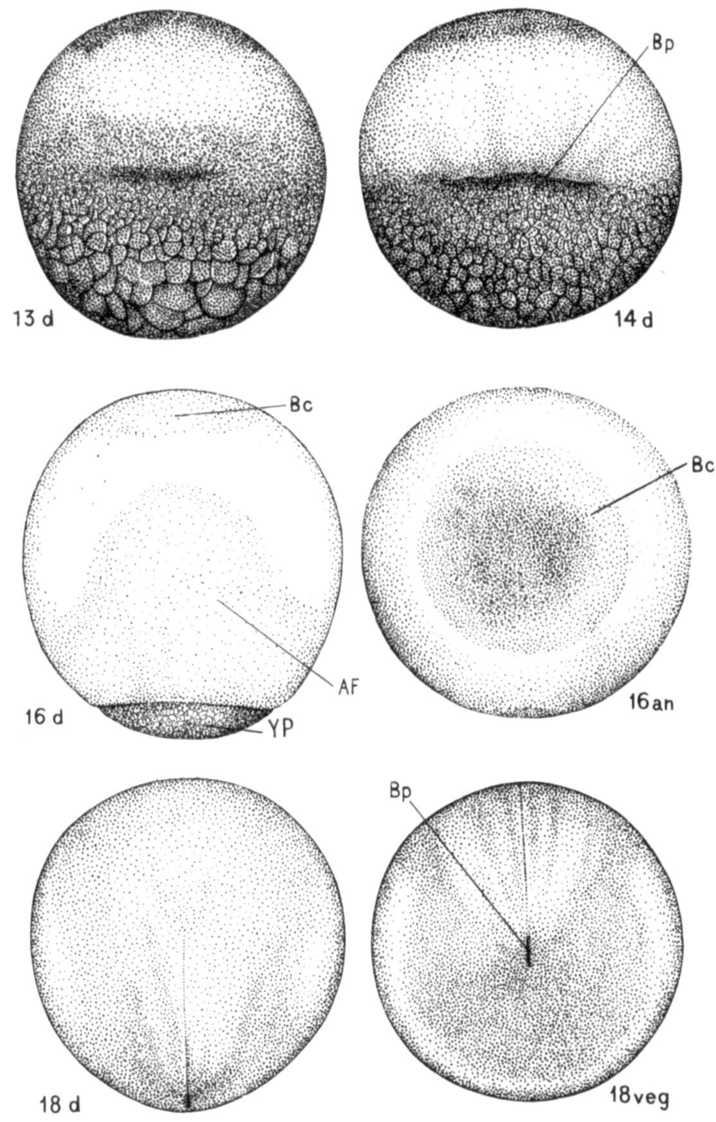

Plate III. Stage of late blastula (stage 12); onset of gastrulation (stage 13); early (stage 14) and middle (stage 15) gastrula; stages of large (stage 16) and small (stage 17) yolk plug; stage of slit-like blastopore (stage 18)

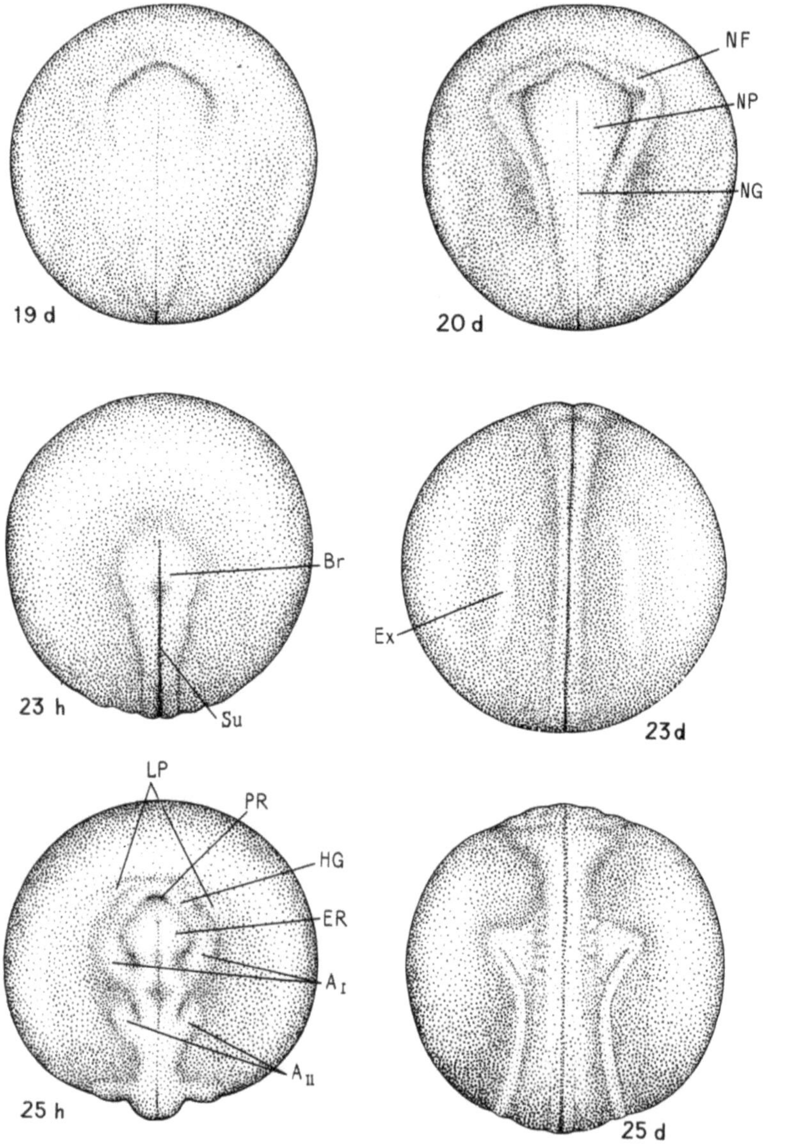
NF
NP
NG
19 d
20 d
Br
23 h
Su
Ex
23d
LP
PR
HG
ER
A_I
A_{II}
25 h
25 d

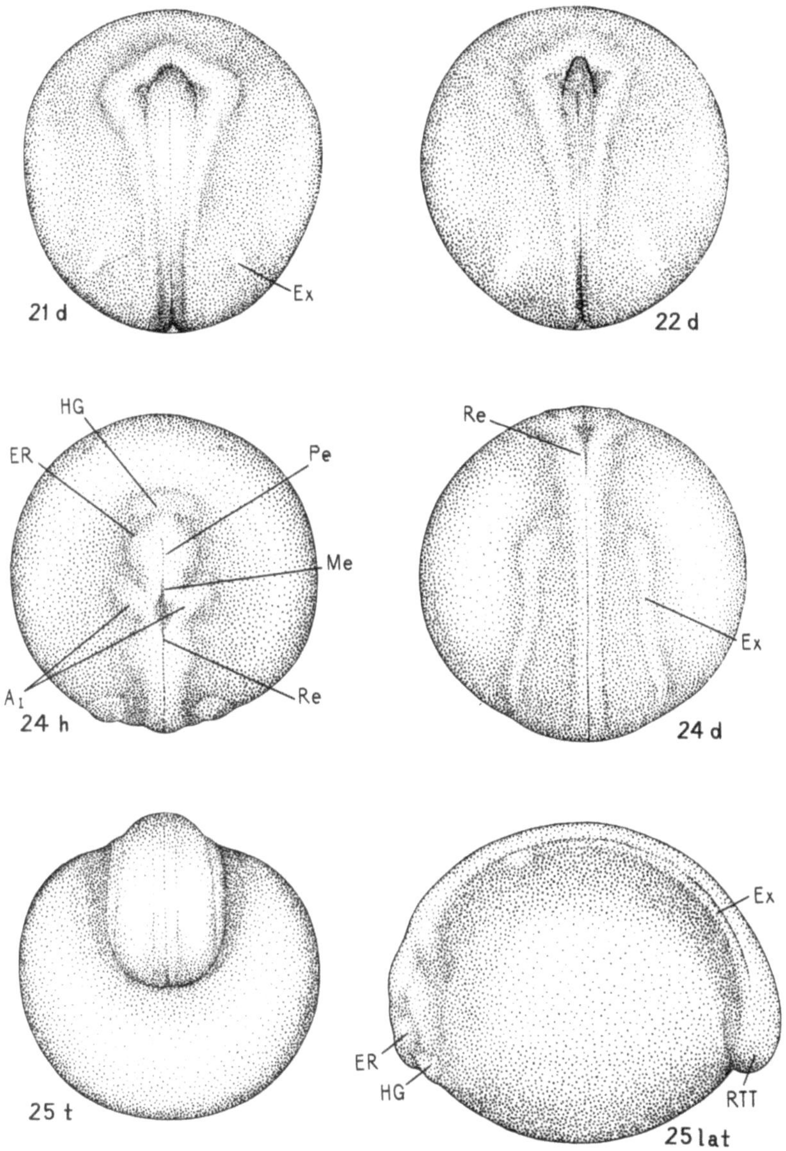

Plate IV. Early neurula (stage 19); stage of broad neural plate (stage 20); stage of the onset of drawing together of the neural folds (stage 21); late neurula (stage 22); stage of closed neural tube (stage 23); stage of appearance of eye protrusions and thickening of the anterior end of the excretory system rudiments (stage 24); stage of drawing together of the lateral plates and appearance of the common rudiment of posterior trunk and tail regions (stage 25)

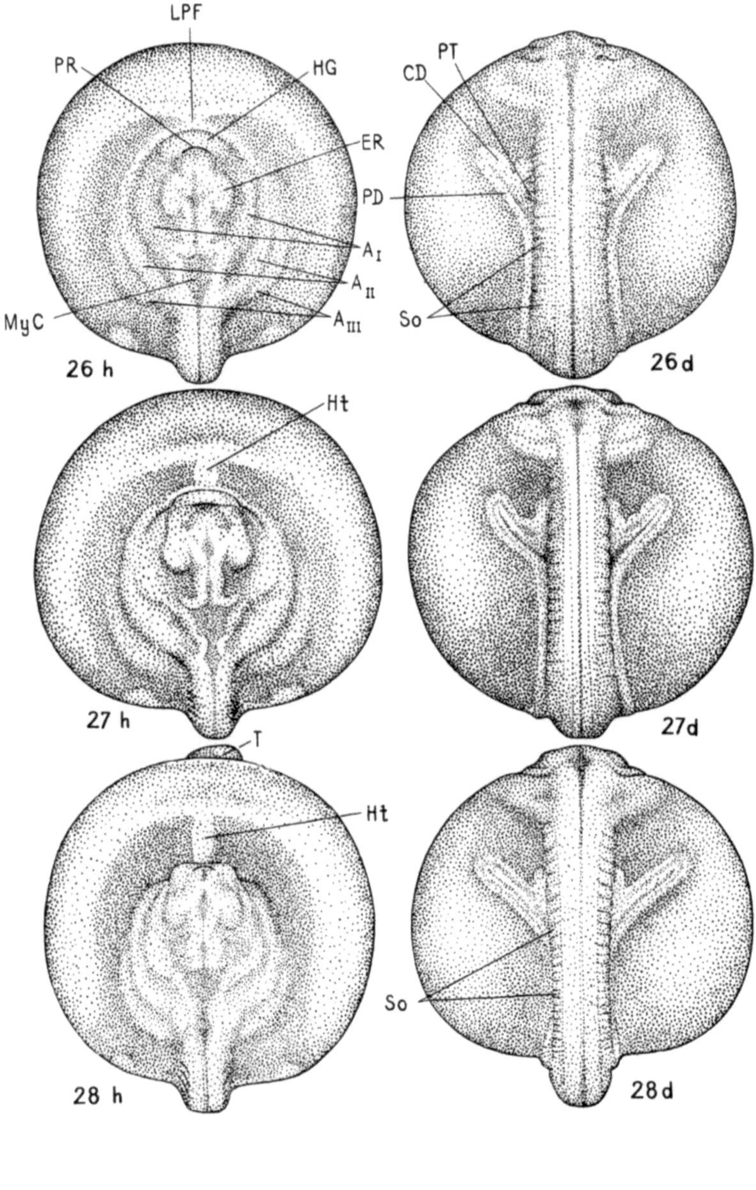
LPF
PR
HG
ER
A I
A II
A III
MyC
26 h
PT
CD
PD
So
26 d
Ht
27 h
27 d
T
Ht
28 h
So
28 d

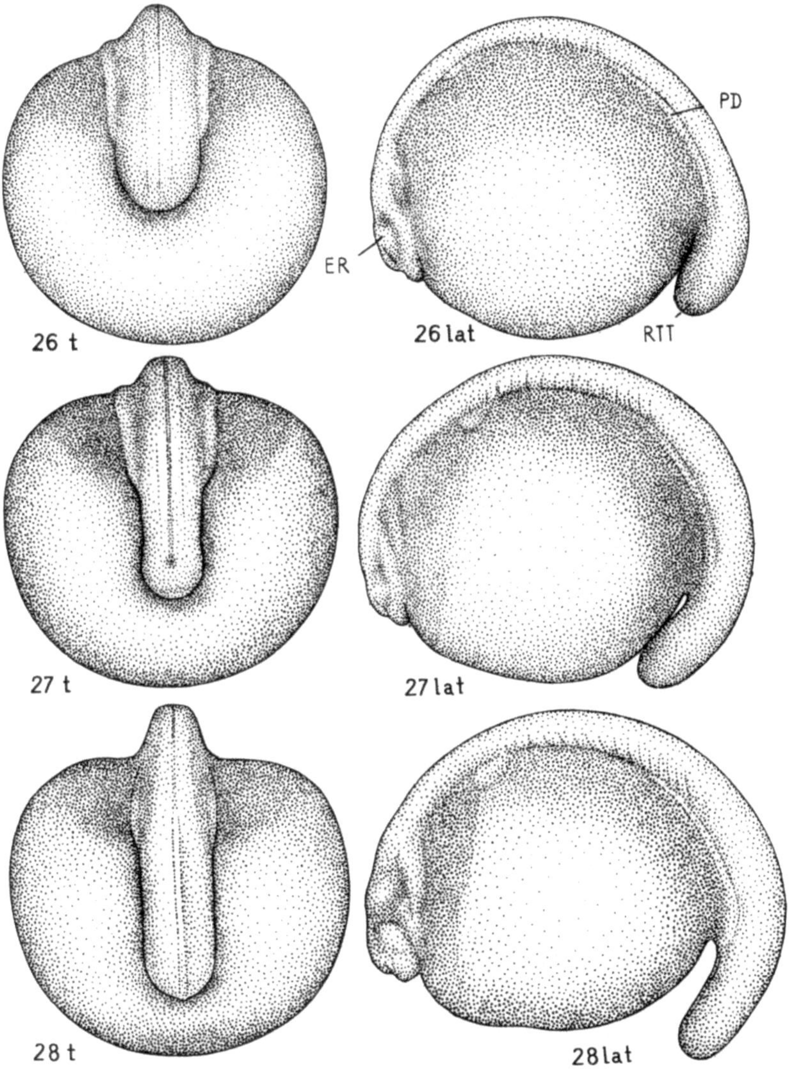

Plate V. Stage of fusion of the lateral plates and onset of separation of a rudiment of the posterior trunk and tail regions (stage 26); stages of short (stage 27) and straight elongated cardiac tube (stage 28)

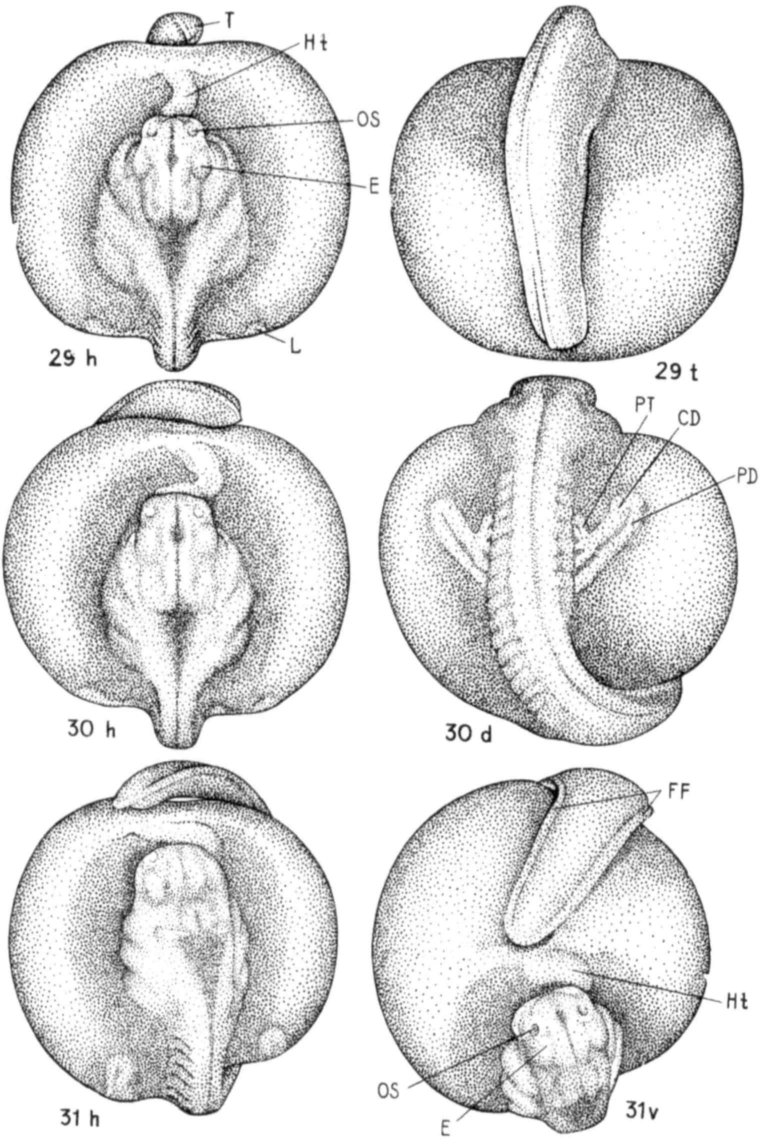
T
Ht
OS
E
L
29 h
29 t
PT
CD
PD
30 h
30 d
FF
Ht
OS
E
31 h
31v

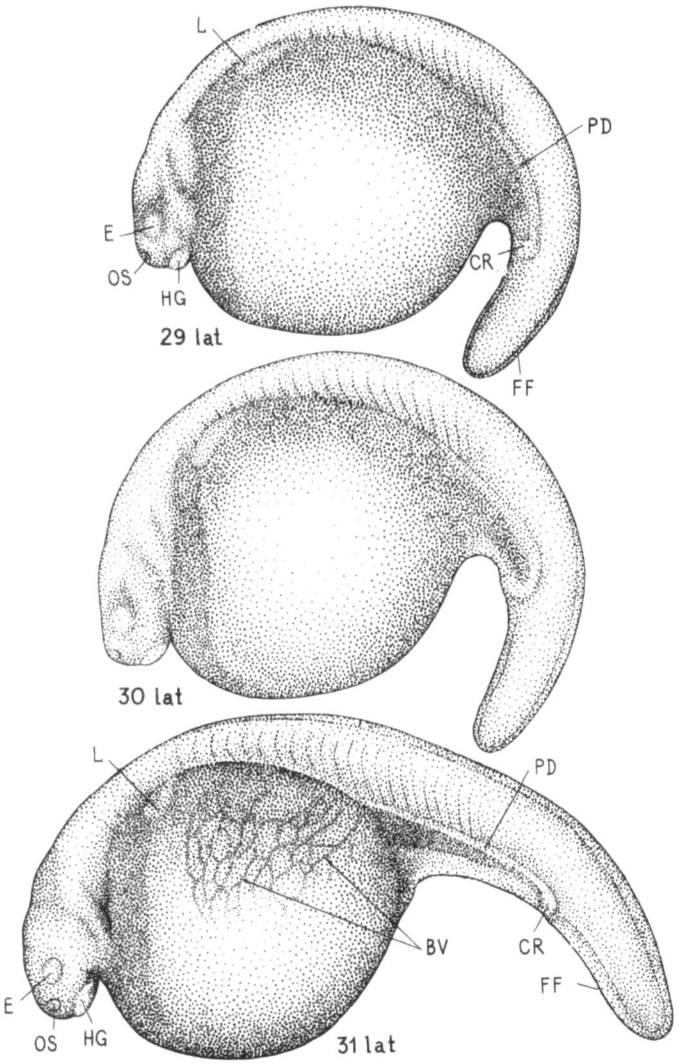

Plate VI. Stage of formation of the cardiac tube curvature (stage 29); stages when the tail tip approaches the heart (stage 30) and reaches the heart (stage 31) in embryos within envelopes

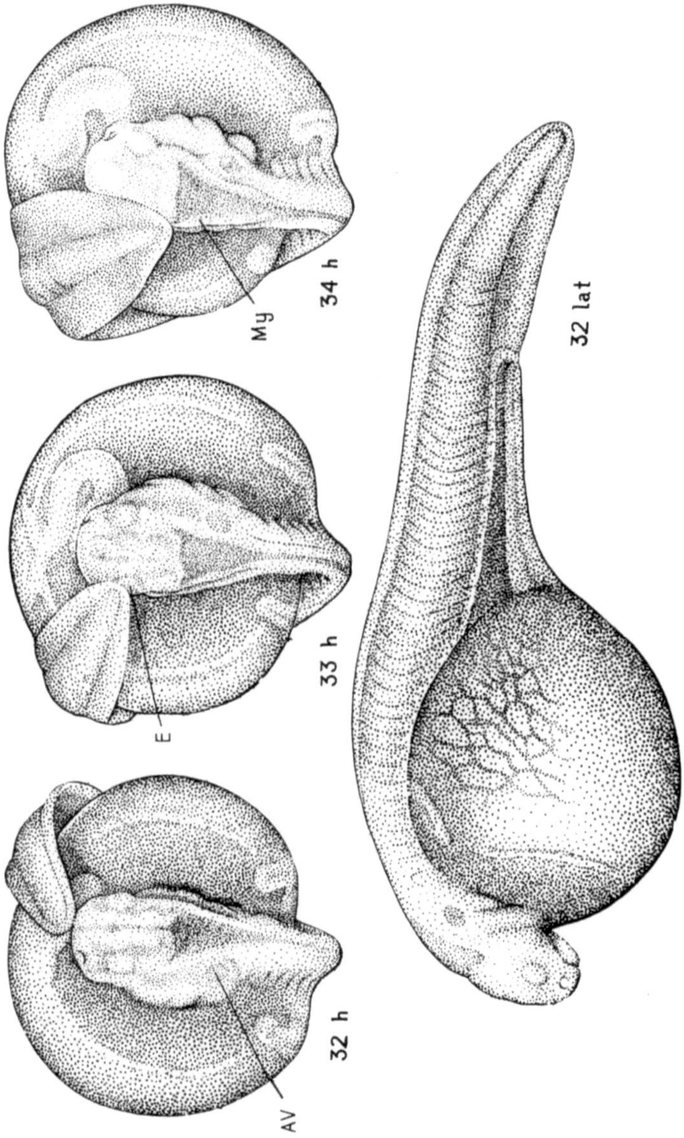

34 h
My
33 h
E
32 h
AV
32 lat

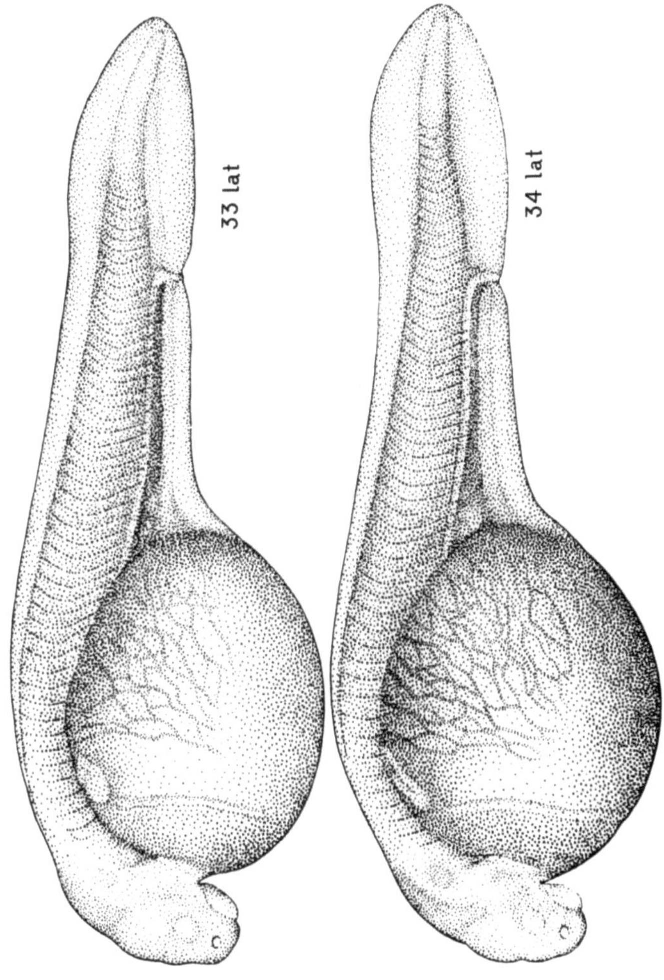

Plate VII. Stages (in embryos within envelopes) when the tail tip touches the head (stage 32), goes slightly beyond the head end (stage 33) and reaches (in *A. gueldenstaedti*) the front part of the myelencephalon (stage 34)

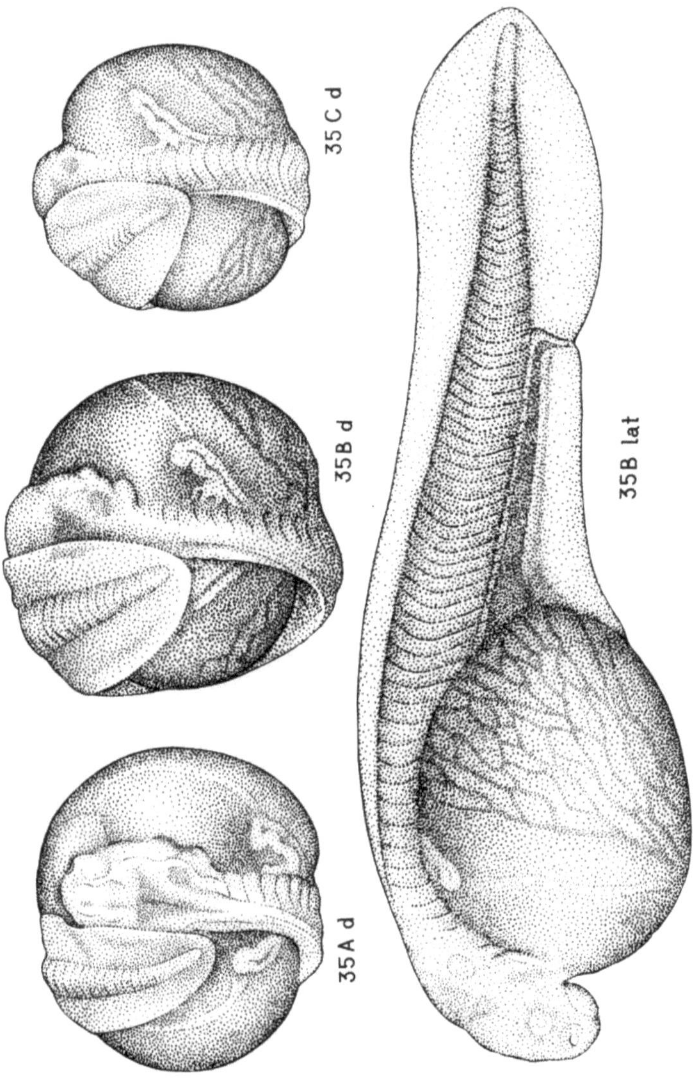

Plate VIII. Stage of hatching of single prelarvae (stage 35) in *A. stellatus* (*A*), *A. gueldenstaedti* (*B*), and *A. ruthenus* (*C*)

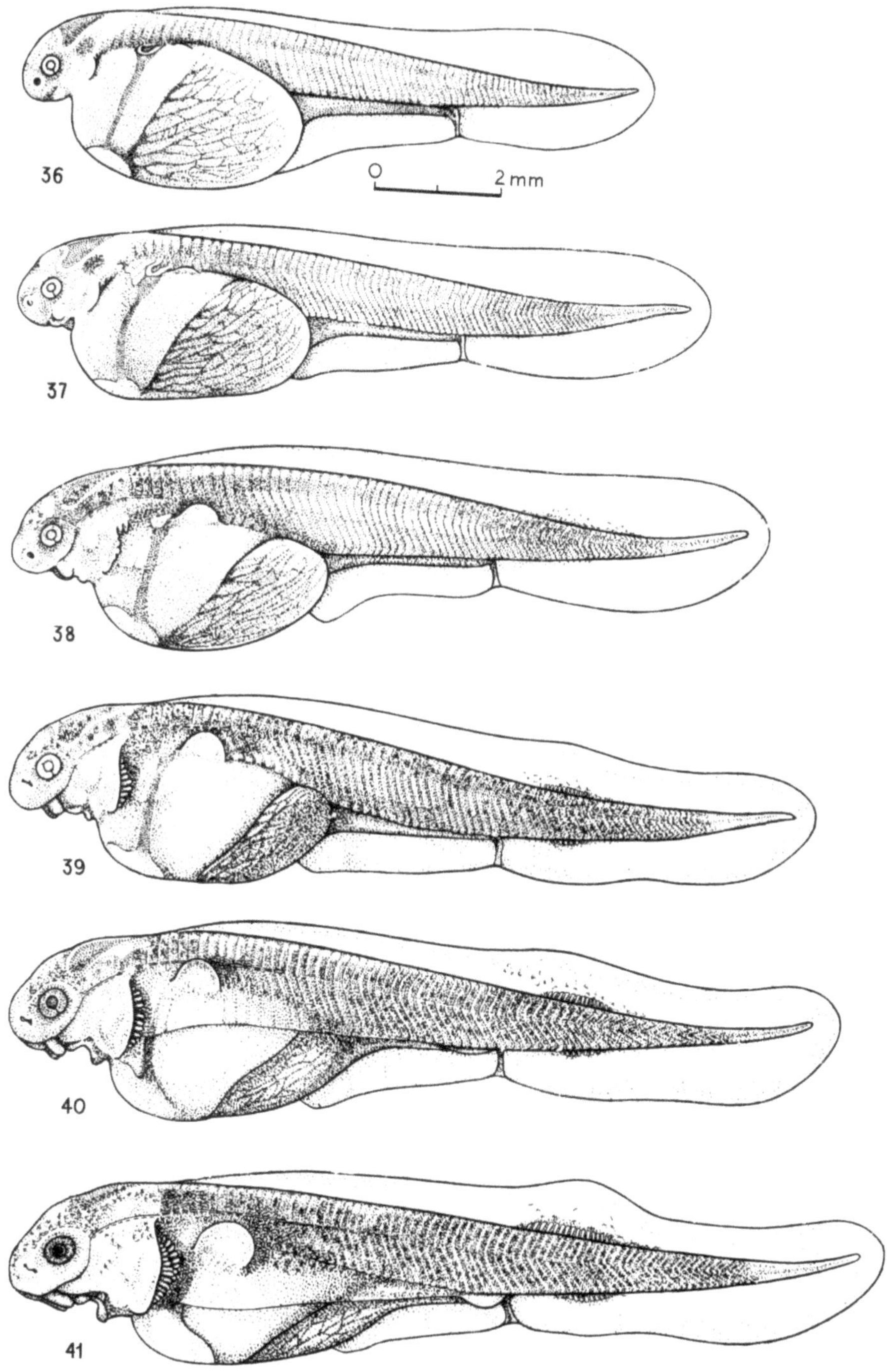

Plate IX. Prelarvae of *A. gueldenstaedti colchicus* at stages 36–41

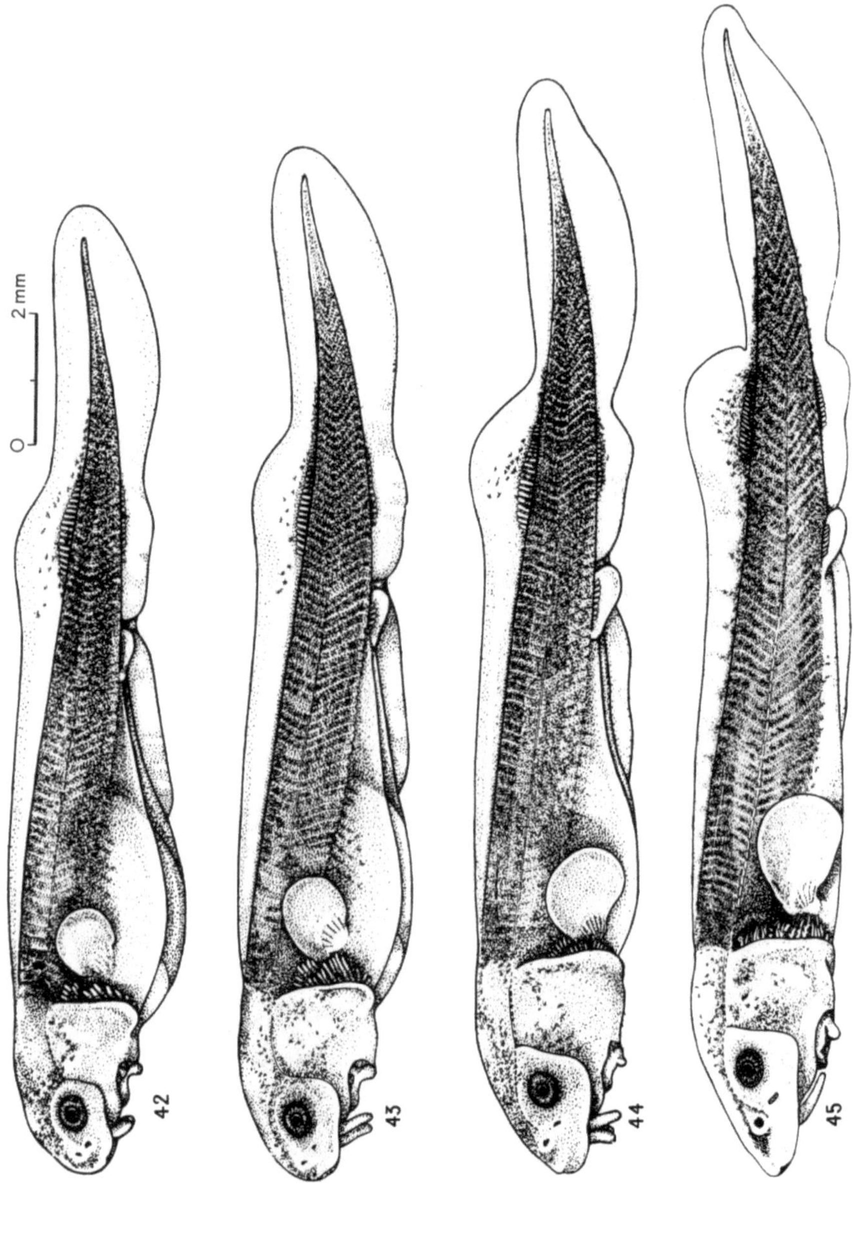

Plate X. Prelarvae of *A. gueldenstaedti colchicus* at stages 42–45

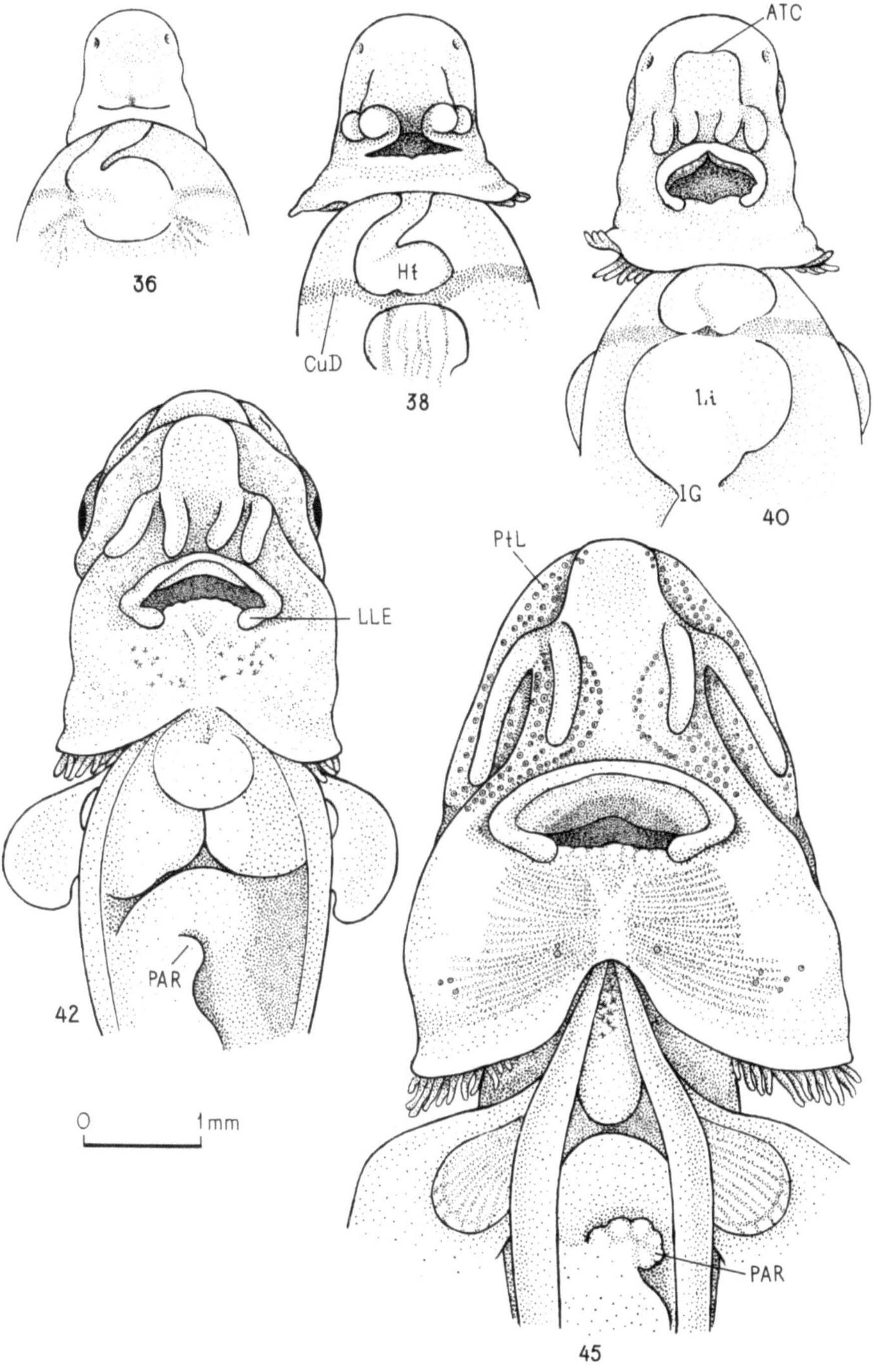

Plate XI. Heads of prelarvae of *A. gueldenstaedti colchicus* at stages 36, 38, 40, 42 and 45. View from the ventral side

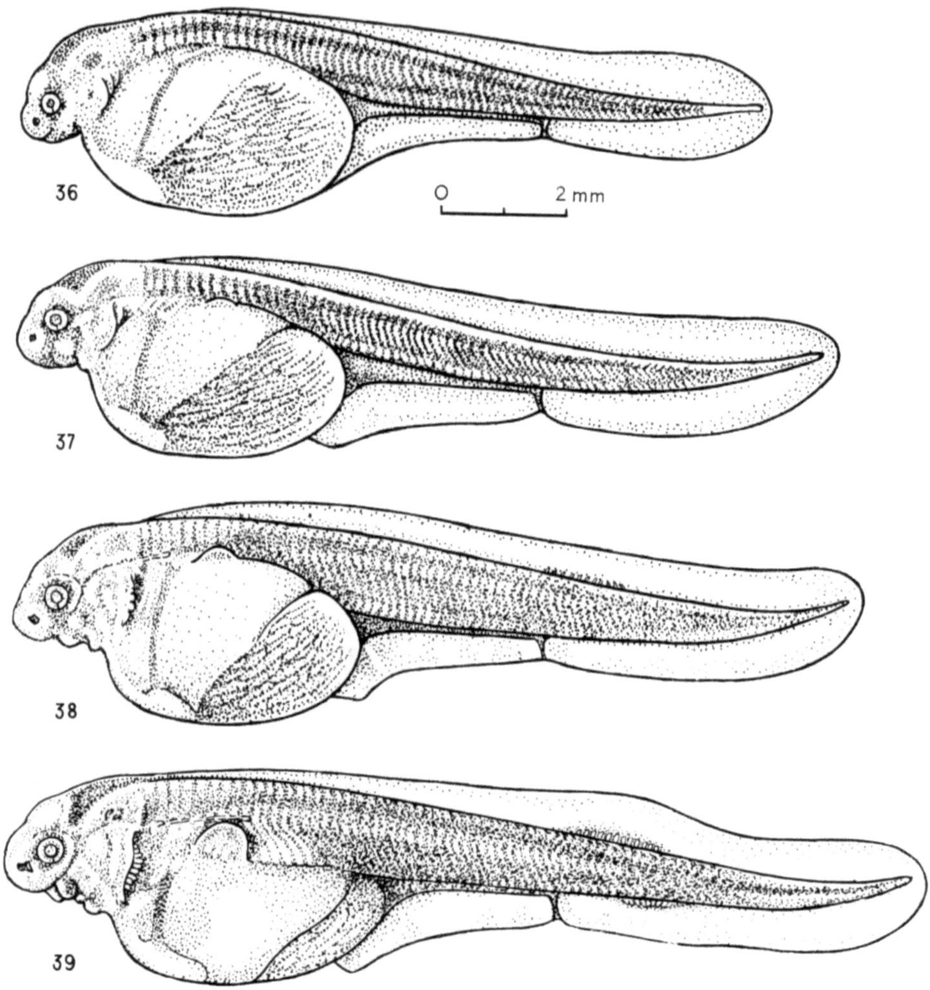

Plate XII. Prelarvae of *H. huso* at stages 36–39

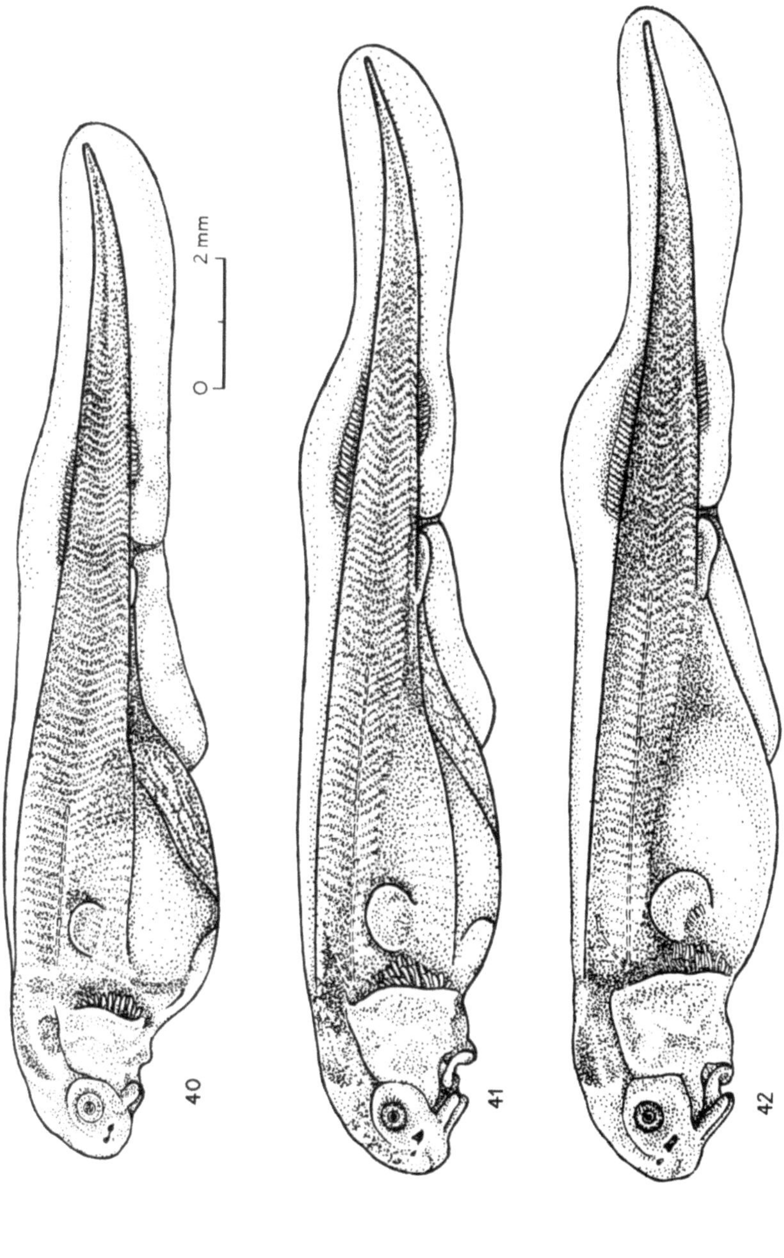

Plate XIII. Prelarvae of *H. huso* at stages 40–42

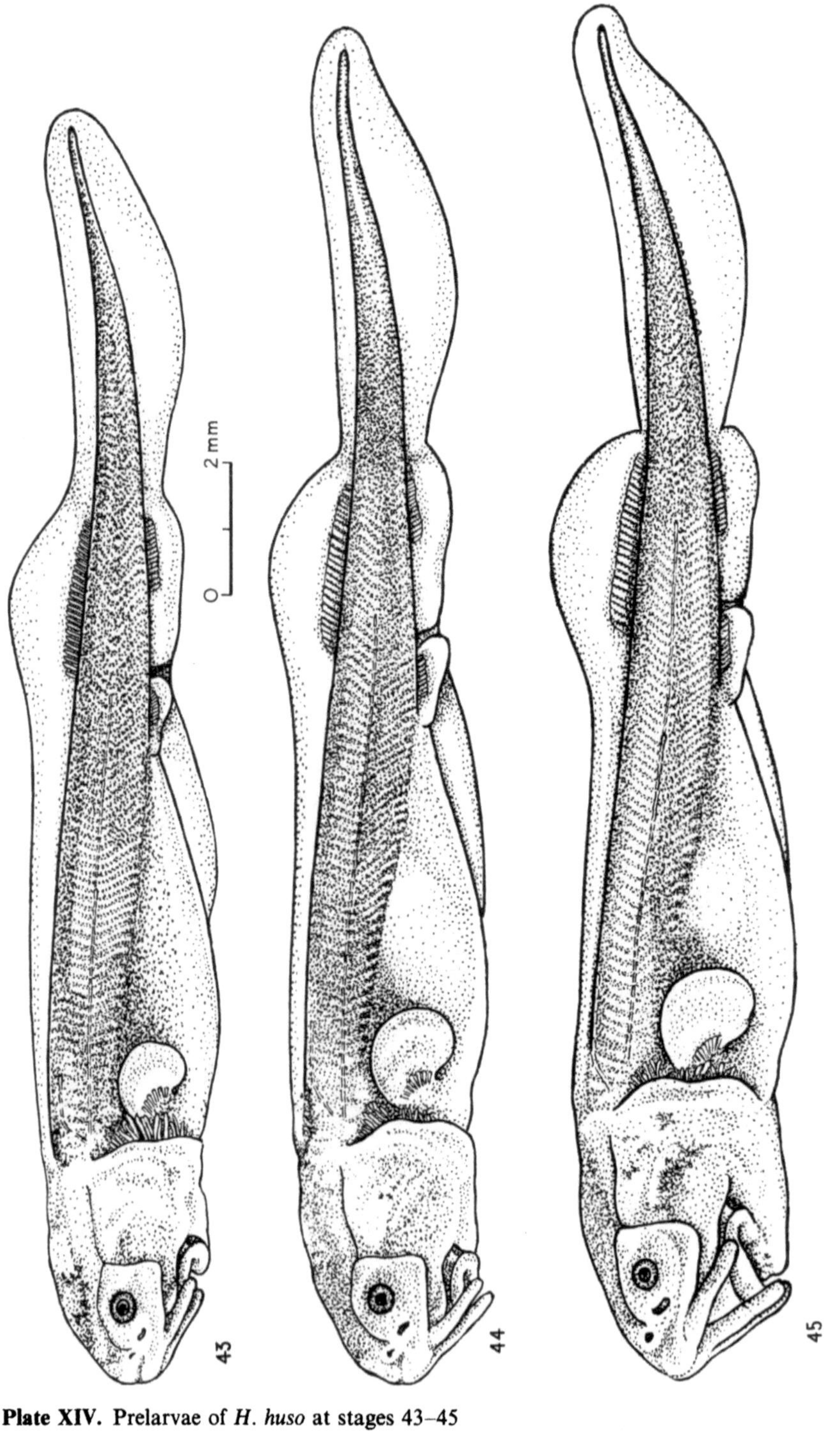

Plate XIV. Prelarvae of *H. huso* at stages 43–45

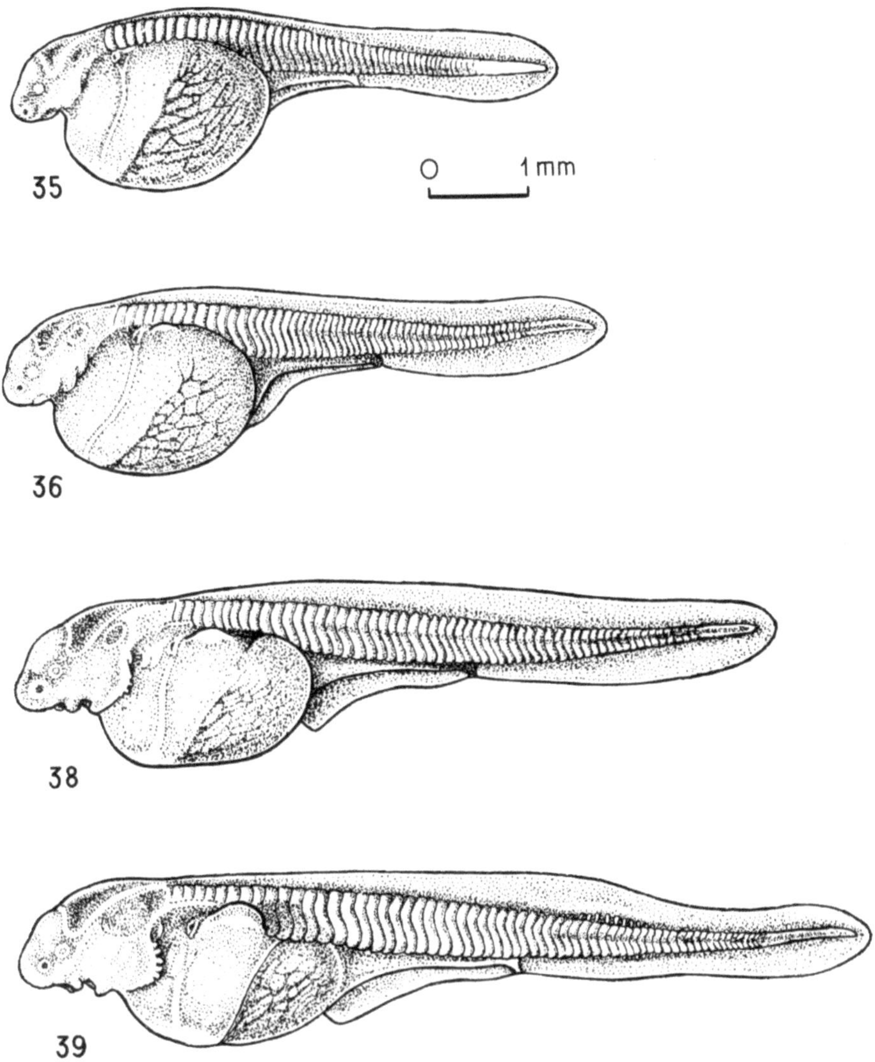

Plate XV. Prelarvae of *P. kaufmanni* at stages 35, 36, 38 and 39

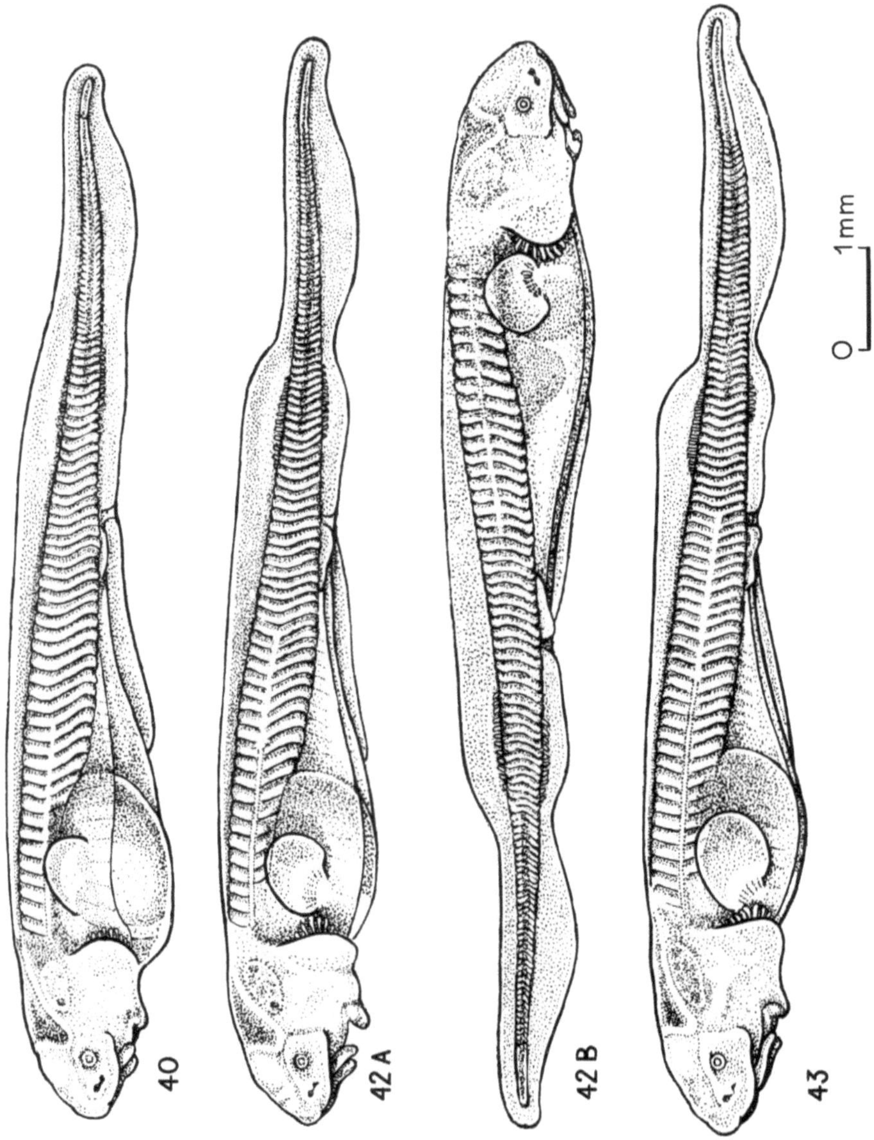

Plate XVI. Prelarvae of *P. kaufmanni* at stages 40 and 42 (*A* view from the left; *B* view from the right side) and 43

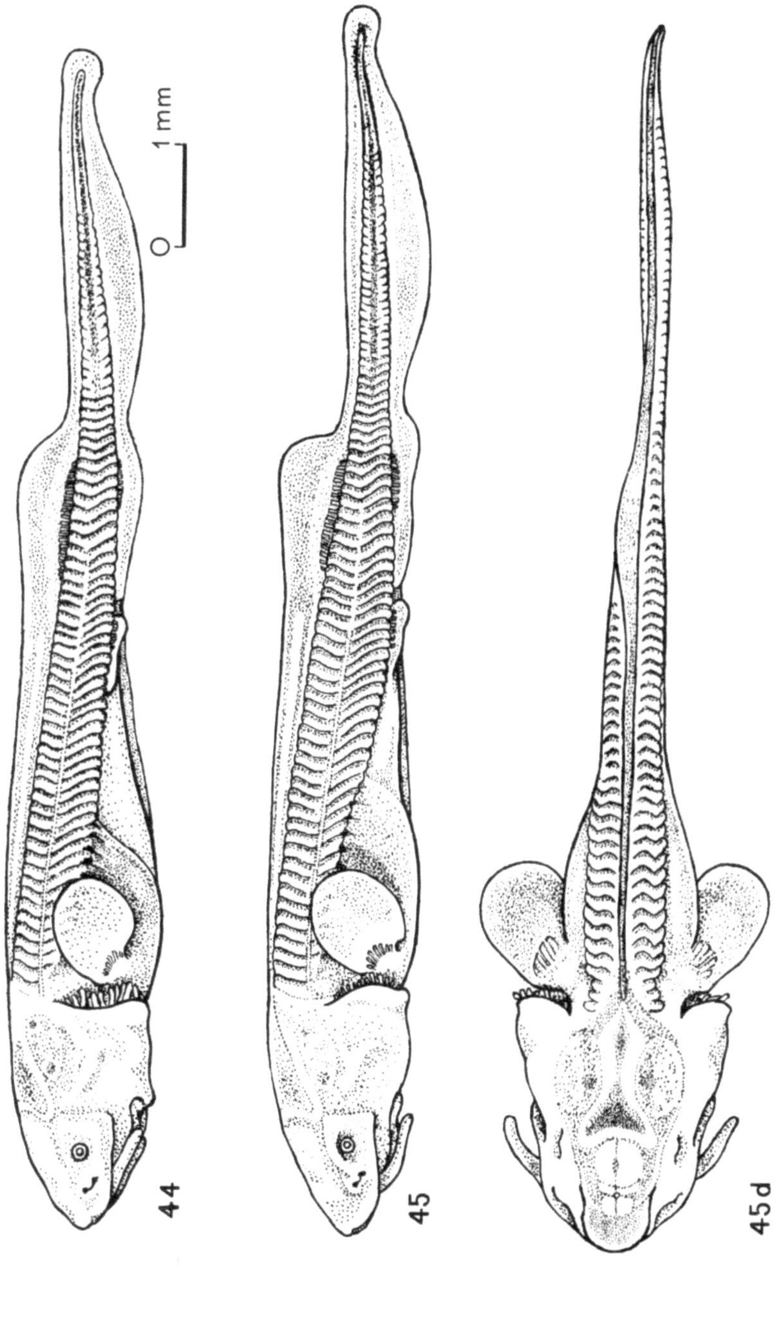

Plate XVII. Prelarvae of *P. kaufmanni* at stages 44 and 45 (*d* view from the dorsal side)

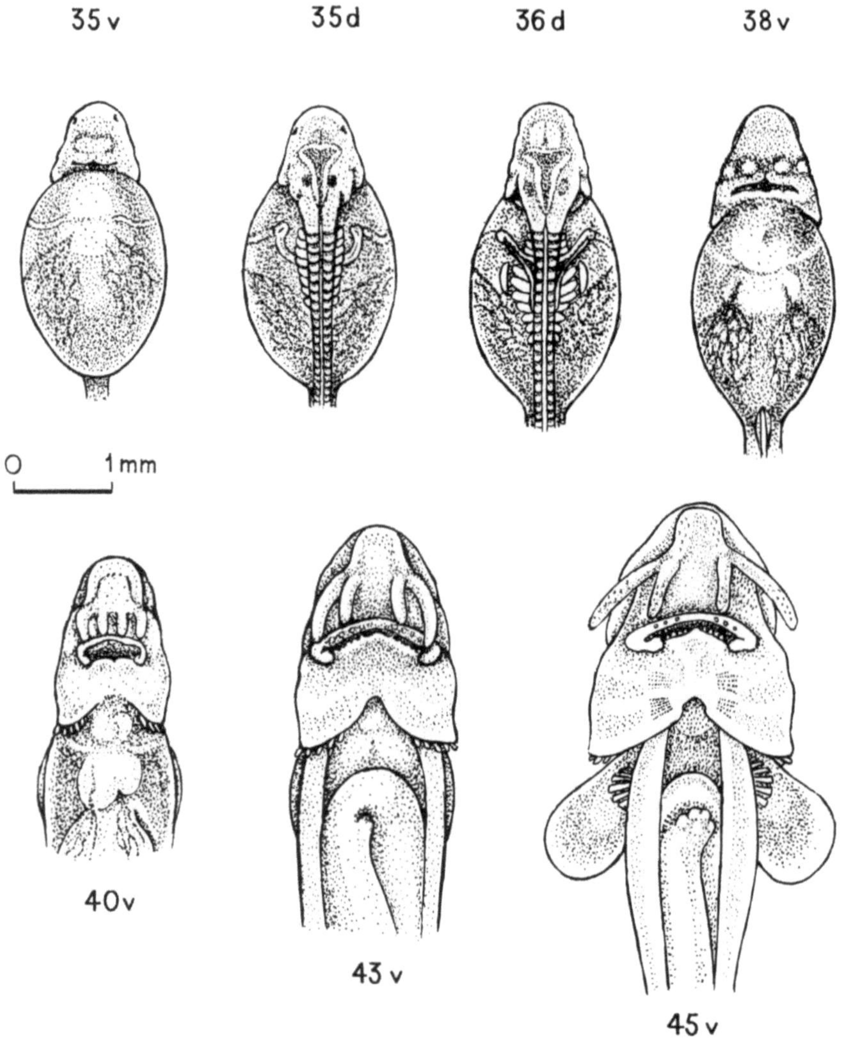

Plate XVIII. Prelarvae of *P. kaufmanni* at stages 35, 36, 38, 40, 43 and 45. View from the dorsal (*d*) and ventral (*v*) side

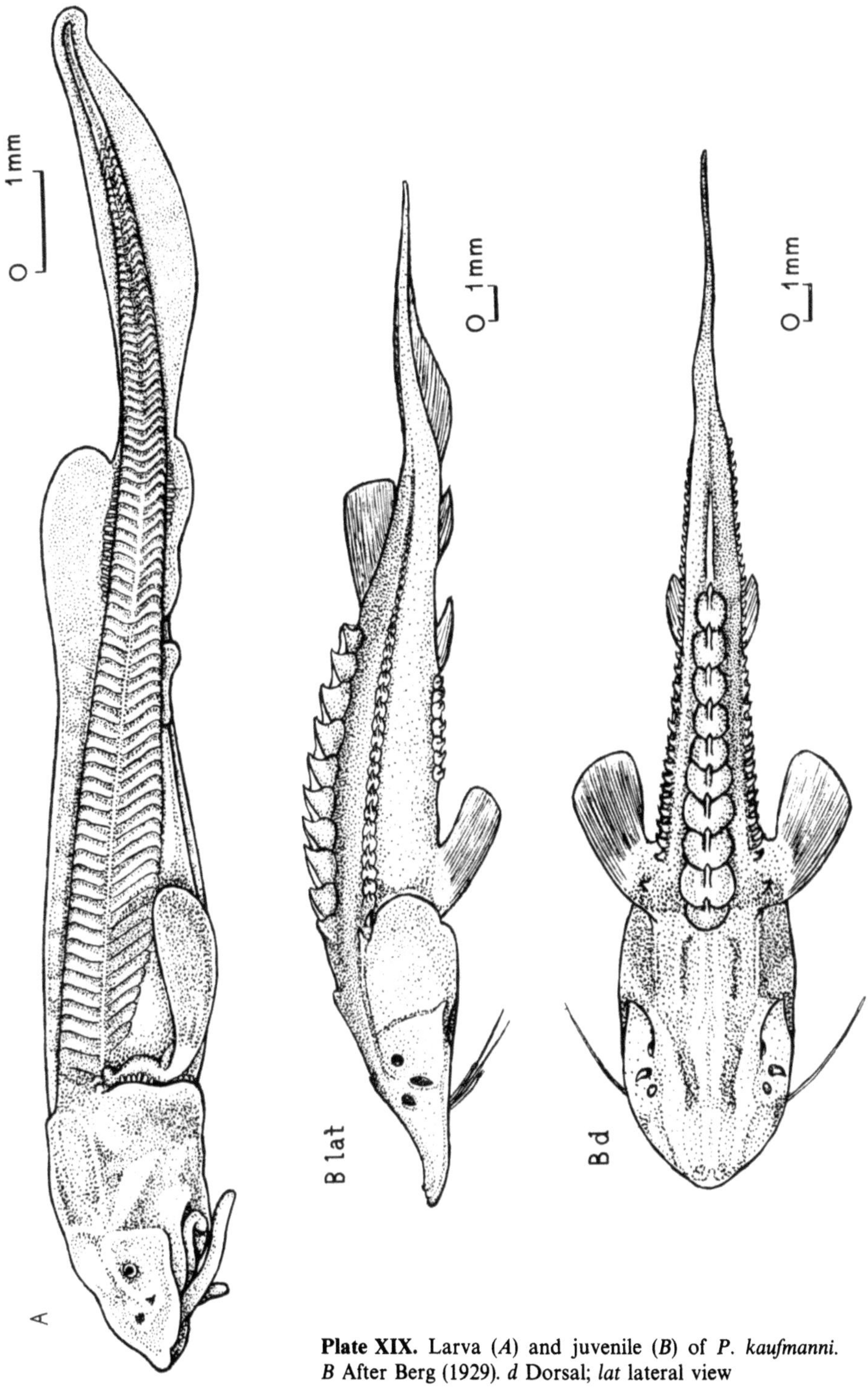

Plate XIX. Larva (*A*) and juvenile (*B*) of *P. kaufmanni*. *B* After Berg (1929). *d* Dorsal; *lat* lateral view

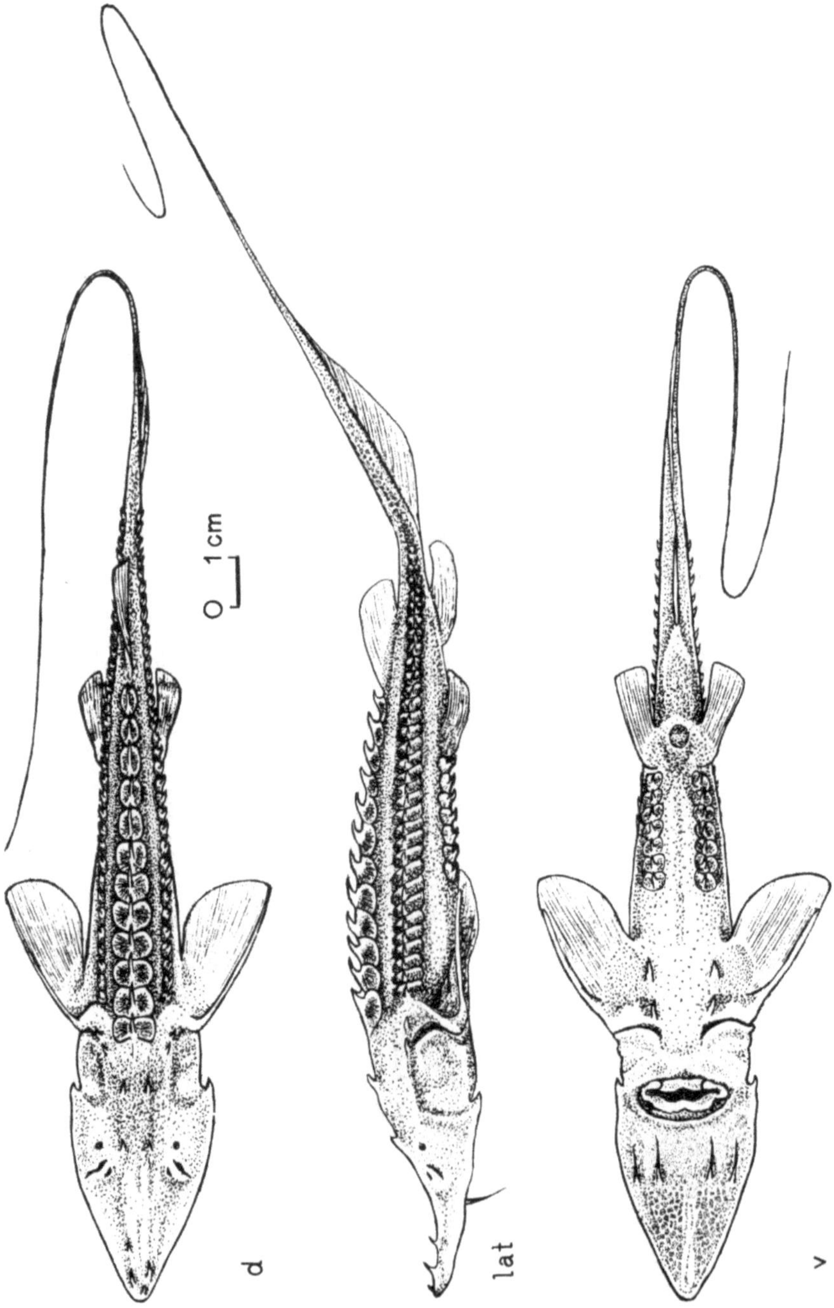

Plate XX. Sexually immature adult *P. kaufmanni*. Dorsal (*d*), lateral (*lat*) and ventral (*v*) views

Abbreviations for Figures and Plates

A_I, A_{II}	rudiments of 1st and 2nd pair of visceral arches, respectively
A_{III}	rudiments of the branchial arches
AB	small blastomeres of the animal area
AcP	acrosomal process
AE	activated egg
AF	pigmented floor of the archenteron seen through the archenteron roof on the dorsal side
an	view from the animal pole
an-veg	animal-vegetal axis of the egg
An/Veg	boundary between the animal and vegetal areas
AP	animal pole
APt	animal part of the egg
Ar	archenteron
AR	animal region
ATC	anterior transversal commissure
AV	auditory vesicle
BA	branchial aortic arch
BA_{IV}	4th branchial arch
BB	rudiment of branchial blood vessels
Bc	blastocoel; blastocoel seen through the outer cell layer
BC	brain cavity
BG	branchial grooves
Bo	boundary of the exposed area (after removal of epithelium and neural tube)
Bp	blastopore
BP	endodermal branchial pouch
Br	brain, brain rudiment
BV	network of blood vessels of the yolk sac
BW	brain wall
C	choroid
Cb	ceratobranchialia of the 1st–3rd branchial arches
CD	collecting part of the pronephric duct
CG	cortical granules
Ch	chromosomes
ChV	chromosomal vesicles
CL	cortical layer of the oocyte

Co	columns forming in the sites of extrusion of the cortical granule contents as a result of fixation
Cp	copula
CR	cloaca rudiment
CuD	duct of Cuvier
d	view from the dorsal side
D	dorsal side
DBp	dorsal lip of blastopore
De	diencephalon
DPB	defective posterior region of the brain
E	eye
EM	egg membranes, individual layers indiscernible
En	yolk-rich endodermal cells
ENL	external nuclear layer
Ep	bilayered covering epithelium, covering epithelium
EP	equatorial plate
EPL	external plexiform layer
Eq	equator
ER	eye rudiment
Es	esophagus
ES	eye stalk
ET	external theca
EV	eye vesicle
Ex	rudiment of the excretory system
Fd	food
FE	follicular epithelium; follicle cell
$\mathbf{F_I}$, $\mathbf{F_{II}}$, $\mathbf{F_{III}}$, $\mathbf{F_{IV}}$	furrows of the 1st, 2nd, 3rd, and 4th cleavage divisions, respectively
FF	fin fold rudiment
FO	large full-grown oocytes in the follicles
FP	process of a follicle cell filling the future micropylar canal
FW	follicle wall
GB	gall bladder
GC	gut cavity
GL	ganglionic cell layer
GO	small oocytes in the beginning of rapid growth
GV	germinal vesicle
GW	gut wall
h	view from the side of head region
H	head
HA	hyoid aortic arch
Hb	hypobranchialia of the 1st–3rd branchial arches
HC	hyomandibular commissure
HCa	hyoid cartilage
Hch	hypochord (subchorda)

Hg	hindgut
HG	hatching gland; hatching gland rudiment
Hh	hypohyale
HHb	rudiment of hyoid hemibranch blood vessels
HLC	hollow in place of lysed cells
HmCa	hyomandibular cartilage
Ht	heart, site of the heart rudiment formation
Hy	hypothalamus
IG	intermediate gut
Ih	interhyale
INL	inner nuclear layer
Inv	invaginated cell sheet
IPL	internal plexiform layer
IPR	intensely pigmented ridge
IT	internal theca
JC	jelly coat
K	karyoplasm; karyoplasm oozed into the cytoplasm
L	loop formed by the anterior part of the pronephric duct
lat	lateral view (at stages 25–35 membranes were removed from the embryo while it was still alive)
LC	light crescent
LD	liver diverticulum
LDr	lipid droplets
LEx	lateral extensions of the cell membrane
Lg	loosening of the cytoplasm in the region of future furrow
Li	liver
LK	lacunae of the karyoplasm in the cytoplasm
LLE	lower lip edge at the site where it is interrupted
LP	lateral plate
LPF	region of fusion of the lateral plates; region of fusion of the lateral plates where the heart rudiment forms
LPS	light polar spot
LR	liver rudiment
Ls	lens
M	mesenchyme; mesenchyme forming the endocardium rudiment
MII	metaphase of the II meiotic division
MA	mandibular aortic arch
MC	micropylar canal; micropylar canal occupied by the process of the closing cell
MCa	Meckel's cartilage
Me	mesencephalon
MeC	mesenteric cord
MF	maturation funnel
Mi	micropyle
Mp	midpiece of the spermatozoon

MS I, MS II spindles of the I and II meiotic divisions
MT main part of the sperm tail
MuS muscle segments
Mv microvilli
My myelencephalon
MyC myelencephalon cavity
MyR myocardium rudiment
MZ marginal zone
N notochord
NC neurenteric canal
NE normal egg
NF neural fold
NG neural groove
NP neural plate
NT neural tube
Nu nucleoli
ON optic nerve
OP olfactory pit
OS olfactory sac
Out outer cell sheet
P pituitary
Pa pancreas
PAR rudiment of the pyloric appendage
PB polar body
PC pericardial cavity
PCa palatoquadrate cartilage
PD pronephric duct; rudiment of the pronephric duct
PDC cavity of pronephric duct
Pe prosencephalon
PE pigment epithelium
PF pectoral fin rudiment
PFE process of the follicular epithelium cell
PG pigment granules
PiS pigment spot
PL parietal layer of the lateral plate
PoE polyspermic egg
PR deepening at the site of formation of the pituitary rudiment; ectodermal part of the pituitary rudiment
PS perivitelline space
PT pronephric tubules
PtL pit lines
Re rhombencephalon
RNF retinal nerve fibers
RTT common rudiment of the posterior trunk and tail regions
Sa spermaster

SB	superbranchial blood vessel
Sc	sclera
SC	spinal cord
SCG	secreted contents of the cortical granules
SG	spiral gut
SH	head of the spermatozoon
SLC	site of prospective light crescent
So	somites
SRL	split between retina and lens
St	stomach
ST	sperm trail
Su	suture along the line of fusion of the neural plate edges
Sy	symplecticum
t	view from the side of tail
T	tail
TeR	teeth rudiments
Th	theca
TR	tail region of the spermatozoon
TT	terminal tubule of the micropyle
UE	unactivated egg
v	view from the ventral side
V	ventral side
VA	ventral aorta
VB	large balstomeres of the vegetal area
veg	view from the vegetal pole
ViB	vitreous body
VL	visceral layer of the lateral plate which forms the heart rudiment here
VPT	vegetal part of the egg
VR	vegetal region; vegetal pole region
YG	yolk granules
YP	yolk plug
ZR	zona radiata
ZR I	zona radiata externa
ZR II	zona radiata interna

References

Afonich RV, Gordienko OL, Soldatova EV (1971) Hatchery estimation of the Russian sturgeon spawners of different weight and age according to the condition of eggs, larvae, and early juveniles. Tr VNIRO 81: 92–111
Aizenshtadt TB (1984) Cytology of oogenesis. Nauka, Moscow (in Russian).
Aizenshtadt TB, Dettlaff TA (1972) Ultrastructure of stellate sturgeon oocytes during maturation. I. Annulate lamellae and Golgi complex. Sov J Dev Biol 3: 220–229
Aizenshtadt TB, Skoblina MN (1970) Fine structure of frog oocytes maturing in vitro under the influence of hypophysial hormones after the removal of their germinal vesicles. Tsitologiya 12: 713–717
Alekseeva TA, Ozernyuk ND (1987) Energy exchange and temperature optimum of loach development. Zh Obshch Biol 48: 525–531
Alpatov VV, Stroganov NS (1950) A new unit for measuring the pituitary activity in fish. Dokl Akad Nauk SSSR 74: 405–407
Alyavdina LA (1951a) Condition and distribution of spawning grounds of the Russian and stellate sturgeons in the Volga region from the Kamyshin to Saratov. Tr Sarat Otd Kaspii Fil VNIRO 1: 14–32
Alyavdina LA (1951b) On biology and taxonomy of sturgeon at early developmental stages. Tr Sarat Otd Kaspii Fil VNIRO 1: 33–73
Alyavdina LA (1953) On ecology of the Russian sturgeon reproduction in the Volga. Tr Sarat Otd Kaspii Fil VNIRO 2: 3–27
Ancel P (1945) Sur l'action tératogène élective de certaines substances chimiques. C R Seances Soc Biol 139 (21/22): 983–984
Ancel P, Vintemberger P (1948) Recherches sur le déterminisme de la symétrie bilatérale dans l'oeuf des Amphibiens. Bull Biol Fr Belg (Suppl) 31: 1–182
Andreev AA, Kolupaev BI, Karpovich TA (1979) Effects of phenol on oxygen tension in the muscle tissue of *Rutilus rutilus lacustris* (Pallas). Vopr Ikhtiol 19: 564–565
Androsov SA, Nepomnyashchii LI, Bondarenko NV (1990) Results of rearing sturgeon in systems with the closed cycle water supply. Rybn Khoz (6): 69–70
Arnol'd IN (1915) Experiments of the Kazan' Division of the Imperial Russian Society of Fish Culture and Fishery on artificial fertilization of the sterlet eggs and rearing fry in 1911–1914. Vestn Rybopromsti 30 (1): 2–19; No. 2: 62–94
Artyukhin EN (1979) The Persian sturgeon in the North Caspian rivers and perspective of its use in sturgeon culture. In: Berdichevskii LS (ed) Biological foundations of sturgeon culture development in the water bodies of the USSR. Nauka, Moscow, pp 105–115 (in Russian)
Babaskin AV (1930) Influence of the concentration of hydrogen ions on the survival rate of *Acipenser ruthenus* eggs and fry. Uch Zap Kazan Univ 90 (3/4): 497–543
Baburina EA (1972) Eye development in *Cyclostomata* and *Pisces* with reference to their ecology. Nauka, Moscow, (in Russian)
Babushkin NYa (1947) Fertility of the Caspian beluga sturgeon. Zool Zh 26: 339–345
Badenko LV (1967) Application of morphophysiological indices for selecting the Don stellate sturgeons for hatchery purposes. Tr TsNIORKh 1: 222–228
Badenko LV, Golovanenko LF, Fedorova LS, Meleshko AA, Androsyuk LYa, Grudanova SD (1971) Temporary instruction for selecting spawners and gametes of the Don sturgeons

for hatchery reproduction. Moscow, Glavrybvod of the USSR Ministry of Fisheries, AzNIIRKh (in Russian)
Badenko LV, Golovanenko LF, Grudanova SD (1972) On the influence of the physiological condition of the stellate sturgeon females on the quality of eggs and the progeny. Tr TsNIORKh 4: 191–199
Badenko LV, Kornienko GG, Gashchenko LA, Shchigel'skaya VP (1976) Condition of the gonads in the Kuban' *Acipenser stellatus* Pallas at the dam of the Fedorovo hydrocenter and its use in fish culture. Vopr Ikhtiol 16 (4): 644–653
Balabai PP (1939) Studies of the respiratory system functioning in the *Acipenseridae*. Tr Inst Zool Biol Akad Nauk Ukr SSR 5: 111–145 (Collected Works on Animal Morphology (in Ukrainian))
Ballard WW (1981) Morphogenetic movements and fate maps of vertebrates. Am Zool 21: 391–399
Ballard WW, Ginsburg AS (1980) Morphogenetic movements in Acipenserid embryos. J Exp Zool 213: 69–103
Ballard WW, Needham RG (1964) Normal embryonic stages of *Polyodon spathula* (Walbaum). J Morphol 114: 465–477
Balon EK (1975) Reproductive guilds of fishes: a proposal and definition. J Fish Res Board Can 32: 821–864
Barannikova IA (1954) Completion of the process of transition to the spawning condition in the hiemal Russian sturgeon females and males of the autumn migration after the river period of the spawning migration was abolished. Dokl Akad Nauk SSSR 99: 641–644
Barannikova IA (1957) Biological differentiation of the brood stock of the Volga-Caspian Russian sturgeon (with reference to the tasks of commercial sturgeon culture in the Volga delta). Uch Zap LGU 228 Ser Biol (44): 54–71
Barannikova IA (1968) Functional foundations of the migratory behavior of anadromous fish. Author's Abstract of the Doctoral Dissertation. Leningrad LGU Press, (in Russian)
Barannikova IA (1970) New data about the response of sturgeon population on disturbance of the migration and reproduction conditions. Tr TsNIORKh 2: 12–19
Barannikova IA (1972a) Pathways of development and functional foundations of intrapopulation differentiation in sturgeons. In: Marti Yu Yu, Barannikova IA (eds) Sturgeons and the problems of sturgeon culture. Pishchevaya Promyshlennost', Moscow, pp 167–179 (in Russian)
Barannikova IA (1972b) Functional foundations of the migration in sturgeons. In: Marti Yu Yu, Barannikova IA (eds) Sturgeons and the problems of sturgeon culture. Pishchevaya Promyshlennost', Moscow, pp 180–204 (in Russian)
Barannikova IA (1975) Functional foundations of migrations in fish. Nauka, Moscow, (in Russian)
Barannikova IA (1978) Histophysiological foundations on application of repeated and single pituitary injections in sturgeon culture. Tr VNIRO 130: 85–92
Barannikova IA (1979) Condition and current tasks of sturgeon culture. In: Berdichevskii LS (ed) Biological foundations of sturgeon culture development in the water bodies of the USSR, Nauka, Moscow, pp 49–58 (in Russian)
Barannikova IA (1987) The study of sturgeons and the problems of sturgeon hatcheries in the Soviet Union. Vopr Ikhtiol 27: 735–746
Barannikova IA, Boev AA (1977) Methodical instructions for application of the method of pituitary injections in fish culture. Glavrybvod of the USSR Ministry of Fisheries, Moscow (in Russian)
Barannikova IA, Bukovskaya OS (1984) Condition of the pituitary and kinetics of gonadotropin in the blood serum of the stellate sturgeon after treatment with LH–RH. In: Sturgeon culture in water bodies of the USSR. Astrakhan' pp 36–37
Barannikova IA, Burenin OK (1971) An experience of application of fractional pituitary injections in artificial reproduction of the Kuban' stellate sturgeon. In: Materials of the Joint Scientific Session of TsNIORKH and AzNIIRKh. Astrakhan' pp 9–11

Barannikova IA, Berdichevskii LS, Sokolov LI (1979) Scientific foundations of sturgeon culture and directions of its further development in water bodies of the USSR. In: Berdichevskii LS (ed) Biological foundations of sturgeon culture development in the water bodies of the USSR, Nauka, Moscow, pp 5–22 (in Russian)

Barannikova IA, Boev AA, Zenkevich GA, Latse ZM, Soltitskaya LP (1981) Results of application of the purified sturgeon gonadotropin for stimulation of maturation of the Russian sturgeon (*Acipenser gueldenstaedti* Brandt) females, as compared with the effect of other pituitary preparations. Vopr Ikhtiol 21: 719–725

Barannikova IA, Boev AA, Bukovskaya OS, Efimova NA (1983) Hormonal regulation of the reproductive function in sturgeons and biotechnics of stimulation of the spawners maturation in sturgeon culture. In: Barannikova IA, Berdichevskii LS (eds) Biological foundations of sturgeon culture. Nauka, Moscow, pp 22–42

Barannikova IA, Boev AA, Efimova NA, Presnov DI (1984) Comparative analysis of the hatchery production obtained as a result of stimulation of the Russian sturgeon maturation by synthetic LH–RH and glycerol pituitary preparation. In: Sturgeon Culture in Water Bodies of the USSR (in Russian). Astrakhan', pp 34–36

Barannikova IA, Boev AA, Bukovskaya OS, Efimova NA (1985) Specific features of hypophysial gonadotropic function and kinetics of gonadotropin in the Russian sturgeon blood serum after stimulation with LH–RH versus pituitary preparation. In: Virbitskas YuB (ed) VI All-Union Conference on Ecological physiology and biochemistry of fishes. abstracts, Vilnius, pp 368–369 (in Russian)

Barannikova IA, Boev AA, Bukovskaya OS, Efimova NA, Presnov DI (1989) Analysis of the Russian sturgeon maturation under the influence of LH–RH with reference to possible application of this preparation in fish culture. Vopr Ikhtiol 29: 439–447

Bělehrádek J (1934) Temperature and living matter. Protoplasma-Monographien 8. Borntraeger, Berlin

Belen'kii ML (1963) Elements of quantitative estimation of the pharmacological effect. Medgiz, Leningrad, (in Russian)

Berdichevskii LS, Sokolov LI, Malyutin VS, Smol'yanov II (1979) *Acipenser baeri* of the Lena river as a valuable object of commercial sturgeon culture and acclimatization in the inland water bodies of the USSR. In: Berdichevskii LS (ed) Biological foundations of sturgeon culture development in the water bodies of the USSR, Nauka, Moscow, pp 74–81 (in Russian)

Berg LS (1929) Fish juveniles in the lower Amu Darja flow. Izv Otd Prikladnoi Ikhtiol 9: 225–230

Berg LS (1948) Fish of fresh waters of the USSR and adjacent countries. Izdatel'stvo Akad Nauk SSSR, Moscow, (in Russian)

Berg LS, Bogdanov AS, Kozhin NI, Rass TS (eds) (1949) Commercial fish of the USSR. Pishchepromizdat, Moscow, (in Russian)

Berman DI, Gorgolyuk SI, Neifakh AA (1987) Temperature dependence of rate of embryonic development in Siberian newt and common triton. Sov J Dev Biol 18: 154–161

Boev AA, Artyukhin EN (1978) Selecting of the optimal doses of the tested preparation of acetone-dried pituitaries for stimulation of the spawners' maturation in the lower Volga flow. Tr VNIRO 130: 93–97

Bogdanova LS (1967) Comparative characteristic of the process of transition to active feeding in larvae of various sturgeon species and ecological forms. Tr TsNIORKh 1: 196–201

Bogdanova LS (1967) Ability of the larvae of the Russian and Siberian sturgeons for long-term starvation. In: Peseridi NE, Popova AA (eds) Materials of the scientific session of TsNIORKh dedicated to 100 years of sturgeon culture, Astrakhan', pp 24–26 (in Russian)

Bogdanova LS (1972a) Ecological plasticity of sturgeon larvae and juveniles. In: Marti Yu Yu, Barannikova IA (eds) Sturgeons and problems of sturgeon culture, Pishchevaya Promyshlennost', Moscow, pp 244–250 (in Russian)

Bogdanova LS (1972b) Comparison of the transition of larvae to active feeding at different temperatures in the Russian and Siberian sturgeons. Tr TsNIORKh 4: 217–223

Bordzilovskaya NP, Dettlaff TA, Duhon ST, Malacinski GM (1989) Developmental-stage series of axolotl embryos. In: Armstrong JB, Malacinski GM (eds) Developmental biology of the axolotl. Oxford Univ Press, New York, pp 201–219

Borovik IA (1916) On breeding of sturgeons. Report of the activity of a temporary hatchery station on the Ural river in 1915. Mater Poznaniyu Russ Rybovod 5 (9): 3–65

Brachet A (1910) Recherches sur l'influence de la polyspermie expérimentale dans le développement de l'oeuf de *Rana fusca*. Arch Zool Exp Gén Ser 5 (6): 1–100

Bratsenyuk GN (1974) On breeding of the sterlet in the Saratov water reservoir. Rybn Khoz (8): 13–14

Burtsev IA (1967) Vitellogenesis in the oocytes of a hybrid *Huso huso* L. × *Acipenser ruthenus* L. Dokl Akad Nauk SSSR 172: 464–467

Burtsev IA (1969) Obtaining of progeny from an intergeneric hybrid between the beluga sturgeon and sterlet. In: Cherfas BI (ed) Genetics, selection, and hybridization of fish. Nauka, Moscow, pp 232–242 (in Russian)

Burtsev IA (1971) Gametogenesis and reproductive capacity of interspecific hybrids between sturgeons. Author's Abstract of the Candidate Dissertation. VNIRO, Moscow, (in Russian)

Burtsev IA (1983) Hybridization and selection of sturgeon during full cycle breeding and domestication. In: Kirpichnikov VS (ed) Biological foundations of fish culture: problems of genetics and selection. Nauka, Leningrad, pp 102–113 (in Russian)

Burtsev IA, Serebryakova EV (1969) First experiments of deep freezing of the sturgeon sperm. In: Vladimirskaya EV (ed) Works of young scientists. VNIRO Moscow, 1: pp 94–100 (in Russian)

Burtsev IA, Serebryakova EV (1980) Estimation of the bester (hybrids between *Huso huso* and *Acipenser ruthenus*) spawners according to the cytological indices and the progeny viability. In: Kirpichnikov VS (ed) Karyological variation, mutagenesis, and gynogenesis in fish. Institute of Cytology, USSR Academy of Sciences, Leningrad, pp 63–70 (in Russian)

Burzawa-Gerard E, Goncharov BF, Fontaine YA (1975a) L'hormone gonadotrope hypophysaire d'un poisson chondrostéen, l'Esturgeon (*Acipenser stellatus* Pall.). 1. Purification. Gen Comp Endocrinol 27: 289–295

Burzawa-Gerard E, Goncharov BF, Fontaine YA (1975b) L'hormone gonadotrope hypophysaire d'un poisson chondrostéen, l'Esturgeon (*Acipenser stellatus* Pall.). 2. Propriétés biochimiques. Gen Comp Endocrinol 27: 296–304

Burzawa-Gerard E, Goncharov BF, Dumas A, Fontaine YA (1976) Further biochemical studies on carp gonadotropin (c-6 TH); biochemical and biological comparison of c-6 TH and a gonadotropin from *Acipenser stellatus* Pall. (Chondrostei). Gen Comp Endocrinol 29: 498–505

Bykov NE, Shilov VI (1974) Some data about biotechnical standards of the sterlet breeding. In: Astakhova TV (ed) Abstracts of the accounting session of TsNIORKh. Astrakhan', pp 24–25 (in Russian)

Cherr GN, Clark WH Jr (1982) Fine structure of the envelope and micropyles in the eggs of the white sturgeon, *Acipenser transmontanus* Richardson. Dev Growth Differ 24: 341–352

Cherr GN, Clark WH Jr (1984a) Jelly release in the eggs of the white sturgeon, *Acipenser transmontanus*: an enzymatically mediated event. J Exp Zool 230: 145–149

Cherr GN, Clark WH Jr (1984b) An acrosome reaction in sperm from the white sturgeon, *Acipenser transmontanus*. J Exp Zool 232: 129–139

Cherr GN, Clark WH Jr (1985a) An egg envelope component induces the acrosome reaction in sturgeon sperm. J Exp Zool 234: 75–85

Cherr GN, Clark WH Jr (1985b) Gamete interaction in the white sturgeon *Acipenser transmontanus*: a morphological and physiological review. In: Environ Biol Fish 14: 11–22

Chulitskaya EV (1961a) Latent differentiation of the auditory vesicle material in the Russian and stellate sturgeons. Dokl Akad Nauk SSSR 138: 718–721

Chulitskaya EV (1961b) Correlation between the latent differentiation of the auditory vesicle material and the cell generations at the same developmental stages in sturgeon embryos (beluga, Russian, and stellate sturgeons). Dokl Akad Nauk SSSR 139: 506–509

Chulitskaya EV (1962) A study of the inducing effect of mesoderm on the auditory vesicle material in sturgeons and amphibians. Dokl Akad Nauk SSSR 144: 245–247
Chulitskaya EV (1965) Relative length of cleavage period and gastrulation and latent differentiation of the labyrinth rudiment in embryos of *Rana temporaria* at different temperatures. Dokl Akad Nauk SSSR 160: 489–492
Chulitskaya EV (1970) Desynchronization of cell divisions in the course of egg cleavage and an attempt at experimental shift of its onset. J Embryol Exp Morphol 25: 359–374
Chusovitina LS (1963) Postembryonic development of the Siberian sturgeon (*Acipenser baeri* Brandt). Tr Ob'-Tazov Otd GosNIORKh 3: 103–114
Dan JC (1967) Acrosome reaction and lysins. In: Metz CB, Monroy A (eds) Fertilization, Vol 1, Academic Press, New York, pp 237–293
Dan JC (1970) Morphogenetic aspects of acrosome formation and reaction. Adv Morphog 8: 1–39
Danil'chenko OP (1977) Sensitivity of fish embryos to toxic agents. Vopr Ikhtiol 17: 518–526
Davydova SI (1968) Maturation of Acipenserid oocytes outside the body of a female. Dokl Akad Nauk SSSR 179: 750–752
Davydova SI (1972) Effect of temperature and keeping time of female sturgeons in captivity on maturation of oocytes under the influence of hormones in vitro. Sov J Dev Biol 3: 339–344
Dean B (1895) The early development of gar-pike and sturgeon. J Morphol 11: 1–62
Dean B (1896) On the early development of ganoids. In: CR 3 Congr Int Zool. Leyde, 1895, Hoek P.P.C. (ed), pp 336–346
Derzhavin AN (1922) The stellate sturgeon (*Acipenser stellatus* Pallas), Izv Bakinskoi Ikhtiol Lab 1: 1–393
Derzhavin AN (1938) Experiments on intensive breeding of sturgeons at the Kura Experimental Hatchery in 1936–1937. Rybn Khoz 18, 2: 26–31
Derzhavin AN (1939) Fish culture on the Kura River and the problem of Mingechaur. Izvestiya Azerbaidzhanskoi Nauchno-issledovatel'skoi Rybokhozyaistvennoi Stantsii, pt 2
Derzhavin AN (1947) Reproduction of reserves of Acipenserid fishes. Izdatel'stvo Akad Nauk AzSSR, Baku, (in Russian)
Derzhavin AN, Digurova TM (1951) Toxic effects of naphthene acids on the eggs and fry of Acipenserid fishes. Izv Akad Nauk AzSSR 17: 39–45
Dettlaff TA (1946) Exogastrulation in the embryos of stellate sturgeon *Acipenser stellatus*. Dokl Akad Nauk SSSR 52: 557–558
Dettlaff TA (1953) Dependence of the rate of cleavage of sturgeon eggs on temperature. Dokl Akad Nauk SSSR 91: 695–698
Dettlaff TA (1956) Species-specific differences in morphogenetic properties of the germ material and the shift of gastrulation with reference to the cleavage stages (the role of relationship between developmental stages and cell generations). Dokl Akad Nauk SSSR 111: 1149–1152
Dettlaff TA (1957) Cortical granules and substances secreted from the animal part of the egg during the period of activation in Acipenserids. Dokl Akad Nauk SSSR 116: 341–344
Dettlaff TA (1961) The ovulation and activation of oocytes of Acipenserid fishes in vitro. In: Symp Germ Cells and Dev, Int Inst Embryol Fond A. Baselli, Milano, pp 141–144
Dettlaff TA (1962) Cortical changes in Acipenserid eggs during fertilization and artificial activation. J Embryol Exp Morphol 10: 1–26
Dettlaff TA (1963) Mitotic dynamics of the first cleavage divisions in the eggs of sturgeon (at various temperatures) and of trout. Exp Cell Res 29: 490–503
Dettlaff TA (1964) Cell divisions, duration of interkinetic states and differentiation in early stages of embryonic development. Adv Morphog 3: 323–362
Dettlaff TA (1966) Action of actinomycin and puromycin upon frog oocyte maturation. J Embryol Exp Morphol 16: 183–195
Dettlaff TA (1970a) Influence of environmental temperature during oocyte maturation and ovulation on the quality of sturgeon eggs (on the thermal regime of keeping sturgeons in captivity during the period of obtaining eggs). Tr TsNIORKh 2: 112–126

Dettlaff TA (1970b) Intercellular influences in the course of oocyte maturation in sturgeons. In: Lopashov GV (ed) Intercellular interactions in differentiation and growth. Nauka, Moscow, pp 241–243 (in Russian)

Dettlaff TA (1977) Some thermal-temporal patterns of embryogenesis in poikilothermic animals. In: Belyaev DK (ed) Problems of experimental biology. Nauka, Moscow, pp 269–287 (in Russian)

Dettlaff TA (1981) Adaptation of poikilothermic animals to development in conditions of fluctuating temperatures and the problem of the integrity of the developing organism. Sov J Dev Biol 12: 139–151

Dettlaff TA (1982) Evolution of the structure of the ectoderm and chordamesoderm and their derivatives among *Anamnia*. Sov J Dev Biol 13: 261–269

Dettlaff TA (1983) A study of the properties, morphogenetic potencies and prospective fate of outer and inner layers of ectodermal and chordamesodermal regions during gastrulation in various Anuran amphibians. J Embryol Exp Morphol 75: 67–86

Dettlaff TA (1985) On general principles of the organism integrity in ontogenesis. Zh Obshch Biol 46: 147–152

Dettlaff TA (1986) The rate of development in poikilothermic animals calculated in astronomical and relative time units. J Therm Biol 11: 1–7

Dettlaff TA (1988) Development of the mature egg organization in amphibians, fish and starfish during the concluding stages of oogenesis, in the period of maturation. In: Dettlaff TA, Vassetzky SG (eds) Oocyte growth and maturation. Plenum Press, New York, pp 165–277

Dettlaff TA (1989) A study of temporal patterns of animal development. Ontogenez 20: 647–657

Dettlaff TA (1990) Biological time units τ_0 and τ_s as applied when using the method of relative characteristic of development duration. Ontogenez 21: 646–652

Dettlaff TA (1991) Introduction: temperature and timing in developmental biology. In: Dettlaff TA, Vassetzky SG (eds) Animal species for developmental studies, vol 2. Vertebrates. Consultants Bureau, New York, pp 1–14

Dettlaff TA, Davydova SI (1974) Effect of triiodothyronine on oocyte maturation in the stellate sturgeon after the action of low temperatures and reservation of females. Ontogenez 5: 402–410

Dettlaff TA, Davydova SI (1978) Thermosensitivity of follicles of the stellate sturgeon at different stages of maturation. Sov J Dev Biol 9: 200–204

Dettlaff TA, Davydova SI (1979a) Influence of triiodothyronine on oocyte maturation in the stellate sturgeon *Acipenser stellatus* Pallas under the influence of gonadotropic hypophysial hormones in the hatchery conditions. Vopr Ikhtiol 19: 503–508

Dettlaff TA, Davydova SI (1979b) Differential sensitivity of cells of follicular epithelium and oocyte in the stellate sturgeon to unfavorable conditions and correcting influence of diiodothyronine. Gen Comp Endocrinol 39: 236–243

Dettlaff TA, Davydova SI (1981) Differential sensitivity of the follicular epithelium cells and oocytes of *Acipenser stellatus* to unfavorable conditions and the correcting effect of triiodothyronine. In: Mitskevich MS (ed) Mechanisms of hormonal regulations and the role of feedback in development and homeostasis. Nauka, Moscow, pp 322–331 (in Russian)

Dettlaff TA, Dettlaff AA (1961) On relative dimensionless characteristics of the development duration in embryology. Arch Biol (Liège) 72: 1–16

Dettlaff TA, Dettlaff AA (1982) Dimensionless criteria as a method of quantitative characterization of animal development. In: Zotin Al, Presnov EV (eds) Mathematical developmental biology. Nauka, Moscow, pp 25–39 (in Russian)

Dettlaff TA, Felgenhauer PE (1980) Maturation of the stellate sturgeon oocytes following the injection of the cytoplasm of mature stellate sturgeon eggs and maturing *Xenopus laevis* oocytes and the effect of cycloheximide on this process. Ontogenez 11: 223–228

Dettlaff TA, Ginsburg AS (1950) A case of parthenogenetic cleavage of non-ovulated oocytes in the ovary of *Acipenser stellatus*. Dokl Akad Nauk SSSR 72: 429–432

Dettlaff TA, Ginsburg AS (1954) The embryonic development of Acipenserid fishes (stellate, Russian and giant sturgeon) with reference to the problems of their breeding. Izdatel'stvo Akad Nauk SSR, Moscow (in Russian)

Dettlaff TA, Ginsburg AS (1963) Acrosomal reaction in Acipenserid fishes and the role of calcium ions in gamete fusion. Dokl Akad Nauk SSSR 153: 1461–1464

Dettlaff TA, Rudneva TB (1973) Dimensionless characteristic of the duration of embryonic development of *Xenopus laevis*. Sov J Dev Biol 4: 423–432

Dettlaff TA, Ryabova LV (1986) Maturation of *Rana temporaria* and *Acipenser stellatus* oocytes induced by the cytoplasm of embryos at different cell cycle phases and at different stages of cleavage and blastulation. Cell Differ 18: 9–16

Dettlaff TA, Skoblina MN (1969) The role of germinal vesicle in the process of oocyte maturation in *Anura* and *Acipenseridae*. Ann Embryol Exp Morphol (Suppl) 1: 133–151

Dettlaff TA, Vassetzky SG (eds) (1990) Animal species for developmental studies, vol 1. Invertebrates. Consultants Bureau, New York

Dettlaff TA, Vassetzky SG (eds) (1991) Animal species for developmental studies, vol 2. Vertebrates. Consultants Bureau, New York

Dettlaff TA, Zubova CE (1962) The ratio between durations of the periods of maturation and embryogenesis in the Russian and stellate sturgeons. Dokl Akad Nauk SSSR 143: 746–748

Dettlaff TA, Zuichenko SI (1963) Metaphase of the I maturation division in sturgeon oocytes. Dokl Akad Nauk SSSR 152: 246–248

Dettlaff TA, Nikitina LA, Stroeva OG (1964) The role of the germinal vesicle in oocyte maturation in anurans as revealed by the removal and transplantation of nuclei. J Embryol Exp Morphol 12: 851–873

Dettlaff TA, Vassetzky SG, Davydova SI (1965) Recommendations for obtaining eggs in Acipenserid fishes after pituitary injection. Glavrybvod, Moscow (in Russian)

Dettlaff TA, Skoblina MN, Davydova SI (1968) Intercellular influences during oocyte maturation in Acipenserid fishes. In: Tumanishvili GD (ed) Symposium Intercellular interactions in differentiation and growth. Abstracts, Tbilisi, pp 5–6 (in Russian)

Dettlaff TA, Felgenhauer PE, Chulitskaya EV (1977) Effect of cycloheximide on germinal vesicle breakdown and changes in the cortical layer in clawed toad and sturgeon oocytes during their maturation under the influence of active cytoplasm. Sov J Dev Biol 8: 417–423

Dettlaff TA, Felgenhauer PE, Stepanov AS, Chulitskaya EV (1980) Protein synthesis in the oocytes of the stellate sturgeon at different stages of maturation and effect of its inhibition on changes in the oocyte. Sov J Dev Biol 11: 17–22

Dettlaff TA, Ginsburg AS, Schmalhausen OI (1981) Development of Acipenserid fishes. Maturation of oocytes, fertilization, development of embryos and prelarvae. Nauka, Moscow (in Russian)

Dettlaff TA, Ignatieva GM, Vassetzky SG (1987) The problem of time in developmental biology: its study by the use of relative characteristics of development duration. Sov Sci Rev F Physiol Gen Biol 1: 1–88

Deuchar EM (1958) Regional differences in catheptic activity in *Xenopus laevis* embryos. J Embryol Exp Morphol 6: 223–237

Disler NN (1949) Development of cutaneous sense organs in the lateral system of the stellate sturgeon *Acipenser stellatus*. Tr Inst Morfol Zhivotn Akad Nauk SSSR 1: 333–362

Disler NN (1960) Sence organs of the lateral line system and their role in fish behavior. Izdatel'stvo Akad Nauk SSSR, Moscow (in Russian)

Dodswell MJ (1976) Biology of the shovelnose sturgeon *Acipenser brevirostrum* Le Sueur in the St John River estuary, New Brunswick. In: Trans Can Soc Environ Biol, Atl Chapter Annu Meet 1975 (cited from Jones et al. 1978)

Doinikov KG (1936) Materials on biology and estimation of reserves of Acipenserid fishes in the Azov Sea. Raboty Dono-Kubanskoi Rybokhoz Stantsii 4: 1–213

Doroshin GYa, Troitskii SK (1949) Characterization of the breeding conditions of the stellate sturgeon in the Kuban' River in 1944–1947. Tr Rybovod biol Lab Az-CherRybVoda 1: 111–130

Doroshov SI, Clark WH, Lutes PB, Swallow RL, Beer KE, McGuire AN, Cochran MD (1983) Artificial propagation of the white sturgeon, *Acipenser transmontanus* Richardson. Aquaculture 32: 93–104

Doudoroff P, Katz M (1953) Critical review of literature on the toxicity of industrial wastes and their components to fish 2. The metals, as salts. Sewage Ind Wastes 25: 802–839

Dragomirov NI (1953a) Species specific features of sturgeon larvae at the stage of hatching. Dokl Akad Nauk SSSR 93: 551–554

Dragomirov NI (1953b) Development of the stellate sturgeon larvae during the period of yolk feeding. Tr Inst Morfol Zhivotn Akad Nauk SSSR 10: 244–263

Dragomirov NI (1954) Development of cutaneous receptors on the lower head side in the Russian sturgeon larvae during their transition to the benthic way of life. Dokl Akad Nauk SSSR 97: 173–176

Dragomirov NI (1957) Larval development of the Russian sturgeon *Acipenser gueldenstaedti* Brandt. Tr Inst Morfol Zhivotn Akad Nauk SSSR 20: 187–231

Dragomirov NI (1961a) Development of the lateral system of organs in the giant sturgeon larvae. Zh Obshch Biol 22: 273–280

Dragomirov NI (1961b) Ecological-morphological features of larval development in the giant sturgeon *Huso huso* L. Tr Inst Morfol Zhivotn Akad Nauk SSSR 33: 72–93

Dragomirov NI (1968) Gradients of development of the temporal follicle rudiments in the Russian sturgeon. Dokl Akad Nauk SSSR 181: 762–764

Dragomirov NI (1971) Pathways and patterns of integration of organogenesis in the supraorbital region of lateral system in the Russian sturgeon larvae. Dokl Akad Nauk SSSR 196: 478–481

Dragomirov NI, Schmalhausen OI (1952) Ecological-morphological features of larvae in *Pseudoscaphirhynchus*. Dokl Akad Nauk SSSR 85: 1399–1402

Emel'yanov SV (1951) Influence of demucilaging of the Russian and stellate sturgeon eggs on hatching. Dokl Akad Nauk SSSR 78: 371–374

Emel'yanov SV (1953) Influence of demucilaging of the Russian and stellate sturgeon eggs on their development. In: Pavlovskii EN (ed) Proc All-Union Conf on the Problems of fisheries, 1951. Izdatel'stvo Akad Nauk SSSR, Moscow, pp 325–339 (in Russian)

Emel'yanov SV (1961) Influence of different condition of the egg membranes (sticky or demucilaged) on embryogenesis in the Russian and stellate sturgeon. Tr Inst Morfol Zhivotn Akad Nauk SSSR 37: 67–219

Emel'yanov SV, Sytina LA (1967) Larvae of Acipenserid fishes at hatching, their structure and variability. Tr TsNIORKh 1: 152–162

Engel'gardt VA, Tatarskaya RI, Kafiani KA, Kanopkaite SI (1957) Demucilaging of sturgeon eggs with chalk. Rybn Khoz 10: 52–55

Faleeva TI (1965) Analysis of atresia of the fish oocytes with reference to the adaptive importance of this phenomenon. Vopr Ikhtiol 5: 455–470

Faleeva TI (1970) Some data about the nature of so-called "broken" eggs in sturgeon culture. Tr TsNIORKh 2: 132–136

Faleeva TI (1979) Comparative and experimental analysis of defects of oogenesis in fishes. Author's Abstract of Candidate Dissertation. LGU, Leningrad (in Russian)

Flerov BA (1965) An experimental study of phenol poisoning of fishes (in Russian). Author's Abstract of Candidate Dissertation KGU, Moscow

Flerov BA (1973) Experimental studies of phenol poisoning in fishes. Tr Inst Biol Vnutrennikh Vod 24: 5–38

Flickinger RA (1956) The relation of phosphoprotein phosphatase activity to yolk platelet utilization in the amphibian embryo. J Exp Zool 131: 307–332

Fontana F, Colombo G (1974) The chromosomes of Italian sturgeons. Experientia 30: 739–742

Gabaeva NS (1974) Development of the follicle and formation of the egg membrane during oogenesis in the Russian sturgeon. Biol Nauki (12): 15–21

Gerbil'skii NL (1941) Method of pituitary injections and its role in fish culture. In: Gerbil'skii

NL (ed) Method of pituitary injections and its role in reproduction of fish resources. LGU Press, Leningrad, pp 5–35 (in Russian)
Gerbil'skii NL (1947) Current state and perspectives of the method of pituitary injections in fish culture. Tr Lab Osnov Rybovod 1: 5–24
Gerbil'skii NL (1949) Experimental and methodical bases of development of the sturgeon culture in the lower Kuban flow. Tr Lab Osnov Rybovod 2: 5–28
Gerbil'skii NL (1951) Intraspecific biological groups of sturgeons and their reproduction in the lower flow of regulated rivers. Rybn Khoz 27 (4): 24–27
Gerbil'skii NL (1957) Histophysiological analysis of the digestive system in sturgeons and teleosteans during the early developmental period and the method of rearing larvae in fish culture. In: Proc Conf on Fish culture, 1954, Izdatel'stvo Akad Nauk SSSR, Moscow, pp 89–94 (in Russian)
Gerbil'skii NL (1962) Theory of biological progress of the Acipenserid fishes and its use in sturgeon culture. Uch Zap LGU No 311 Ser Biol Sci (48): 5–18
Gerbil'skii NL, Barannikova IA, Kazanskii BN (1951) Stock material for rearing sturgeon fry. Rybn Khoz 27(9): 46–49
Gerhart J, Wu M, Kirschner M (1984) Cell cycle dynamics of an M-phase specific cytoplasmic factor in *Xenopus laevis* oocytes and eggs. J Cell Biol 98: 1247–1255
Gilbert SF (1988) Developmental biology, 2nd edn. Sinauer Associates, Sunderland, MA
Ginsburg AS (1950) Species specific features of the early stages of the labyrinth development in amphibians. Dokl Akad Nauk SSSR 73: 229–232
Ginsburg AS (1953a) Establishment of bilateral symmetry in the eggs of Acipenserid fishes. Dokl Akad Nauk SSSR 90: 477–480
Ginsburg AS (1953b) Disturbances in development of Acipenserid fishes related to the conditions of insemination. Dokl Akad Nauk SSSR 92: 1097–1100
Ginsburg AS (1954) Variations of the furrowing pattern in Acipenserid fishes. Dokl Akad Nauk SSSR 95: 1117–1120
Ginsburg AS (1957) The time of establishment of the contact between the egg and spermatozoon in sturgeons. Dokl Akad Nauk SSSR 115: 845–848
Ginsburg AS (1959) Fertilization in Acipenserid fishes. 1. The fusion of gametes. Tsitologiya 1: 510–526
Ginsburg AS (1961) The block to polyspermy in sturgeon and trout with special reference to the role of cortical granules (alveoli). J Embryol Exp Morphol 9: 173–190
Ginsburg AS (1963a) Sperm-egg association and its relationship to the activation of the egg in Salmonid fishes. J Embryol Exp Morphol 11: 13–33
Ginsburg AS (1963b) Instruction for artificial insemination of the eggs of Acipenserid fishes. Glavrybvod, Moscow (in Russian)
Ginsburg AS (1968) Fertilization in fishes and the problem of polyspermy. Nauka, Moscow (in Russian)
Ginsburg AS (1972) Fertilization in fishes and the problem of polyspermy. Israel Program for Scientific Translations, Jerusalem
Ginsburg AS (1977) Ultrastructure of the spermatozoon and the acrosomal reaction in the stellate sturgeon. In: Belyaev DK (ed) Problems of experimental biology. Nauka, Moscow, pp 246–256 (in Russian)
Ginsburg AS (1987) Egg cortical reaction during fertilization and its role in block to polyspermy. Sov Sci Rev F Physiol Gen Biol 1: 307–375
Ginsburg AS (1989) Determination of ectoderm derivatives in various amphibian species. Sov J Dev Biol 20: 18–27
Ginsburg AS, Dettlaff TA (1944) Experiments on transplantation and removal of organ rudiments in embryos of *Acipenser stellatus* in early developmental stages. C R Acad Sci URSS 44: 209–212
Ginsburg AS, Dettlaff TA (1955) Development of the embryos of Acipenserid fishes. Izdatel'stvo Akad Nauk SSSR, Moscow (in Russian)
Ginsburg AS, Dettlaff TA (1969) Development of Acipenserid fishes. Oocyte maturation, fertilization, and embryogenesis. Nauka, Moscow (in Russian)

Ginsburg AS, Dettlaff TA (1975) The Russian sturgeon *Acipenser gueldenstaedti*. In: Dettlaff TA (ed) Objects of developmental biology. Nauka, Moscow, pp 217–263 (in Russian)

Ginsburg AS, Dettlaff TA (1991) The Russian sturgeon *Acipenser gueldenstaedti*. Part I. gametes and early development up to time of hatching. In: Dettlaff TA, Vassetzky SG (eds) Animal species for developmental studies. Vol 2. Vertebrates. Consultants Bureau, New York, pp 15–65

Ginsburg AS, Nikiforova GP (1978) Cortical reaction in the stellate sturgeon egg during fertilization and artificial activation. Sov J Dev Biol 9: 189–199

Ginsburg AS, Nikiforova GP (1979) Development of the cortical reaction during oocyte maturation in *Acipenser stellatus*. Sov J Dev Biol 10: 385–392

Ginsburg AS, Zubova SE, Filatova LA (1963) An efficient method of artificial insemination of Acipenserid eggs. In: Pavlovskii EN (ed) Sturgeon culture in water bodies of the USSR. Izdatel'stvo Akad Nauk SSSR, Moscow, pp 47–55 (in Russian)

Ginzburg YaI (1939) Materials on taxonomy of sturgeon fry in the South Caspian and a key to their larvae and juveniles. Tr Zool Inst Azerb Fil Akad Nauk SSSR 10: 77–108

Ginzburg YaI (1967) Role of larval consumers in decreasing the efficiency of spawning of sturgeons after the Volga flow regulation at Volgograd. Vopr Ikhtiol 7: 480–490

Goncharov BF (1971a) Functional relationship between length of the hormone-dependent period of follicle maturation in the common frog and the concentration of pituitary suspension. Sov J Dev Biol 2: 50–55

Goncharov BF (1971b) Studies of the pattern of transition of amphibian and sturgeon oocytes from growth to maturation. Author's Abstract of the Candidate Dissertation. Inst Biol Razv Akad Nauk SSSR, Moscow (in Russian)

Goncharov BF (1972) Estimation of gonadotropic activity of sturgeon pituitary according to the reaction of in vitro oocyte maturation. In: Marti Yu Yu, Barannikova IA (eds) Sturgeons and the problems of sturgeon culture. Pishchevaya Promyshlennost', Moscow, pp 257–262 (in Russian)

Goncharov BF (1976) The physiological condition of the elder generation of the stellate sturgeon follicles during spawning migration. In: Abstracts III All-Union Conf on Ecological physiology of fishes. Naukova Dumka, Kiev, pt 2, pp 139–140 (in Russian)

Goncharov BF (1977) Hormonal regulation of vitellogenesis and maturation in fishes and amphibians. In: Dettlaff TA (ed) Current problems of oogenesis. Nauka, Moscow, pp 173–199 (in Russian)

Goncharov BF (1978) Influence of the composition of culture medium on the capacity of sturgeon follicles to respond to the effect of gonadotropic hormones by maturation. In: Abstracts II All-Union Conf on the Problems of early ontogenesis of fishes. Naukova Dumka, Kiev, pp 77–78 (in Russian)

Goncharov BF (1981) The use of the method of pituitary injection in fish culture. Some results and perspectives. In: Koshelev BV, Gulidov MV (eds) Studies of fish reproduction and development (methodical recommendations). Nauka, Moscow, pp 16–48 (in Russian)

Goncharov BF (1984a) Application of synthetic LH–RH analogs for inducing maturation of gametes in Acipenserid fishes. In: Abstracts All-Union Conf Sturgeon culture in water bodies of the USSR, Dec 11–14, 1984. Astrakhan', pp 94–95 (in Russian)

Goncharov BF (1984b) A synthetic analog of LH–RH as a new promising stimulus of maturation of the gametes in Acipenserid fishes. Dokl Akad Nauk SSSR 276: 1002–1006

Goncharov BF (1985) Perspectives of utilization of synthetic LH–RH analogs when obtaining mature gametes from sturgeon spawners. In: Ecological physiology and biochemistry of fishes, Sept 1985, Abstracts, Vil'nyus, pp 384–385 (in Russian)

Goncharov BF (1990) A new method of obtaining mature gametes from spawners in Acipenserid fishes. In: Orlov VM, Chernyak LE (eds) Fundamental sciences for the national economy. Nauka, Moscow, pp 297–300 (in Russian)

Goncharov BF, Kuznetsov AA (1981) Differences in the biological action of isoforms of the gonadotropic hormone in the stellate sturgeon (*Acipenser stellatus*) and identify of their spectra in female and male pituitaries. Dokl Akad Nauk SSSR 258: 243–246

Goncharov BF, Burzawa-Gerard E, Fontaine YA (1976) Comparative study of the biological action of purified gonadotropins of the starred sturgeon (*Acipenser stellatus*) and the carp (*Cyprinus carpio*). Sov J Dev Biol 7: 68–72

Goncharov BF, Kuznetsov AA, Burzawa-Gerard E (1980) Analysis of heterogeneity of the gonadotropic hypophysial hormone in the stellate sturgeon. Biokhimiya 45: 455–462.

Goncharov BF, Kuznetsov AA, Burzawa-Gerard E (1983) Pituitary gonadotropic hormone from a chondrostean fish, starred sturgeon (*Acipenser stellatus* Pall.). 4. Differences in biological action of isoforms and absence of sex-specific forms. Gen Comp Endocrinol 49: 375–382.

Goncharov BF, Shubravyi OI, Uteshev VK (1986) On the problem of conservation of the big Amu-dar shovelnose by creating populations maintained in captivity. In: Formation of sturgeon resources under conditions of complex use of aquatic resources. Abstracts, TsNIORKh, Astrakhan', pp 75–76 (in Russian)

Goncharov BF, Igumnova LV, Polupan IS, Savelieva EA (1989) Induced oocyte maturation, ovulation and spermiation in sturgeons (*Acipenseridae*) using synthetic analogue of gonadotropin-releasing hormone. In: Summaries of oral communications, posters, films of the 1st Int Sym on the Sturgeon, 3–6 Oct 1989, Bordeaux. CEMAGREF, Bordeaux, p 34

Goncharov BF, Igumnova LV, Polupan IS, Savelieva EA (1991) Induced oocyte maturation, ovulation and spermiation in sturgeons (*Acipenseridae*) using synthetic analogue of gonadotropin-releasing hormone. In: *Acipencer*. Aetes du premier colloque international sur l'esturgeon, Bordeaux 3–6 Octobre 1989, Bordeaux, CEMAGREF, pp 354–364

Gordienko OL (1953) Rearing of the giant sturgeon fry. Gizlegpishcheprom, Moscow (in Russian)

Grodziński Z (1963) The yolk morphology of *Acipenser gueldenstaedti colchicus* Brandt. Acta Biol Cracov Ser Zool 6: 97–110

Gross PR (1952) An enzymatic study of yolk platelet lysis. Biol Bull 103: 281

Gross PR (1954) On the mechanism of the yolk-lysis reaction. Protoplasma 43: 416–428

Haldane JBS (1930) Enzymes. Lougmans & Green, London

Harris DL (1946) Phosphoprotein phosphatase, a new enzyme from the frog egg. J Biol. Chem 165: 541–550

Hayes FR (1949) The growth, general chemistry and temperature reactions of salmonid eggs. Q Rev Biol 24: 281–308

Herlant M (1911) Recherches sur les oeufs di- et trispermiques de grenouille. Arch Biol 26: 103–336

Hewitt EJ, Nicholas DJD (1963) Cations and anions: inhibitions and interactions in metabolism and in enzyme activity. In: Hochster RM, Quastel JH (eds) Metabolic inhibitors vol 2. A comprehensive treatise. Academic Press, New York, pp 311–436

Hörstadius S (1950) The neural crest. Oxford Univ Press, London

Horton HF, Ott AG (1976) Cryopreservation of fish spermatozoa and ova. J Fish Res Board Can 33: 995–1000

Ignatieva GM (1957a) A morphophysiological study of the hatching gland in the giant sturgeon- *Huso huso* (L.). Dokl Akad Nauk SSSR 114: 908–911

Ignatieva GM (1957b) A morphophysiological study of the hatching gland in the stellate sturgeon at different temperatures of incubation. Dokl Akad Nauk SSSR 114: 1132–1135

Ignatieva GM (1958) On conditions of synchronous hatching during incubation of sturgeon eggs. Rybn Khoz (3): 20–25

Ignatieva GM (1979) Early embryogenesis of fishes and amphibians (comparative analysis of the temporal patterns of development). Nauka, Moscow (in Russian)

Igumnova LV (1974) A characteristic of the duration of maturation of the giant sturgeon (*Huso huso*) females after pituitary injection at different temperatures. Vopr Ikhtiol 14: 1029–1033

Igumnova LV (1975a) Chronological patterns of embryonic development of the beluga. Sov J Dev Biol 6: 38–43

Igumnova LV (1975b) Instruction for biotechnics of hatchery reproduction of the giant sturgeon. Glavrybvod, Moscow (in Russian)
Igumnova LV (1979) Temporal patterns of maturation and embryogenesis of the giant sturgeon with reference to the improvement of biotechnics of its breeding. Author's Abstract of the Candidate Dissertation. Inst Biol Rasv Acad Nauk SSR, Moscow (in Russian)
Igumnova LV (1985a) Temporal patterns of embryogenesis in the sterlet. Ontogenez 16: 67–73
Igumnova LV (1985b) Effects of temperature drops on embryogenesis of *Huso huso* (L.) and *Acipenser stellatus* Pallas (*Acipenseridae*). Vopr Ikhtiol 25: 478–482
Igumnova LV, Dettlaff TA (1983) The temporal patterns of maturation in the stellate sturgeon *Acipenser stellatus* Pallas (*Acipenseridae*) from the Volga spawning population. Vopr Ikhtiol 23: 76–80
Igumnova LV, Dubinin VI (1987) Effect of changes and short-term drops in temperature during embryogenesis of *Huso huso* on the structure and survival of prelarvae. Vopr Ikhtiol 27: 515–518
Igumnova LV, Goncharov BF (1986) Obtaining of mature gametes from the giant sturgeon females through the use of a synthetic LH–RH analog. In: Formation of sturgeon reserves under the conditions of complex utilization of aquatic resources. Abstracts, TsNIORKh, Astrakhan', p 111 (in Russian)
Ivanova AL (1954) Thyroid gland of the Russian and stellate sturgeon during spawning migration and spawning. Dokl Akad Nauk SSSR 98: 693–696
Jaana H, Yamamoto TS (1981) The ultrastructure of spermatozoa with a note on the formation of the acrosomal filament in the lamprey, *Lampetra japonica*. Jpn J Ichthyol 28: 135–147
Jacobson AG (1966) Inductive processes in embryonic development. Science 152: 25–34
Jones JRE (1938) The relative toxicity of salts of lead, zinc and copper to stickleback (*Gasterosteus aculeatus* L.) and the effect of calcium on the toxicity of lead and zinc salts. J Exp Biol 15: 394–407
Jones JRE (1939) Antagonism between salts of the heavy and alkaline-earth metals in their toxic action on the tadpole of the toad, *Bufo bufo* (L.). J Exp Biol 16: 313–333
Jones PW, Martin FD, Hardy JD Jr (1978) Development of fishes of the Mid-Atlantic Bight. An atlas of eggs, larval and juvenile stages, vol 1. *Acipenseridae* through *Ictaluridae*. Fish and Wildlife Service, US Department of Interior
Kalashnikov GN, Skadovskii SN (1940) Observations of the physiology of Acipenserid fishes during reproduction with reference to the problem of artificial propagation. Zool Zh. 19: 671–679
Kasimov RYu, Kasimov MA, Guseinov MSh, Sidorov PA (1964) Biotechnics of breeding of Acipenserid fishes at the Kura experimental hatchery. Tr VNIRO 56: 25–37
Kasimov RYu, Dzhafarov AI, Periligin VV, Gasanov GI, Rustamova ShA (1974) Influence of deep freezing of the sperm of *Acipenser güldenstädti* and *A. stellatus* on their resistance and fertilizing capacity. In: Astakhova TV (ed) Materials of the accounting session of TsNIORKh Astrakhan', pp 60–61 (in Russian)
Kazanskii BN (1951) New data about introduction of the Russian and stellate sturgeons of the Kura River in culture. Rybn Khoz 27(1): 31–36
Kazanskii BN (1952) Completion of ovulation outside the female body in sturgeons. Dokl Akad Nauk SSSR 83: 965–968
Kazanskii BN (1953a) On maturation and fertilization of the Russian sturgeon eggs. Dokl Akad Nauk SSSR 89: 757–760
Kazanskii BN (1953b) Reproduction and breeding of the Kura River sturgeon in autumn. Dokl Akad Nauk SSSR 89: 957–960
Kazanskii BN (1957a) Analysis of phenomena occurring in the sturgeon oocytes upon application of pituitary injections. In: Proc Conf on Fish breeding. Izdatel'stvo Akad Nauk SSSR, Moscow, pp 130–138 (in Russian)
Kazanskii BN (1957b) Rationalization of the sturgeon culture on the Kura River based on the analysis of intraspecific biological groups. Uch Zap LGU No 228 Ser Biol Nauk 44: 33–53

Kazanskii BN (1962) Experimental analysis of the seasonal pattern in reproduction of the Volga sturgeons with reference to intraspecific biological differentiation. Uch Zap LGU No 311 Ser Biol Nauk (48): 19–45

Kazanskii BN (1963) Obtaining of different season progeny of fishes in order to provide for repeated fish culture cycles (sturgeons taken as an example). In: Pavlovskii SN (ed) Sturgeon culture in water bodies of the USSR. Izdatel'stvo Akad Nauk SSSR, Moscow, pp 56–64 (in Russian)

Kazanskii BN (1975) Patterns of gametogenesis and ecological plasticity of fish reproduction. In: Kazanskii BN (ed) Ecological plasticity of gonadal cycles and reproduction in fishes. LGU Press, Leningrad, pp 3–32 (in Russian)

Kazanskii BN (1979) Ecological-evolutionary principles of the sturgeon culture organization in the south sea basin of the USSR. In: Berdichevskii LS (ed) Biological foundations of the sturgeon culture development in water bodies of the USSR. Nauka, Moscow, pp 22–33 (in Russian)

Kazanskii BN, Molodtsov AN (1973) Methods of handling sturgeon spawners in production lines with regulated water temperature. Tr VNIRO 92: 21–33

Kazanskii BN, Molodtsov AN (1974) Recommendations for handling sturgeon spawners, migrants of different types, according to the continuous regime in production lines with regulated water temperature. Glavrybvod, Moscow (in Russian)

Kazanskii BN, Nusenbaum LM (1947) The loach as an object for estimating the gonadotropic activity of the pituitary preparations in fish. Tr Lab Osnov Rybovod 1: 111–120

Kazanskii BN, Feklov YuA, Podushka SB, Molodtsov AN (1978) Express method for determining the degree of gonad maturity in sturgeon spawners. Rybn Khoz (2): 24–27

Khakimullin AA (1979) Duration of maturation of the Siberian sturgeon after pituitary injection. Rybn Khoz (8): 20

Khakimullin AA (1987) Oxypreferendum of hatchery produced juvenile Siberian sturgeon *Acipenser baeri*. Vopr Ikhtiol 27: 693–696

Khakimullin AA, Molodykh LG (1985) Metabolic rates of the developing eggs of *Acipenser baeri* Brandt. Vopr Ikhtiol 25: 1001–1007

Khoroshko PN (1968) Ecology and efficiency of reproduction of sturgeons in the lower Volga flow. The Author's Abstract of the Candidate Dissertation. GosNIORKh, Astrakhan' (in Russian)

Khoroshko PN, Pashkin LM, Vlasenko AD (1974) Disturbance of the hydrological regime and the yield of Russian sturgeons in 1973. In: Astakhova TV (ed) Abstracts, TsNIORKh Session, Astrakhan', p 165 (in Russian)

Kokhanskaya EM (1965) Instruction for using the trough incubator of the Sadov and Kokhanskaya system, for sturgeon eggs. Glavrybvod, Moscow (in Russian)

Kokhanskaya EM (1980) The method of reproduction of fish with sticky egg membranes without demucilaging. Nauka, Moscow (in Russian)

Kolodkova LG, Shevchenko VN (1976) The result of rearing sturgeon larvae until the transition to active feeding in ponds with different types of nets. In: Abstracts of the accounting session of the TsNIORKh (1971–1975), TsNIORKh, Gur'ev, p 84 (in Russian)

Korzhuev PA (1941) Oxygen consumption by the eggs and fry of the Russian sturgeon (*Acipenser gueldenstaedti*) and stellate sturgeon (*Acipenser stellatus*). Izv Akad Nauk SSSR Ser Biol (2): 291–302

Korzhuev PA, Sharkova LB (1967) On peculiarities of digestion of the Russian sturgeon in the Caspian Sea. In: Karzinkin GS (ed) Metabolism and biochemistry of fishes, Nauka, Moscow, pp 205–209 (in Russian)

Korzhuev PA, Nikol'skaya IS, Radzinskaya LI (1960) Respiration of sturgeon eggs during their incubation. Vopr Ikhtiol 14: 113–118

Kowalewsky A, Owsjannikow Ph, Wagner N (1870) Die Entwicklungsgeschichte der Störe. Vorläufige Mitteilung. Bull Acad Sci St Petersb 14: 317–325

Kozhin NI (1964) Acipenserid fishes of the USSR and their reproduction. Tr VNIRO 52: 21–58

Kozhin NI (1970) The sterlet and reproduction of its resources. Tr TsNIORKh (2): 28–33
Kozlov VI (1990) The situation is catastrophic. Rybn Khoz (6): 60–63
Krönig (Chulitskaya) EV (1960) On the correlation between the processes of cleavage and gastrulation in *Acipenser gueldenstaedti* and *A. stellatus*. Dokl Akad Nauk SSSR 134: 984–986
Krykhtin ML (1979) Current state and perspectives of the sturgeon culture in the Amur river basin. In: Berdichevskii LS (ed) Biological foundations of the sturgeon culture development in the water bodies of the USSR, Nauka, Moscow, pp 68–73 (in Russian)
Krylova VD (1970) Early stages of development of a second generation hybrid between the giant sturgeon and sterlet. TrVNIRO 76: 231–237
Kryzhanovskii SG (1934) Die Atmungsorgane der Fischlarven (Teleostomi). Zool Jahrb Abt Anat Ontogenie 58:21–60
Kryzhanovskii SG (1949) Ecological-morphological patterns of development of the *Cyprinoidei* and *Siluroidei*. Tr Inst Morfol Zhivot Akad Nauk SSSR 1: 5–332
Kulinchenko PM (1939) Ecology of spawning of Acipenserid fishes. Author's Abstract of the Candidate Dissertation. VNIRO, Moscow (in Russian)
Kusa M (1956) Studies on cortical alveoli in some teleostean eggs. Embryologie 3: 105–129
Kuznetsov AA, Goncharov BF, Burzawa-Gerard E (1983) Pituitary gonadotropic hormone from a chondrostean fish, starred sturgeon (*Acipenser stellatus* Pall.). 3. Polymorphism. Gen Comp Endocrinol 49: 364–374
Landauer N (1947) Insulin-induced abnormalities of beak, extremities and eyes in chickens. J Exp Zool 105: 145–172
Lloyd R (1961) Effect of dissolved oxygen concentrations on the toxicity of concentrations of several poisons to rainbow trout (*Salmo gairdneri* Richardson). J Exp Biol 38: 447–455
Lukin AV (1947) The main features of the ecology of sturgeons in the middle Volga flow. Tr Ova Estestvoz Pri Kazanskom Univ 57 (3/4): 1–143
Lukin AV (1979) The sterlet in the Kuibyshev water reservoir. In: Berdichevskii LS (ed) Biological foundations of the sturgeon culture development in the water bodies of the USSR. Nauka, Moscow, pp 146–154 (in Russian)
Luk'yanenko VI (1965) General characteristics and the phase pattern of the course of phenol intoxication of the crucian carp in the light of "stress" theory. Vopr Ikhtiol 5: 540–547
Luk'yanenko VI (1967a) Physiological aspects of the problem of the waste poison influence on fishes. Vopr Ikhtiol 7: 547–565
Luk'yanenko VI (1967b) Toxicology of fishes, Pishchevaya Promyshlennost', Moscow (in Russian)
Luk'yanenko VI, Flerov BA (1965) Species-specific features of fish sensitivity and resistance to phenol. Gidrobiol Zh 1 (2): 48–53
L'vov LF (1976) Breeding and rearing of the sterlet in the conditions of the Volga delta. In: Abstracts of the accounting session TsNIORKh (1971–1975), TsNIORKh Gur' ev, pp 85–86
Magnin E (1959) Répartition actuelle des Acipenseridés. Rev Trav Inst Pêches Marit 23: 277–285
Mailyan RA, Alekperov AP (1977) An experience of using the stellate sturgeon spawners caught in the sea in fish culture. Rybn Khoz (9): 28–29
Mamedov TF (1977) Hatchery reproduction of all spawning populations of the Kura stellate sturgeon. Rybn Khoz (5): 20–21
Markov KP (1975) A study of the egg membrane microstructure in the Russian sturgeon *Acipenser gueldenstaedti* Brandt using scanning electron microscopy. Vopr Ikhtiol 15: 822–832
Markov KP (1978) On the stickiness of egg envelopes in the *Acipenseridae*. Vopr Ikhtiol 18: 483–493
Marti YuYu (1964) Preface. In: Acipenserid fishes of the southern seas of the USSR (biology, fishery, reproduction). Tr VNIRO 52: 7–19
Maslov PA (1919) Instruction for artificial propagation of sturgeons. Astrakhan Ikhtiol Lab, Astrakhan'

Masui Y (1967) Relative roles of the pituitary, follicle cells, and progesterone in the induction of oocyte maturation in *Rana pipiens*. J Exp Zool 166: 365–367
Masui Y, Markert CL (1971) Cytoplasmic control of nuclear behavior during meiotic maturation of frog oocytes. J Exp Zool 177: 129–146
Matveev BS (1951) On tasks of studying the developmental biology of Acipenserid fishes under the conditions of artificial breeding. Tr Inst Morfol Zhivot Akad Nauk SSSR 5: 123–129
Matveev BS (1953) On biological stages in postembryonic development of Acipenserid fishes. Zool Zh 32: 249–255
Meshcheryakov AI, Bikbaev NKh (1973) A device for demucilaging and washing of fish eggs. Rybn Khoz (12): 15–17
Metelev VV, Kanaev AI, Dzasokhova NG (1971) Aquatic toxicology. Kolos, Moscow (in Russian)
Mil'shtein VV (1972) Sturgeon culture, Pishchevaya Promyshlennost', Moscow (in Russian)
Mironov OG (1973) Oil pollutions and life of the sea. Naukova Dumka, Kiev (in Russian)
Molodtsov AN (1971) Kepping of sturgeon spawners at low temperatures. Rybn Khov (6): 17–18
Molodtsov AN (1972) Physiological stimulation of maturation of the hiemal Russian sturgeon spawners. Rybn Khoz (6): 20–21
Molodtsov AN (1975) On new possibilities of the stellate sturgeon breeding at the hatcheries. Rybn Khoz (10): 21–22
Molodtsov AN (1978) An experience of introduction of the hiemal stellate sturgeon in culture. Rybn Khoz (3): 26–28
Molodtsov AN (1979) Biotechnics of the hatchery utilization of the hiemal Russian sturgeon of summer migration. In: Berdichevskii LS (ed) Biological foundations of the sturgeon culture development in water bodies of the USSR. Nauka, Moscow, pp 85–92 (in Russian)
Molodtsov AN, Meshcheryakov AI (1972) The use of the hiemal Russian sturgeon of autumn migration for hatchery purposes. Rybn Khoz (2): 15–16
Montalembert de J, Jalabert B, Bry C (1978) Precocious induction of maturation and ovulation in the northern pike (*Esox lucius*). Ann Biol Anim Biochim Biophys 18: 969–975
Motlokh NN, Poponov SYu, Podolsky IYa (1989) Enhancement of maturation and releasing of gametes in fish induced by combined administration of LH–RH superactive analogue and sulpiride, a D-2 antagonist. In: Abstracts of the symp The Signal molecules and mechanisms of animal behavior Biol Res Centre USSR Acad Sci, Pushchino, p 41 (in Russian)
Newport J, Kirschner M (1982a) A major developmental transition in early *Xenopus* embryos. 1. Characterization and timing of cellular changes at the midblastula stage. Cell 30: 675–686
Newport J, Kirschner M (1982b) A major developmental transition in early *Xenopus* embryos. 2. Control of the onset of transcription. Cell 30: 687–696
Nicander L, Sjöden I (1971) An electron microscopical study of the acrosomal complex and its role in fertilization in the river lamprey, *Lampetra fluviatilis*. J Submicrosc Cytol 3: 309–317
Nikiforov ID (1949) The influence of temperature on embryogenesis in the stellate sturgeon. Izv VNIORKh 29: 156–164
Nikitina LA (1974) Transplatation of nuclei from growing oocytes into fully grown enucleated sturgeon oocytes. Ontogenez 5: 251–255
Nikitina LA (1978) Presence of cytotomy factor in the karyoplasm of *Acipenser* oocytes at the onset of the period of rapid growth. Wilhelm Roux's Arch Dev Biol 184: 109–113
Nikitina LA, Dettlaff TA (1986) Appearance of a maturation-stimulating factor in the karyoplasm of *Bufo viridis*, *Xenopus laevis*, and *Acipenser stellatus* oocytes. Sov J Dev Biol 17: 148–151
Nikol'skaya MP (1985) Development of the system of ampullar receptors in the *Acipenseriformes* (*Chondrostei*) and its functional role. In: Akoev GN (ed) Signalization and Behavior of Fish. Murmanskii Morskoi Biologicheskii Institut Kol'skogo Filiala Akad Nauk SSSR Apatity, pp 46–56 (in Russian)
Nikol'skaya MP (1989) Development of the system of ampullar organs in *Pseudoscaphirhync-*

hus kaufmanni (Bogd.). In: Pavlov DS, Sbikin YuN (eds) Morphology, ecology and behavior of Acipenserid fishes, Nauka, Moscow, pp 71–88 (in Russian)
Nikol'skaya NG, Sytina LA (1978a) Comparative analysis of the effect of constant temperatures on embryogenesis of various acipenserid species. Vopr Ikhtiol 18: 101–116
Nikol'skaya NG, Sytina LA (1978b) Temperature conditions essential for development of the eggs of Siberian sturgeon from the Lena River. Rybn Khoz (9): 17–20
Nikolyukin NI (1952) Interspecific hybridization of fishes. Obl Gos Izdatel'stvo, Saratov (in Russian)
Nikolyukin NI, Burtsev IA (1969) Instruction for breeding and industrial rearing of the hybrids between the giant sturgeon and sterlet, VNIRO, Moscow (in Russian)
Ninua NSh (1976) The Atlantic sea sturgeon of the Rioni River. Metsniereba, Tbilisi, (in Russian)
Nuccitelli R (1980) The electrical changes accompanying fertilization and cortical vesicle secretion in the medaka egg. Dev Biol 76: 483–498
Owsjannikow PhV (1871) On artificial breeding of sterlets. In: Proc II Congr Russian naturalists, Div Zool Anat Physiol, pp 191–200 (in Russian)
Ozernyuk ND (1985) Energy metabolism in early ontogenesis of fishes. Nauka, Moscow (in Russian)
Ozernyuk ND (1988) The principle of energy minimum in ontogenesis and steady state of development. Zh Obshch Biol 49: 552–562
Pavlenko N (1916) Development of the pronephros in the sterlet. Rus Zool Zh 1: 239–242
Persov GM (1941a) Some data about survival of the stellate sturgeon (*Acipenser stellatus*) spermatozoa. Dokl Akad Nauk SSSR 33: 327–329
Persov GM (1941b) An account of sturgeon culture work with reference to the use of the method of pituitary injections. In: Gerbil'skii NL (ed) The method of pituitary injections and its role in reproduction of fish resources. LGU Press, Leningrad, pp 42–50 (in Russian)
Persov GM (1947) Sexual function of Acipenserid males (a histological and experimental study). Candidate Dissertation. LGU, Leningrad (in Russian)
Persov GM (1953) Dosage of spermatozoa as a method of control of egg fertilization in sturgeons. Dokl Akad Nauk SSSR 90: 1184–1185
Persov GM (1957) The method of handling the sterlet spawners. Uch Zap LGU 228 Ser Biol Nauk (44): 72–86
Persov GM (1963) The sterlet as an object of fish culture, acclimatization and industrial rearing. In: Pavlovskii EN (ed) Sturgeon culture in water bodies of the USSR. Izdatel'stvo Akad Nauk SSSR, Moscow, pp 40–43 (in Russian)
Petru A (1965) Industrial waste. Stroiizdat, Moscow (in Russian)
Piri A, Van Geiningen R (1968) Biochemistry of the eye. Mir, Moscow (in Russian)
Popov OP, Budarin VV (1976) The use of progesterone for stimulating maturation of the carp and wild carp females. Rybn Khoz (2): 18–19
Popova AA (1973) Analysis of work with the Russian sturgeon spawners at the hatcheries in the Volga delta. Tr VNIRO 92: 15–20
Papova AA (1978) Breeding of the hiemal giant sturgeon at the hatcheries in the Volga delta. Rybn Khoz (6): 21–22
Privol'nev TI (1947) Changes in the respiration during ontogenesis of the fish at different partial oxygen pressure. Izv VNIORKh 25: 57–112
Pushkar' NS, Belous AM, Kopeika EF, Gordienko EA (1979) Low temperature conservation of sturgeon sperm. In: Luk'yanenko VI (ed) Abstracts II All-Union Conf Sturgeon culture of the inland water bodies of the USSR. Astrakhan', p 220 (in Russian)
Pushkar' NS, Itkin YuA, Kopeika EF, Gordienko EA, Bronshtein VL Glushko TA (1978) A method of conservation of fish sperm. Ofits Byull Gos Komiteta Soveta Ministrov SSSR po Delam Izobretenii i Otkrytii 5: 3
Rass TS (1941) Geographical parallelisms in the structure and development of the *Teleostei* in the northern seas, MOIP, Moscow (in Russian)
Rass TS (1946) Stages of ontogenesis in *Teleostei*. Zool Zh 25: 137–146

Romanov AA (1979) Hatchery reproduction of Acipenserid fishes in the Volga delta (1971–1976). In: Berdichevskii LS (ed) Biological foundations of the sturgeon culture development in water bodies of the USSR. Nauka, Moscow, pp 92–105 (in Russian)
Romanycheva OD (1976) Instruction for rearing the bester in fish tanks. VNIRO, Moscow (in Russian)
Romanycheva OD (1979) On rearing the bester and the giant sturgeon juveniles in marine cages. In: Berdichevskii LS (ed) Biological foundations of the sturgeon culture development in water bodies of the USSR, Nauka, Moscow, pp 81–84 (in Russian)
Rott NN (1973) Correlation between karyo- and cytokinesis during the first cell divisions in the axolotl (*Ambystoma mexicanum* Cope). Sov J Dev Biol 4: 175–177
Rott NN (1980) Cell divisions during pregastrulation period of development. Sov J Dev Biol 11: 1–17
Rott NN (1987) Cell cycles in early development of animals. Nauka, Moscow, (in Russian)
Rudneva TB (1972) Duration of karyokinesis and cell division in cleavage divisions II–IV in the clawed frog, *Xenopus laevis*. Sov J Dev Biol 3: 526–530
Rustamova ShA (1974) Experimental data about the influence of naphthene acids on embryogenesis of Acipenserid fishes. In: Astakhova TV (ed) Abstracts of the accounting session TsNIORKh. Astrakhan', pp 135–136 (in Russian)
Ryabova LV, Dettlaff TA (1981) A study of activity of the factor of germinal vesicle membrane disintegration at different phases of the two first cleavage divisions in the ooplasm of *Acipenser stellatus*, *Rana temporaria*, and *Xenopus laevis*. Ontogenez 12: 517–520
Ryabova LV, Trubnikova OB, Sleptsova LA (1984) A contribution to the chronology of embryogenesis in the common frog. Ontogenez 15: 91–93
Sadov IA (1941) Morphobiological characteristic of developmental stages in sturgeons. Rybn Khoz (5): 23–25
Sadov IA (1948a) Breeding and rearing of the Russian and stellate sturgeon juveniles. Rybn Khoz (1): 38–42
Sadov IA (1948b) Influence of water oversaturated with oxygen on morphology of the Russian and stellate sturgeon juveniles. Rybn Khoz (1): 43–44
Sadov IA (1950) Dependence of the egg mortality in the Russian and stellate sturgeons on the methods of their obtaining and conditions of incubation. Tr Inst Morfol Zhivot Akad Nauk SSSR 3: 3–18
Sadov IA (1951) Influence of the conditions of egg incubation on development of juveniles in the Russian and stellate sturgeons. Tr Inst Morfol Zhivot Akad Nauk SSSR 5: 129–146
Sadov IA (1957) About the role of the male in spawning of sturgeon and the methods of egg insemination. In: Cherfas BI (ed) Proc Conf on Fish culture Izdatel'stvo Akad Nauk SSSR, Moscow, pp 150–159 (in Russian)
Sadov IA, Kokhanskaya EM (1953) On dependence between the cleavage pattern of the sturgeon eggs during the first three divisions and structure of hatching larvae. Dokl Akad Nauk SSSR 93: 1135–1138
Sadov IA, Kokhanskaya EM (1961) Incubation of sturgeon eggs in troughs. Tr Inst Morfol Zhivot Akad Nauk SSSR 37: 5–66
Sadov IA, Kokhanskaya EM (1963) Trough incubator for sturgeon eggs. Rybn Khoz (6): 23–27
Salensky VV (1878) Developmental history of the sterlet (*Acipenser ruthenus*). 1. Embryogenesis. Tr Ova Estestvoispytatelei pri Kazanskom Univ 7, pt 3: 1–226
Salensky VV (1880) Developmental history of the sterlet (*Acipenser ruthenus*). 2. Postembryonic development and development of organs. Tr Ova Estestvoispytatelei pri Kazanskom Univ 10, pt 2: 227–545
Salensky VV (1881) Recherches sur le développement du sterlet, *Acipenser ruthenus*. Arch Biol 2: 233–241
Sawadski AM (1911) Development of rudimentary larval organ of attachment in the sterlet. Tr Ova Estestvoispytatelei pri Kazanskom Univ 44: 3–27
Sawadski AM (1926a) Untersuchungen zur Entwicklungsgeschichte des Sterlets (*Acipenser*

ruthenus L.). Veränderungen vor der Gastrulation, Bildung der Keimblätter und die damit verbundenen Prozesse. Z Anat Entwicklungsgesch 78: 26–65

Sawadski AM (1926b) Untersuchungen zur Entwicklungsgeschichte des Sterlets (*Acipenser ruthenus* L.). Entwicklung des Pronephros. Z Anat Entwicklungsgeschi 78: 682–713

Schmalhausen OI (1951) Development of the oro-branchial system in the stellate sturgeon. Dokl Akad Nauk SSSR 80: 681–684

Schmalhausen OI (1952) Development of gills and blood vessels of the visceral apparatus in the stellate sturgeon. Dokl Akad Nauk SSSR 86: 193–196

Schmalhausen OI (1955a) Development of gills in the larvae of the Russian sturgeon from the Volga River. Dokl Akad Nauk SSSR 100: 397–400

Schmalhausen OI (1955b) Development of gill vessels in the larvae of the Russian sturgeon from the Volga River. Dokl Akad Nauk SSSR 100: 605–608

Schmalhausen OI (1955c) Ecological-morphological features in the branchial apparatus development in the larvae of the Russian sturgeon from the Volga River. Dokl Akad Nauk SSSR 100: 837–839

Schmalhausen OI (1957) Formation of defective olfactory organs in Acipenserid fishes upon their artificial breeding. Dokl Akad Nauk SSSR 114: 216–219

Schmalhausen OI (1962) Disturbance of the olfactory organ development in Acipenserid fishes Under certain conditions of rearing. Tr Inst Morfol Zhivot Akad Nauk SSSR 40: 188–218

Schmalhausen OI (1968) Development of the digestive system in Acipenserid fishes. In: Disler NN (ed) Morphoecological studies of fish development Nauka, Moscow, pp 40–70 (in Russian)

Schmalhausen OI (1971) Effect of phenol on development of prelarvae of Acipenserid fish. 1. External structure and behaviour. Sov J Dev Biol 2: 481–487

Schmalhausen OI (1972) Effect of phenol on development of prelarvae of Acipenserid fish. 2. The digestive system. Sov J Dev Biol 3: 407–413

Schmalhausen OI (1973) Effects of phenol on development of prelarvae of Acipenserid fish. 3. Pigmentation of the skin. Structure of the eyes. Sov J Dev Biol 4: 28–35

Schmalhausen OI (1975) The Russian sturgeon *Acipenser gueldenstaedti*. Development of prelarvae. In: Dettlaff TA (ed) Objects of developmental biology. Nauka, Moscow, pp 264–277 (in Russian)

Schmalhausen OI (1983) Duration and patterns of development of the giant and stellate sturgeon prelarvae at different temperatures. In: Barannikova IA, Berdichevskii LS (eds Biological foundations of sturgeon culture, Nauka, Moscow, pp 94–103 (in Russian)

Schmalhausen OI (1991) The Russian sturgeon *Acipenser gueldenstaedti* Part II. Later prelarval development. In: Dettlaff TA, Vassetzky SG (eds) Animal species for developmental studies, vol 2. Vertebrates. Consultants Bureau, New York, pp 67–88

Schuetz AW (1967) Action of hormones on germinal vesicle break-down in frog (*Rana pipiens*) oocytes. J Exp Zool 166: 347–354

Schuetz AW (1972) Hormones and follicular functions. In: Biggers JD, Schuetz AW (eds) Oogenesis. Univ Park Press, Baltimore; Butterworths, London, pp 479–512

Semenov KI (1957) On influence of various illumination conditions on the Russian sturgeon larval development in the period from hatching until transition to active feeding upon artificial breeding. Dokl Akad Nauk SSSR 113: 937–940

Semenov KI (1958) Morphological and biological features of the Russian sturgeon larval development under different conditions of illumination. Vidavnitstvo Akad Nauk USSR, Kiev (in Ukrainian)

Semenov KI (1965) Influence of the conditions of keeping the Russian sturgeon spawners on their maturation after pituitary injection and the quality of progeny at the early developmental stages. In: Vladimirov VI (ed) Influence of the quality of spawners on the progeny in fishes. Naukova Dumka, Kiev, pp 123–142 (in Russian)

Severtsov AN (1948) Origin and evolution of lower vertebrates. Part II. Visceral system of fish. In: Severtsov AN Collected works vol 4. Izdatel'stvo Akad Nauk SSSR, Moscow-Leningrad, pp 201–397 (in Russian)

Shelukhin GK, Metallov GF, Geraskin PP (1990) Influence of the thermal-salt regime of Caspian water on juveniles of the Russian sturgeon *Acipenser gueldenstaedti*. Vopr Ikhtiol 30: 296–304

Shilov VI, Khazov YuK (1971) Reproduction of Acipenserid fishes in the Volgograd and Saratov water reservoirs after the construction of the Saratov hydroelectric power station. In: Belyaeva VN et al. (eds) Materials of Joint Session of TsNIORKh and AzNIIRKh. Astrakhan', pp 123–124 (in Russian)

Shilov VI, Khazov YuK (1979) On the necessity of artificial breeding of the sterlet in the Volga River. In: Luk'yanenko VI (ed) Abstracts II All-Union Conf Sturgeon culture of the inland water bodies of the USSR, Astrakhan', pp 272–279 (in Russian)

Shmidtov AI (1936) On survival of sturgeon sperm under different environmental conditions. Dokl Akad Nauk SSSR 3: 89–91

Shmidtov AI (1939) The sterlet (*Acipenser ruthenus* L.). Uch Zap Kazan Univ Zool 99, pt 4/5, (6/7): 1–279

Shubravyi OI, Goncharov BF, Uteshev VK (1989) An experience of keeping and breeding in captivity of *Pseudoscaphirhynchus kaufmanni* Bogdanov, 1874. In: Davletova LV (ed) Problems of domestication of animals. Institute of Evolutionary Morphology and Animal Ecology, USSR Academy of Sciences, Moscow, pp 202–206 (in Russian)

Skadovskii SN (1949) Perspectives of sturgeon culture according to the data of ecological-physiological studies of the Kura stellate sturgeon. Rybn Khoz (2): 99–104

Skoblina MN (1963) Characteristics of the duration of the main phases of embryogenesis in *Ambystoma mexicanum* as expressed in dimensionless units. Abstracts IV Conf of Embryologists (Leningrad, June 19–27, 1963), LGU Press, Leningrad, p 172 (in Russian)

Skoblina MN (1968) Maturation of the cortex of enucleated oocytes of frog and stellate sturgeon induced by gonadotropic pituitary hormones. Dokl Akad Nauk SSSR 183: 643–645

Skoblina MN (1969) Independence of the cortex maturation from germinal vesicle material during the maturation of amphibian and sturgeon oocytes. Exp Cell Res 55: 142–144

Skoblina MN (1976) Role of karyoplasm in the emergence of capacity of egg cytoplasm to induce DNA synthesis in transplanted sperm nuclei. J Embryol Exp Morphol 36: 67–72

Soin SG (1968) Adaptive properties of fish development. MGU Press, Moscow (in Russian)

Soin SG (1977) On some specific features of the carp *Cyprinus carpio* L. development with reference to the incubation of its eggs under hatchery conditions. Vopr Ikhtiol 5: 900–911

Sokolov LI (1965) Maturation and fertility of the Siberian sturgeon *Acipenser baeri* Br. from the Lena River. Vopr Ikhtiol 5: 70–81

Sokolov LI, Malyutin VS (1977) Specific features of the population structure and characteristics of the spawners of the Siberian sturgeon *Acipenser baeri* Brandt in the region of spawning grounds on the Lena River. Vopr Ikhtiol 17: 237–246

Solmann T (1949) Correlation of the aquarium goldfish toxicities of some phenols, quinones, and other benzene derivatives with their inhibition of autooxidative reactions. J Gen Physiol 32: 671–679

Stroganov NS (1938) Studies of spawning grounds of the *Acipenseridae* and *Clupeideae* in the Volga in summer 1934. In: Proc I All-Caspian Fishery Conf VNIRO Moscow. vol 2 pp 101–112 (in Russian)

Svetlov PG (1978) Developmental physiology (mechanics), vol 2 Nauka, Moscow (in Russian)

Svirskii VG (1967) The Amur and Kaluga sturgeons (taxonomy, biology, perspectives of reproduction). Author's Abstract of the Candidate Dissertation. DGU, Vladivostok (in Russian)

Svirskii VG (1970) The first positive experiments on breeding of the Amur and kaluga sturgeons. Tr TsNIORKh 2: 197–198

Svirskii VG (1971) The Amur and kaluga sturgeons (state of reserves, some features of biology, perspectives of reproduction). Uch Zap Dal'nevost Univ 15 (3): 19–33

Sytina LA (1971a) Some peculiarities of the rate of larval development in different Acipenserid species during the early postembryonic period. Tr TsNIORKh 3: 282–297

Sytina LA (1971b) Comparative study of the development of sense organs and specific features of the rate of their formation in Acipenserid larvae. Author's Abstract of the Candidate Dissertation, Institute of Animal Evolutionary Morphology and Ecology, USSR Academy of Sciences, Moscow (in Russian)

Sytina LA (1979) Morphological aspects in studying the patterns of variability in early fish ontogenesis. In: Sokolov VE (ed) State and perspectives of development of morphology. Materials of All-Union Conf. Nauka, Moscow, pp 88–91 (in Russian)

Sytina LA, Timofeev OB (1973) Periodization of development in the *Acipenseridae* and the problem of variability of organisms. Vopr Ikhtiol 13: 275–289

Sytina LA, Nikol'skaya NG, Shagaeva VG (1985) On relationship between hyperthermal resistance of the stellate sturgeon embryos and the time of spawning migration. Dokl Akad Nauk SSSR 280: 247–250

Tatarko KI (1936) Apparatus of the operculum and its connection with the hyoid and maxillar arches in the *Acipenseridae*. Tr Inst Zool Biol Akad Nauk USSR 10: 5–53 (in Ukrainian)

Tatarskaya RI, Kafiani KA, Kanopkaite SI (1958) Some enzymes of phosphorus metabolism and the intensity of respiration and aerobic glycolysis in embryogenesis of Acipenserid fishes. Biokhimiya 23: 527–539

Tchou-Su, Wang Yu-lan (1958) Etudes comparatives sur l'ovulation et la maturation *in vivo* et *in vitro* chez le Crapaud asiatique (*Bufo bufo asiaticus*). Acta Biol Exp Sin 6: 129–180

Teichmann H (1957) Defektstufen von Nase und Auge bei fehlgebiltdeten Larven der Regenbogenforelle (*Salmo irideus* W. Gibb.) Wilkelm Roux' Arch Entwicklungsmech Org 149: 365–386

Timofeev OB (1971) Circulatory system of sturgeons at the early stages of ontogenesis. Tr TsNIORKh 3: 306–316

Tleuov R, Sagitov NI (1973) Acipenserid fishes of the Amu Darja River. Izdatel'stvo Akad Nauk UzSSR, Tashkent (in Russian)

Trubnikova OB, Ryabova LV (1989) Preovulatory changes in the follicle of the stellate sturgeon *Acipenser stellatus* Pall. Ontogenez 20: 532–542

Trusov VZ (1964a) Some specific features of maturation and the maturity scale of the Russian sturgeon gonads. Tr VNIRO 56: 69–78

Trusov VZ (1964b) Method of estimation of the degree of the gonad maturity in sturgeon females. Rybn Khoz (1): 26–28

Trusov VZ (1967) Biological foundations for handling summer-spawning (late vernal) Russian sturgeons. Tr TsNIORKh 1: 168–180

Trusov VZ (1975) Maturation of gonads in the stellate sturgeon *Acipenser stellatus* females during the marine period of life. Vopr Ikhtiol 15: 71–82

Ugolev AM (1963) Parietal (contact) digestion. Izdatel'stvo Akad Nauk SSSR, Moscow-Leningrad (in Russian)

Usynin VF (1978) Biology of the sterlet *Acipenser ruthenus* L, from the Chulym River. Vopr Ikhtiol 18: 624–635

Van Asselt LAC, Goos HJT, Smit-Van Dijk W, Speetjens PAM, van Oordt PGWJ (1988) Evidence for the involvement of D_2 receptors in the dopaminergic inhibition of gonadotropin release in the African catfish, *Clarias gariepinus*. Aquaculture 72: 369–378

Vasil'ev VP (1980) Chromosome numbers of the *Agnatha* and *Pisces*. Vopr Ikhtiol 20: 387–342

Vassetzky SG (1966) The thermosensitivity of sturgeon eggs at the stages of their maturation and cleavage. Zh Obshch Biol 27: 583–595

Vassetzky SG (1970) Dynamics of the I maturation division in sturgeon oocytes. Zh Obshch Biol 31: 84–93

Vassetzky SG (1971) Fishes of the family *Polyodontidae*. Vopr Ikhtiol 11: 26–42

Vassetzky SG (1988) Meiotic divisions. In: Dettlaff TA, Vassetzky SG (eds) Oocyte growth and maturation. Plenum Press, New York, pp 279–339

Vernidub MF (1951) Morphophysiological stages in development of Acipenserid eggs and larvae and their importance for fish culture. Uch Zap LGU 142 Ser Biol Nauk 29:75–106

Vernidub MF (1952) Instruction for determining the hatchery quality, obtaining, fertilizing,

rational demucilaging and incubating eggs and rearing larvae of Acipenserid fishes, Biological Foundation for Line II of Sturgeon Hatchery (in Russian)
Vernidub MF (1957) Biological foundation of the methods for obtaining sturgeon larvae from demucilaged eggs under the conditions of hatchery. In: Cherfas BI (ed) Proc Conf on Fish culture, 1954, Izdatel'stvo Akad Nauk SSSR, Moscow, pp 160–175 (in Russian)
Vernidub MF, Kiseleva EV (1953) Influence of "overmaturation" on the pattern of initial cleavage in the Acipenserid eggs and their mortality at the early developmental stages. Dokl Akad Nauk SSSR 92: 1093–1095
Vernidub MF, Kudryashova EN, Nishchaeva GI, Ratnikova GI (1971) Age changes in the structure and function of the digestive system in the Russian sturgeon (*Acipenser gueldenstaedti* Brandt) and stellate sturgeon (*Acipenser stellatus* Pall.) in the early period of ontogenesis. Tr TsNIORKh 3: 77–113
Veshchev PD (1979) Biological characteristics of the Russian and stellate sturgeon spawners on the Volga spawning grounds. In: Berdichevskii LS (ed) Biological foundations of the sturgeon culture development in the water bodies of the USSR. Nauka, Moscow, pp 115–122 (in Russian)
Vincent JP, Oster GF, Gerhart JC (1986) Kinematics of gray crescent formation in *Xenopus* eggs. The displacement of subcortical cytoplasm relative to the egg surface. Dev Biol 113: 484–500
Vladykov VD, Greeley JR (1963) Order *Acipenseridae*. In: Fishes of the western North Atlantic. Mem Sears Found Mar Res 1, pt 3 (cited from Jones et al. 1978)
Vlasenko AD (1990) Formation of the numbers of Acipenserid fishes in the Caspian Sea. Rybn Khoz (7): 53–56
Vogt W (1925) Gestaltungsanalyse am Amphibienkeim mit örtlicher Vitafärbung. 1. Methodik und Wirkungmethode. Wilhelm Roux Arch Entwicklunsmech Org 106: 542–610
Volodin VM (1973) Study of the effects of various phenol concentrations on embryogenesis of *Danio rerio*. Tr Inst Biol Vnutrennikh Vod Akad Nauk SSSR 24: 57–71
Volodin VM, Luk'yanenko VI, Flerov BA (1965) Dynamics of changes in resistance of fishes to phenol at the early stages of ontogenesis. In: Abstracts I Congr All-Union Hydrobiological Society. Nauka, Moscow, p 82 (in Russian)
Volodin VM, Luk'yanenko VI, Flerov BA (1966) Comparative characteristics of the resistance of fishes to phenol at the early stages of ontogenesis. Tr Inst Biol Vnutrennikh Vod Akad Nauk SSSR 10: 300–310
Vorob'eva EI, Rubtsov VV, Markov KP (1986) Influence of environmental factors on the microstructure of egg membranes in fishes, Nauka, Moscow (in Russian)
Votinov NP (1947) The ovary of the stellate sturgeon during the spawning migration, spawning and downstream migration. Tr Lab Osnov Rybovod 1: 139–154
Votinov NP (1963) Biological foundations of artificial reproduction of the Siberian sturgeon in the Ob' River. Tr Ob'-Tazov Otd GosNIORKh 3: 5–102
Votinov NP, Kas'yanov VP (1978) Ecology and efficiency of reproduction of the Siberian sturgeon *Acipenser baeri* Brandt in the Ob'River under the conditions of hydroconstruction. Vopr Ikhtiol 18: 25–35
Votinov NP, Kas'yanov VP (1979) Current state of the sturgeon culture in water bodies of Siberia and perspectives of its development. In: Berdichevskii LS (ed) Biological foundations of the sturgeon culture development in water bodies of the USSR, Nauka, Moscow, pp 59–67 (in Russian)
Wasserman WJ, Smith LD (1978) The cyclic behavior of a cytoplasmic factor controlling nuclear membrane breakdown. J Cell Biol 78: 15–22
Waters W (1966) Mechanisms of oxidation of organic compounds. Mir, Moscow (in Russian)
Wilson EB (1925) The cell in development and heredity, 3rd edn, Macmillan, New York
Woskoboinikoff MM (1928) Apparatus of respiration in fishes (an experience of synthesis in morphology). In: Deryagin KM (ed) Proc III Congr Russian zoologists anatomists and histologists, 1927. Izdanie Glavn Upr Nauchn Uchrezhd Leningrad, pp 103–105 (in Russian)

Woskoboinikoff MM (1932) Der Apparat der Kiemenatmung bei den Fischen. Zool Jahrb Abt Allg Zool Phys 55: 315–488

Wright PA (1945) Factors affecting in vitro ovulation in the frog. J Exp Zool 100: 565–575

Yaroslavtseva AM, Sergeeva ZP, Zotin AI (1991) Influence of temperature and salinity on cleavage of the sea urchin *Strongylocentrotus intermedius* eggs (in press)

Yntema CL (1955) Ear and nose. In: Willier BH, Weiss PA, Hamburger V (eds) Analysis of development. Saunders Philadelphia, pp 415–428

Yurovitskii YuG, Reznichenko PN (1961) On some specific features of development of the sturgeon from the Kura River. Vopr Ikhtiol 1: 314–320

Yurovitskii YuG, Reznichenko PN (1963) Morphophysiological features of the Russian sturgeon embryos during incubation in conditions of varying oxygen regime. In: Pavlovskii EN (ed) Sturgeon culture in water bodies of the USSR. Izdatel'stvo Akad Nauk SSSR, Moscow, pp 77–82 (in Russian)

Yushchenko PC (1957) A device for incubation of the eggs of Acipenserid fishes. USSR Ministry of Fisheries, Moscow (in Russian)

Zenkevich GA, Latse ZM (1979) Isolation and characterization of the gonadotropic hormone preparation from pituitaries of Acipenserid fishes *Acipenser stellatus* Pallas and *Acipenser gueldenstaedti* Brandt. Vopr Ikhtiol 19: 883–890

Zenkevich GA, Latse ZM, Slanke VP (1981) Structural-functional closeness of pituitary gonadotropins of the Russian sturgeon *Acipenser gueldenstaedti* Br. and stellate sturgeon *Acipenser stellatus* Pall. Vopr Ikhtiol 21: 366–372

Zinichev VV, Zotin AI (1988) The total oxygen consumption as a function of temperature at different developmental stages in *Oncorhynchus keta.* Ontogenez 19: 217–220

Zotin AI (1953a) Consumption of water by developing Acipenserid eggs from the environment. Dokl Akad Nauk SSSR 89: 377–380

Zotin AI (1953b) Hatching enzyme in sturgeon embryos. Dokl Akad Nauk SSSR 92: 685–687

Zotin AI (1953c) Changes in the strength of the egg membranes of sturgeon embryos during development. Dokl Akad Nauk SSSR 92: 443–446

Zotin AI (1961) Physiology of water exchange in the embryos of *Pisces* and *Cyclostomata,* Izdatel'stvo Akad Nauk SSSR, Moscow (in Russian)

Zotin AI, Ozernyuk ND (1966) The effect of temperature on respiration and ATP level in the cleaving *Misgurnus fossilis* eggs. Dokl Akad Nauk SSSR 171: 1002–1004

Zotin AI, Alekseeva TA, Ozernyuk ND (1989) Method of determination of the optimal developmental conditions according to the total oxygen consumption. Ontogenez 20: 103–107

Subject Index

MIX
Papier aus verantwortungsvollen Quellen
Paper from responsible sources
FSC® C105338

If you have any concerns about our products,
you can contact us on
ProductSafety@springernature.com

In case Publisher is established outside the EU,
the EU authorized representative is:
Springer Nature Customer Service Center GmbH
Europaplatz 3, 69115 Heidelberg, Germany

Printed by Libri Plureos GmbH
in Hamburg, Germany